OENOTHERA

CYTOGENETICS AND EVOLUTION

EXPERIMENTAL BOTANY
An International Series of Monographs

CONSULTING EDITORS

J. F. Sutcliffe

School of Biological Sciences, University of Sussex, England

AND

P. Mahlberg

Department of Botany, Indiana University, Bloomington, Indiana, U.S.A.

IN PREPARATION

Frontispiece: Portrait of Hugo de Vries (1848-1935).
From I. Dörfler (1906), "Botaniker-Porträts", No.5. Vienna.

OENOTHERA
CYTOGENETICS AND EVOLUTION

RALPH E. CLELAND

Department of Botany
Indiana University
Bloomington, Indiana, U.S.A.

1972

ACADEMIC PRESS · LONDON AND NEW YORK

ACADEMIC PRESS INC. (LONDON) LTD.

24/28 Oval Road,
London NW1

United States Edition published by
ACADEMIC PRESS INC.
111 Fifth Avenue
New York, New York 10003

Copyright © 1972 by
ACADEMIC PRESS INC. (LONDON) LTD.

Library of Congress Catalog Card Number: 73-189936
ISBN: 0-12-176450-8

PRINTED IN GREAT BRITAIN BY
WILLIAM CLOWES & SONS LIMITED
LONDON, COLCHESTER AND BECCLES

Preface

When Hugo de Vries retired as head of the Botanical Institute and Garden of the University of Amsterdam in 1918 he moved to Lunteren, a small village in central Holland and there, adjoining his residence, erected a rather large laboratory building, with an experimental garden attached. In this institution he planned not only to carry on his own research, but also to make space available for scientists from elsewhere to come and work. Unfortunately, many of his investments had been placed in Germany, and the inflation that occurred in that country following World War I wiped these out, and as a result he was unable to furnish and equip the completed building. When we visited Lunteren in 1927 and 1928, the institute was essentially empty and unoccupied. Nevertheless, de Vries continued his own work there, carrying on his investigation up to the time of his death in 1935, at the age of 87.

When we arrived in Lunteren in 1927, we found de Vries, at the age of 80, a man of distinguished appearance, gracious manner and cultivated taste. He was very happy to make his garden, with its extensive collection of mutants, available to a visiting scientist. My wife and I had the great privilege of working in his garden for a few weeks during the summers of 1927 and 1928. Although not wholly convinced of the correctness of the cytological findings and conclusions of his guest, he was most generous with his time, advice and hospitality. Consequently, I look back upon those days in Lunteren with particular pleasure and gratitude.

Already renowned as a plant physiologist, de Vries turned in the 1880's to a study of evolution which became centered around the genus *Oenothera*. In initiating this work, he opened up a field that for a time became one of central importance in genetics because of the fact that this genus seemed not to follow the conventional Mendelian laws, and so became an apparent exception to normal genetical behavior. The field which de Vries thus opened up has over the years attracted many investigators, and the *Oenothera* literature has become copious, widely scattered and difficult to follow.

The purpose of the present monograph, therefore, is to bring together into a connected account the more important developments that have led to the present state of knowledge of *Oenothera* cytogenetics and evolution. For the most part we will be discussing already published work. Two sections, however, will present hitherto unpublished details of research which has been conducted in our own laboratory. In the latter half of Chapter 12 studies on the missing petal character are reported, and in Chapter 18 the story of the evolution of the subgenus *Oenothera* is spelled out in considerable detail.

I am greatly indebted to a number of persons who have been of assistance in the preparation of this treatise. The following have kindly read and criticized portions of the manuscript—Professors L. C. Dunn, D. G. Frey, F. F. Schötz, W. Stubbe and E. E. Steiner; to them I owe a special debt of gratitude. They, of course, are not responsible for any errors that may have crept into the text. Professor Clarence Flaten of the Audio-visual Center at Indiana University has been of great assistance in the preparation of photographs.

It is also a pleasure to acknowledge the major sources of support that have made my *Oenothera* research possible over the years, especially the Penrose Fund of the American Philosophical Society, the Rockefeller Foundation, and the Indiana University Foundation. The latter two agencies are still supporting my work. Without their assistance, this monograph could not have been written.

June, 1971*   RALPH E. CLELAND

---

* This manuscript, begun in the early 1940's and summarizing a lifetime's work on the genetics and cytogenetics of *Oenothera*, was completed and sent to the publishers during the first week of June, 1971. On June 11th, while working in the laboratory, Ralph Cleland suffered a stroke and died.

Contents

Introduction

In the more than eighty years since de Vries first began the study of the evening primrose, *Oenothera*, the genus has been an object of interest for several different reasons. It first attracted attention as the material upon which de Vries largely based his celebrated Mutation Theory of Evolution. Later it became apparent that in many respects its genetical behavior did not conform to the usual pattern of Mendelian inheritance, and for some time this behavior constituted one of the major puzzles confronting geneticists. When the genetical riddle was finally solved on the basis of an unusual type of chromosome behavior, the genus became an outstanding demonstration of the validity of the chromosome theory of inheritance, in that the exceptions which it showed to the normal process of Mendelian inheritance were found to be based upon a parallel and equally exceptional type of chromosome behavior. There were still biologists in the twenties who were skeptical of the chromosome theory of inheritance!

Later, the genus presented investigators with one of the few opportunities they have had to trace in some detail the course of evolution as it has taken place in a given group. Not many cases are known where it is possible to identify the major factors involved and to show how the interplay of these factors has brought about the evolution of a group.

Finally, *Oenothera* continues to yield important results in studies of certain areas of physiological genetics and of extra-nuclear inheritance.

Oenothera, therefore, has played a distinctive role in the areas of cytogenetics, cytotaxonomy and evolution.

The genus *Oenothera* is an extensive one. According to the recent classification of Munz (1965), it consists of 15 subgenera (Table I), distributed widely throughout North and South America.* Although native only to the New World,† it has been carried to other areas, and in Europe has become established as an important element in the flora. In the words of Renner (1956, p. 253), "für die Oenotheren Europa die neue Welt ist."

The first European record of what has been thought to be an *Oenothera* was by Prosper Alpinus of Padua in a treatise written in 1612 but not published until 1627. His material was grown from seed received from Dr. John More of England. He referred to the plant under the name *Hyosciamus virginianus*. Examination of his plate, however, makes it

*But see footnote to Table I.
†See p. 314.

doubtful if this was an *Oenothera* (see Gates, 1911 a, Plate I). For one thing, its flowers apparently had five petals. Other seventeenth century authors included what were apparently evening primroses under the generic name *Lysimachia*, as shown by the descriptions of Caspar Bauhin (1623), Columna (1628), Parkinson (1629), Morison (1680) and Ray (1686). Tournefort (1694) applied the designation *Onagra*. The term *Oenothera* was first used by Linnaeus (1737). An account of early references to *Oenothera* will be found in "The Mutation Factor in Evolution" (Gates, 1915) and also in two papers by Gates (1911 a, b). This author made an extensive survey of early records of *Oenothera* in Europe because of his interest in ascertaining the origin of *O. lamarckiana* and other races with which de Vries worked. His attempts to relate the early descriptions, drawings and herbarium specimens to the races of *Oenothera* as we know them at present lack validity, but the list of early references which he compiled is valuable for its own sake.

TABLE I

Subgenera of the genus Oenothera
(Arrangement is that of Munz, 1965)[a]

Hartmannia	Bolivia and Peru to southwestern United States
Kneiffia	Texas, northward and eastward over United States and eastern Canada
Gauropsis	Mexico, through the western plains states
Megapterium	Mexico, through the western plains states
Lavauxia	Mexico, United States from California to Kentucky; Chile, Argentina
Pachylophus	Northern Mexico, California to Washington, Saskatchewan and South Dakota
Raimannia	Terra del Fuego to Canada, east to west (probably includes South American races earlier assigned to the subgenus *Oenothera*, including Fischer's *Renneria*)
Anogra	Northern Mexico, western United States and Canada
Oenothera	Widespread over North America, doubtfully in South America
Salpingia	Mexico to the northern plains
Calylophus	Northern Mexico to western Canada
Heterostemon	Western United States
Eulobus	Western United States and lower California
Sphaerostigma	Western United States, lower California
Chylismia	Western United States, northern Mexico

[a] Raven (*Britonia* **16**, 276-288, 1964) differs in his classification in that he divides *Oenothera*, as defined by Munz, into three genera: *Camissonia* including subgenera *Sphaerostigma*, *Eulobus*, *Chylismia* and *Heterostemon* (all with capitate stigmas); *Calylophus* including *Salpingia* (with peltate stigmas); and *Oenothera* including the remaining nine subgenera (with four-lobed stigmas).

To my wife, Elizabeth

PART I GENETICAL BEHAVIOR OF OENOTHERA

CHAPTER 1

de Vries and the Mutation Theory

Most of the research for which *Oenothera* is well known has been done on the subgenus *Oenothera* (commonly referred to in recent genetical literature as *Onagra* or *Euoenothera*). This is the group upon which de Vries worked and it is upon this subgenus that we will focus most of our attention in this monograph. de Vries had long been interested in the problem of the origin of species. He accepted in general the concept of organic evolution as expounded by Darwin, but was dissatisfied with two aspects of Darwin's hypothesis. In the first place, he disliked the assumption that evolutionary steps are so minute that they cannot be observed: he reacted against a theory that could not be subjected to direct experimentation. To quote his own words in the preface to the first volume of his Mutation Theory (the English translation, p. viii, 1909):

"The origin of species has so far been the object of comparative study only. It is generally believed that this highly important phenomenon does not lend itself to direct observation, and, much less, to experimental investigation. This belief has its root in . . . the opinion that the species of animals and plants have originated by imperceptible gradations. These changes are indeed believed to be so slow that the life of a man is not long enough to enable him to witness the origin of a new form. The object of the present book is to show that species arise by saltations and that the individual saltations are occurrences which can be observed like any other physiological process."

The second aspect of Darwin's work that de Vries found unsatisfactory was his theory of pangenesis. de Vries was willing to accept the concept of discrete and separable particles, each responsible for a particular character or quality, but he disagreed with the idea that gemmules are passed from the various cells and tissues of the body to the germ cells, thus determining the traits to be passed on to the next generation. This concept carried with it the implication that modifications in the body would result in the production of modified gemmules and so bring about the inheritance of acquired characters. de Vries was unable to accept such an implication.

He was searching, therefore, on the one hand for concrete and visible evidence of evolutionary change, and on the other hand for an explanation of the mechanism whereby such changes are brought about. In the evening primrose, upon which he began work in 1886, he thought he found evidence for detectable modifications of evolutionary significance. About the same time, he developed a modification of Darwin's theory of pangenesis, which he called the theory of Intracellular Pangenesis, a theory designed to explain how characteristics old and new are determined and transmitted from generation to generation. Since he interpreted his experimental findings in terms of this theory, it will be necessary for us to outline the theory first, in order to follow his reasoning as he undertook experiments with *Oenothera*.

1. The Theory of Intracellular Pangenesis

This theory was in some respects a forerunner of the modern gene theory, in that it visualized unit characters as being dependent upon separable and independent entities, the pangenes. He visualized them, however, as being present in both nucleus and cytoplasm. Those in the nucleus are capable of multiplication and transmission from one cell generation to the next and from parents to offspring. They are present also in the cytoplasm. As long as they reside in the nucleus they remain in passive condition, i.e., not influencing the characters for which they are responsible. Those capable of activity in a given individual or cell will take on such activity only after they have passed into the cytoplasm, where they retain their ability to multiply. The living part of the cytoplasm is composed entirely of active pangenes. Only those pangenes necessary to the activity of a particular cell are transmitted from the nucleus to the cytoplasm of that cell. Pangenes that are not needed in a particular type of cell remain latent in the nucleus, all nuclei of all cells of an individual having all the various kinds of pangenes needed for the whole organism, but the cytoplasm of a given cell having only those pangenes necessary for the life of that particular cell.

In his original treatise (1889) de Vries described pangenes as existing in one of two conditions—active and inactive (latent), active pangenes being found only in the cytoplasm, inactive pangenes being of two kinds, those that are inactive merely by reason of being in the nucleus and those that are incapable of activity in the species, race or individual to which they belong. Active pangenes can mutate to the inactive condition or vice versa.

Later, as a result of his breeding experiments with *Oenothera* and other genera, he expanded his theory to include additional categories, as follows (Muta. Theory II, p. 371 *et seq.*):
Semi-active pangenes are present when two clusters of dissimilar but homologous pangenes compete on more or less equal terms, so that each group is able to obtain the mastery in a considerable proportion of the

cases. This is the situation found in "ever-sporting" varieties which de Vries called "Halbrassen".

Semilatent pangenes are those that under similar circumstances are able to express themselves only rarely. Races showing this situation de Vries termed "Mittelrassen".

These two types of pangenes were later included in the category of "labile pangenes" (de Vries, 1913, p. 14, 282 *et seq.*). Labile pangenes were considered (p. 345) to be indistinguishable from active pangenes so far as their phenotypic expression is concerned, but to differ from the latter in being unstable, easily able to mutate to the inactive or active (stable) condition. The labile condition of a pangene was considered to be the result of a "premutation" by which a stable active or inactive pangene has become unstable.

In his "Gruppenweise Artbilding" (1913, p. 15) de Vries emphasized the idea that pangenes exist in groups or clusters, and when a labile pangene mutates, it is liable to cause associated pangenes to change also. Thus a mutant individual may be altered in respect to more than one character.

From this cursory survey of the concept of Intracellular Pangenesis it is seen that de Vries' ideas, developed long before he knew about Mendel or of the role of chromosomes in inheritance, approached the modern concept of the gene. They differed from the latter principally in three respects: (1) in his introduction of the idea of labile pangenes, upon which he based his explanation of the origin and behavior of the so-called mutants produced by *lamarckiana*; (2) in his assumption that pangenes may exist in varying numbers in the nuclei—the more numerous a given pangene is, the more strongly will the character determined by it be expressed, and vice versa; (3) in the fact that he did not clearly relate the pangenes to the chromosomes, and consequently did not visualize them as being present in pairs in somatic cells, separating at germ cell formation. Even as late as 1913, the year in which his "Gruppenweise Artbildung" appeared, he did not associate the pangenes with the chromosomes, except in a very vague and tentative manner, although by this time the role of the chromosomes as the carriers of the genes was clearly understood. Genetic linkage and independent assortment based on chance segregation of distinct pairs of chromosomes played no role in his thinking.

2. DISCOVERY OF OENOTHERA AND ITS MUTANTS

Let us now turn to the discoveries that seemed to de Vries to support the possibility that evolution can occur in detectable steps. His search for such evidence was apparently rewarded when in 1886 he found in an abandoned potato field at Hilversum near Amsterdam a large assemblage of a race that he identified as *O. lamarckiana* Ser. (Fig. 1.1). The population was quite uniform on the whole, but here and there variants were

FIG. 1.1. *O. lamarckiana* of de Vries. Flowers usually have four stigmatic lobes, but an occasional flower will have an odd number.

present that he thought might be of evolutionary significance. He analyzed the members of this colony with great thoroughness and collected seed from both the normal and the various types of aberrant individuals. These he planted in his experimental garden and for many years grew large colonies of the lines thus started, together with their hybrids. He also grew a number of other species obtained from other sources. At the same time, he followed the fate of the population at Hilversum, collecting additional seed or rosettes from year to year. In the second year (1887) he discovered in the field two small colonies that showed much more striking deviations from the normal than any he had observed in the first year. The two plants comprising one of these had styles that were much reduced in length, and stigmas that were flattened and leaf-like. He called this form *brevistylis*. The other colony had smoother foliage and smaller flowers than typical *lamarckiana* but otherwise was very similar. This he called *laevifolia*. *Laevifolia* bred true to type, but *brevistylis*, because it was female-sterile, had to be crossed to *lamarckiana* and recovered in later generations. Curiously enough, these two mutants never occurred again in his cultures, although most of his other mutant forms appeared on several, or even on many occasions.

It was not long before other aberrant forms were found. In his second garden generation, *nanella*, a dwarf form, and *lata* with short, very crinkled, obtuse leaves, short stature, and stout buds appeared, and these cropped up almost yearly in subsequent generations. *Lata* also appeared at Hilversum in 1888. The third garden generation yielded a third mutant—*rubrinervis*, a brittle-stemmed plant, and this continued to appear in most subsequent generations. In the fourth cultivated generation a burst of new forms was observed. In addition to those that had appeared previously, a single plant of *gigas* was produced, a form which never again appeared in de Vries' cultures of *lamarckiana*. In addition, three other new mutants appeared: *albida*, a weak form; *oblonga*, with thick, narrow, characteristically oblong leaves; and *scintillans* with smooth, shiny, narrow leaves. These were the principal forms that made their appearance prior to 1900. Subsequent to that date, other mutants were found, the number ultimately rising to more than sixty. A breakdown of the forms appearing in the experimental garden through the first eight generations is shown in Table 1. I.

The continual appearance of aberrants in cultures of *O. lamarckiana* furnished the principal evidence upon which de Vries based his Mutation Theory of Evolution. But such behavior required explanation, and for this he fell back upon his theory of Intracellular Pangenesis. He postulated that a species may from time to time enter upon a mutable period and concluded that *O. lamarckiana* is now passing through such a period.

TABLE 1.I

Occurrence in the experimental garden of mutants of *O. lamarckiana* prior to 1900
(after de Vries, 1901)

	gigas	albida	oblonga	rubrinervis	lamarckiana	nanella	lata	scintillans
1886–1887					9			
1888–1889					15,000	5	5	
1890–1891				1	10,000	3	3	
1895	1	15	176	8	14,000	60	73	1
1896		25	135	20	8,000	49	142	6
1897		11	29	3	1,800	9	5	1
1898			9		3,000	11		
1899		5	1		1,700	21	1	

When in such a state, an individual pangene may suffer what he termed a "premutation", i.e., from having been a stable active or inactive pangene it has become labile (de Vries, 1913, p. 335). It is then easily influenced by the environment to undergo true mutation. *O. lamarckiana* has accumulated a number of labile pangenes during the mutable period through which it is passing.

If a pangene that has been active becomes labile, the phenotype is not thereby altered, but in its labile condition it may mutate to the inactive condition, in which case the character which it controls will be lost or modified. Such a change de Vries termed a *retrogressive* mutation. For example, *brevistylis* has lost its ability to produce long, fully functional styles, *rubrinervis* has lost much of the lignin in its bast fibers so that its stems have now become brittle, and in *nanella* a pangene for height has become inactive.

If an inactive pangene has become labile, a character that has long remained latent now appears, but in an unstable state. It may mutate either to the active or the inactive condition; if it mutates to the active condition, the atavistic character becomes fixed, if to the inactive condition it disappears again. The apperance of an atavistic character de Vries called a *degressive* mutation. Examples of degressive mutants were *lata*, *scintillans*, *oblonga* and *albida*. These were unstable, and reverted back to the inactive condition in a percentage of their progeny.

A third kind of mutation was found in one case (that of *gigas*). This was called a *progressive* mutation, and was ascribed to the creation of a new pangene or group of pangenes, which were added to the normal complement. de Vries regarded *gigas* as a new elementary species. The other mutants were considered to be at the varietal, rather than the specific level.

In Chapter 15 we will discuss more fully the nature of de Vries' mutants.

CHAPTER 2

de Vries' Early Genetical Work

In order to understand the process by which mutations appear and persist, it was necessary for de Vries to become acquainted with the mechanism by which hereditary characters are normally transmitted from parents to offspring. He approached this problem both from a theoretical and an experimental point of view. His theory of Intracellular Pangenesis was, however, not formulated on the basis of his experimental findings. Rather, it was published (1889) before his breeding work had advanced beyond the incipient stage (*Oenothera* is not mentioned in this publication), and a major objective of his extensive experimental work became to interpret his findings in terms of this theory. This prevented him from giving proper consideration, even after Mendel's work was rediscovered, to the newly emerging knowledge of the relation between chromosomes and genes.

de Vries used a number of genera in his breeding work, including *Agrostemma, Chelidonium, Coreopsis, Datura, Hyoscyamus, Zea, Solanum* and others, as well as a number of species or races of *Oenothera*. Most of these genera agreed in showing the type of behavior that we now know as Mendelian; only *Oenothera* stood out as an exception to the rule. Except for one character (*brevistylis*), *Oenothera* behaved differently from the rest. He thus obtained in a number of organisms the same results that Mendel had found earlier, without knowing of Mendel's work, but he also found a type of behavior in *Oenothera* that did not fit this scheme. Instead of pure-bred races giving uniform F_1s and splitting F_2s, true-breeding races or species of *Oenothera*, when crossed, often produced splitting progenies in F_1, and the hybrids thus formed often bred true—a result quite the opposite of that obtained with other genera. For example, (de Vries, 1913, pp. 122, 126), *biennis* × *lamarckiana* gives two* classes of F_1, and each class breeds true in succeeding generations. de Vries, however, did not interpret the *Oenothera* behavior as exceptional, although he found it in only one genus. Because evolution was de Vries' primary interest, and he believed that he had found in this genus a clue to the way in which species originate, *Oenothera* took on special importance for him. Consequently he gave special weight to this

---

* Renner (1917c, p. 287) found that this cross yields three classes of F_1 all of which breed true in most of their characters.

genus and concluded that the type of behavior that it showed was of equal significance with that found in organisms that we would call Mendelian. He suggested (1900e) that crosses which behave in Mendelian fashion ("erbgleichen Kreuzungen") involve hereditary units of "equal value", resulting in what he called "true" or "isogonous" hybrids; in *Oenothera*, however, crosses ("erbungleichen Kreuzungen") may involve hereditary units that are quite dissimilar in quality and as a result either fail to segregate, or separate according to other than Mendelian rules, giving rise to "false" (using Millardet's term) or "anisogonous" hybrids. He was inclined to consider the two systems of comparable importance in nature. He states (1900e, p. 437), "Es liegt vorlaufig kein Grund vor, anzunehmen, dass die erbungleichen Kreuzungen in Pflanzenreiche seltener sein wurden als der erbgleichen . . . Nach meinen bisherigen Erfahrungen sind die ersteren jedenfalls nicht weniger zahlreich als die letzteren." In this respect he differed from Correns who regarded "false hybrids" as exceptions to the general rule.

If de Vries had found uniformity of behavior as between the various genera that he studied he would no doubt have published his results before 1900. As it was, only a single paper relating to these researches was published during the 1890's (1895)—an account of his discovery and early study of *O. lamarckiana* and its derivatives. de Vries was in the position that Mendel would have been in, had the latter completed his study of *Hieracium* or of honeybees before he had published on peas. In a private conversation in 1928, de Vries told me that his failure to publish his findings prior to 1900 was because he obtained discordant results, *Oenothera* contrasting with the other plants he had studied; and he was endeavoring to understand more fully the reason for this discrepancy before publishing. His discovery in 1900 of Mendel's paper, however, stimulated him to begin the presentation of his results, and six papers appeared in that year.

There has been some criticism of the fact that the first of these papers to appear in print, although it outlined the essentials of Mendelian behavior, did not mention Mendel, reference to whom was made only in the second paper. Examination of the dates of submission of his several manuscripts, however, shows that Mendel is mentioned in the first two papers which he submitted for publication, but not in the third. The first paper to appear in print was actually the third that he submitted. In order of submission, the papers were as follows:

1. Das Spaltungsgesetz der Bastarde. *Ber. deut. bot. Ges.* **18**, 83–90. Received March 14, 1900.* Calls attention to Mendel's paper.

---

* In the collected works of de Vries (Opera e Periodicis Collata, Vol. VI, p. 208), the date of receipt was omitted although it appeared in the original.

2. Sur les unitès des caractères spécifiques et leur application a l'étude des hybrides. *Rev. gen. Bot.* **12**, 257–271. Submitted March 19, 1900. Mentions Mendel's work.
3. Sur la loi de disjonction des hybrides. *C. R. Acad. Sci. Hebd.*, *Paris*, **130**, 845–847. Submitted March 26, 1900.

Of the other three papers that he published in 1900 dealing with *Oenothera*, two dealt wholly with *lamarckiana* and its "mutants" and were not concerned with breeding behavior. The third, on "false hybrids", referred to Mendel and emphasized the generality of what he called the Mendelian law. It was submitted for publication on November 21.

We may now outline de Vries' major experimental findings with regard to the hereditary behavior of the races of *Oenothera*, results obtained prior to the beginning of Renner's work, and show how he interpreted these results in terms of the theory of Intracellular Pangenesis. Later we shall see how some of these interpretations were modified, partly as a result of Renner's findings.

It should be mentioned parenthetically that de Vries performed an incredible amount of work and amassed a mountain of data on *Oenothera* behavior, the accuracy of which has not been questioned. de Vries was more than an armchair philosopher, and irrespective of the validity of his theoretical conclusions, one of his chief contributions was to emphasize the importance of the experimental approach at a time when there was too much tendency toward *a priori* reasoning. It is true that he developed his theory first and tried to interpret his experimental results in terms of the theory. Unlike many others, however, he was tireless in the accumulation of vast quantities of empirical data.

As we have seen, de Vries classified his mutants on the basis of their mode of origin into progressive, retrogressive and degressive mutants. In addition to this system, however, he developed two others which included both races and mutants and were based on hereditary behavior rather than on mode of origin (Table 2.1).

1. The first of these systems of classification (de Vries, 1913, p. 30) was based upon whether the offspring of reciprocal crosses were phenotypically alike or different: de Vries found that his races tended to fall into two groups in this respect, though some showed certain degrees of intermediacy.

(A) Those in the first group he called *isogamous* races; these were the races which tended to produce similar or identical reciprocals. Among these, he recognized two categories: (1) one produced reciprocals that were not only alike, but also uniform (e.g., *O. hookeri*); (2) the second yielded reciprocal progeny that were alike, but each contained two types, which he called twin hybrids, the twins being similar in the reciprocals. *O. lamarckiana* was the first and chief example of this type of behavior.

TABLE 2.I

Crosses upon which de Vries based his conclusions with regard to the nature of laetas and velutinas. In each case, de Vries' interpretation is given, followed by interpretation in terms of Renner's theory of complexes, showing that the latter theory is able to explain all of de Vries' results without reference to the condition of the pangenes

Crosses using laetas as female		Progeny	
(*hookeri* × *lamarckiana*) *laeta* × *lamarckiana*	→ *laeta*	*velutina*
(^{h}hookeri·gaudens) (velans·gaudens)	(^{h}hookeri·gaudens) (velans·gaudens)	(^{h}hookeri·velans)
(*hookeri* × *lamarckiana*) *laeta* × *hookeri*	→ *laeta*	*velutina*
(^{h}hookeri·gaudens) (^{h}hookeri·^{h}hookeri)	(^{h}hookeri·gaudens)	(^{h}hookeri·^{h}hookeri)
(*hookeri* × *lamarckiana*) *laeta* × (*hookeri* × *lamarckiana*) *velutina*	→ *laeta*	*velutina*
(^{h}hookeri·gaudens) (^{h}hookeri·velans)	(^{h}hookeri·gaudens) (velans·gaudens)	(^{h}hookeri·^{h}hookeri) (^{h}hookeri·velans)
(*biennis* × *lamarckiana*) *laeta* × *lamarckiana*	→ *laeta*	*velutina*
(albicans·gaudens) (velans·gaudens)	(albicans·gaudens) (gaudens·velans)	(albicans·velans)
(*muricata* × *lamarckiana*) *laeta* × *lamarckiana*	→ *laeta*	*velutina*
(rigens·gaudens) (velans·gaudens)	(rigens·gaudens) (gaudens·velans)	(rigens·velans)
(*biennis* × *lamarckiana*) *laeta* × *hookeri*	→ *laeta*	*velutina*
(albicans·gaudens) (^{h}hookeri·^{h}hookeri)	(gaudens·^{h}hookeri)	(albicans·^{h}hookeri)
(*muricata* × *lamarckiana*) *laeta* × *hookeri*	→ *laeta*	*velutina*
(rigens·gaudens) (^{h}hookeri·^{h}hookeri)	(gaudens·^{h}hookeri)	(rigens·^{h}hookeri)
(*lamarckiana* × *chicaginensis*) *laeta* × *hookeri*	→ *laeta*	*velutina*
(gaudens·punctulans) (^{h}hookeri·^{h}hookeri)	(gaudens·^{h}hookeri)	(**punctulans** ·^{h}hookeri)
(*muricata* × *lamarckiana*) *laeta* × *chicaginensis*	→ *laeta*	*velutina*
(rigens·gaudens) (excellens·punctulans)	(gaudens·punctulans)	(rigens·punctulans)

Crosses using laetas as male		Progeny	
lamarckiana × (*hookeri* × *lamarckiana*) *laeta*	→ *laeta*	*velutina*
(velans·gaudens) (^{h}hookeri·gaudens)	(gaudens·^{h}hookeri) (velans·gaudens)	(velans·^{h}hookeri)
biennis × (*biennis* × *lamarckiana*) *laeta*	→ *laeta*	—
(albicans·rubens) (albicans·gaudens)	(albicans·gaudens)	
hookeri × (*hookeri* × *lamarckiana*) *laeta*	→ *laeta*	*velutina*
(^{h}hookeri·^{h}hookeri) (^{h}hookeri·gaudens)	(^{h}hookeri·gaudens)	(^{h}hookeri·^{h}hookeri)
(*hookeri* × *lamarckiana*) *velutina* × (*hookeri* × *lamarckiana*) *laeta*	→ *laeta*	*velutina*
(^{h}hookeri·velans) (^{h}hookeri·gaudens)	(^{h}hookeri·gaudens) (velans·gaudens)	(^{h}hookeri·^{h}hookeri) (velans·^{h}hookeri)

Crosses using velutinas as female		Progeny	
(*hookeri* × *lamarckiana*) *velutina* × *lamarckiana*	→ *laeta*	*velutina*
(^{h}hookeri·velans) (velans·gaudens)	(^{h}hookeri·gaudens) (velans·gaudens)	(^{h}hookeri·velans)
(*hookeri* × *lamarckiana*) *velutina* × *hookeri*	→ —	*velutina*
(^{h}hookeri·velans) (^{h}hookeri·^{h}hookeri)		(^{h}hookeri·^{h}hookeri) (velans·^{h}hookeri)
(*biennis* × *lamarckiana*) *velutina* × *lamarckiana*	→ *laeta*	*velutina*
(albicans·velans) (gaudens·velans)	(albicans·gaudens) (velans·gaudens)	(albicans·velans)
(*muricata* × *lamarckiana*) *velutina* × *lamarckiana*	→ *laeta*	*velutina*
(rigens·velans) (velans·gaudens)	(velans·gaudens) (rigens·gaudens)	(rigens·velans)
(*biennis* × *lamarckiana*) *velutina* × *chicaginensis*	→ —	*velutina*
(albicans·velans) (excellens·punctulans)		(albicans·punctulans) (velans·punctulans)
(*muricata* × *lamarckiana*) *velutina* × *chicaginensis*	→ —	*velutina*
(rigens·velans) (excellens·punctulans)		(rigens·punctulans) (velans·punctulans)

TABLE 2.I—*continued*

Crosses using velutinas as female			Progeny
(*biennis* × *lamarckiana*) *velutina* × *hookeri*		→ —	*velutina*
(albicans · velans)	(^{h}hookeri · ^{h}hookeri)		(albicans · ^{h}hookeri)
			(velans · ^{h}hookeri)
(*muricata* × *lamarckiana*) *velutina* × *hookeri*		→ —	*velutina*
(rigens · velans)	(^{h}hookeri · ^{h}hookeri)		(rigens · ^{h}hookeri)
			(velans · ^{h}hookeri)
(*lamarckiana* × *chicaginensis*) *velutina* × *hookeri*		→ —	*velutina*
(velans · punctulans)	(^{h}hookeri · ^{h}hookeri)		(velans · ^{h}hookeri)
			(punctulans · ^{h}hookeri)

Crosses using velutinas as male			Progeny
lamarckiana × (*hookeri* × *lamarckiana*) *velutina*		→ *laeta*	*velutina*
(velans · gaudens)	(^{h}hookeri · velans)	(gaudens · velans)	(velans · ^{h}hookeri)
		(gaudens · ^{h}hookeri)	
biennis × (*biennis* × *lamarckiana* *velutina*		→ —	*velutina*
(albicans · rubens)	(albicans · velans)		(albican · velans)
muricata × (*muricata* × *lamarckiana*) *velutina*		→ —	*velutina*
(rigens · curvans)	(rigens · velans)		(rigens · velans)

^{a} ^{h}*hookeri* is an alethal complex and can therefore exist in homozygous condition. *Velans* and *gaudens* can be transmitted through either sperm or egg. *Albicans* and *rigens* are transmitted only through the egg. *Punctulans* is transmitted occasionally through the egg, but mostly through the sperm. *Excellens* is almost exclusively transmitted through the egg.

^{b} The segregation of complexes in certain hybrids may not be entirely clear-cut, since these hybrids may possess more than one independent chromosome group. For example, *rigens · gaudens* has ☉6, 4 pairs and may, therefore, produce gametes that contain mixtures of *rigens* and *gaudens* genes.

(B) The second group was made up of the so-called *heterogamous* races, which produced different (often radically different) phenotypes when crossed reciprocally. Examples were *biennis, muricata* and *biennis Chicago.*

2. His second system of classification was based, not on a comparison between reciprocal F_1 progenies, but rather on the genetic behavior of individual hybrid populations—whether splitting occurred in the F_1 or not until the F_2 or whether hybrid classes bred true. We will discuss the first classification at this point, then follow with a consideration of the second.

1. First System of Classification
A. *Isogamous Races*

(1) The first category of isogamous races included those whose reciprocal progenies were not only alike, but also uniform. The races that de Vries thought belonged in this category were *hookeri, cockerelli* and his strain of *strigosa*. He was correct in considering *hookeri* an isogamous race, as defined by him. It is essentially homozygous and therefore transmits the same genome through sperm and egg. He also included

cockerelli and *strigosa* in this category in spite of the fact that their reciprocals showed slight phenotypic differences (de Vries, 1913, pp. 32, 59, 60). However, these two races are in fact both strictly heterogamous, transmitting different genomes when used as male and female. The hybrids which they produce reciprocally, however, are quite similar phenotypically, and de Vries considered the differences of little importance.

FIG. 2.1. *O. lamarckiana* × *O. hookeri* showing twin hybrids (*laeta* to the left, *velutina* to the right).

(2) The second category of isogamous forms included those that produced twin hybrids when outcrossed. The outstanding example was *O. lamarckiana*. Although this race breeds true when selfed, apart from an occasional mutant, it produces two kinds of offspring when crossed in either direction to other forms. To these twin types he applied the terms *"laeta"* and *"velutina"* (Fig. 2.1) (de Vries 1907, 1908).* Laetas have broad, thin, strongly crinkled, clear green foliage, reduced pubescence and relatively slender stems; velutinas have narrower, thicker, less crinkled, greyish-green foliage, heavier velutinous pubescence, and stouter, more woody stems (de Vries, 1913; Figs. 46, 48). All laetas look very much alike, no matter what race has been crossed with *lamarckiana*

* When crossed with certain races, the corresponding twins were *"laxa"* and *"densa"*.

and they resemble *lamarckiana*. Velutinas, however, vary in appearance much more from hybrid to hybrid, showing more influence of the other parent. Laetas and velutinas are as a rule quite unlike phenotypically, and are easily distinguished.

de Vries' explanation for the occurrence of twins was as follows (1913, p. 133): Two things are necessary for their appearance: (a) one parent (*lamarckiana*) has its *laeta* pangene in the labile condition; (b) the other parent must have the ability to induce this labile pangene to mutate to the inactive or *velutina* condition. F_1 plants that show the *laeta* characters have an unmutated pangene; those that have the *velutina* phenotype possess a mutated *laeta* pangene.

de Vries arrived at conclusion (b) because of the fact that one cross (*lamarckiana* × *biennis*) failed to yield twins, and gave only laetas (1913, p. 157). From this he reasoned that *biennis* differs from other species in not being able to bring about the mutation of the *laeta* pangene. He did not know, however, what Renner later showed, that *biennis* produces only one class of functional pollen, which carries a genome very similar to the *laeta*-producing genome of *lamarckiana*—in fact, it has the same zygotic lethal, so that it has to combine with the *velutina*-producing genome, over which it dominates to such a degree that the hybrid is phenotypically a *laeta*.

There was however, an obvious inconsistency in de Vries' reasoning in regard to the origin of twins. It is clear that a pangene from one parent could not be induced to mutate by the other parent prior to fertilization—such a mutation would have to occur in the zygote. de Vries emphasized on more than one occasion, however, his belief that mutations occur before, and not during or subsequent to fertilization (1913, pp. 308, 317). He could have explained twinning by assuming that mutation occurs prior to gamete formation, thus producing two kinds of gametes in *lamarckiana*. What prevented his adoption of this explanation was the fact that when *lamarckiana* is pollinated by *biennis* only laetas are produced.

de Vries also stated that the *laeta* pangene, when it mutates, commonly causes associated pangenes to mutate along with it to the inactive condition (see above, p. 5), so that more than one character is affected by the change, the total result being the production of a *velutina* hybrid.

de Vries concluded that a labile pangene would not necessarily mutate in every case where it came in contact with an inducing genome. Mutation occurred in only a proportion (often approaching 50%) of the zygotes. Mutated zygotes would then develop into velutinas, whereas those in which mutation failed to occur would develop into laetas.

A further conclusion that de Vries reached in order to explain the behavior of twin hybrids of *lamarckiana* seems quite bizarre. In those zygotes that do not mutate and thus develop into laetas, he suggested that the labile *laeta* pangene remains in the labile condition in the eggs, but is converted to the active condition in the pollen (de Vries, 1913, pp. 138–139). This conversion may be thought of as a reverse mutation, a return to the condition before the original "premutation" occurred that gave to *lamarckiana* a labile *laeta* pangene. This reverse mutation, however, affects only the pollen, so that a chimeral situation with regard to stability of the pangene is set up. This condition is of course not reflected in the phenotypes since labile pangenes do not differ from active ones in their phenotypic effect. The mechanism by which such a differentiation could be effected was not discussed. The assumption of such a differentiation was made in order to account for the behavior in outcrosses of certain laetas (de Vries, 1913, pp. 135, 139, 162). de Vries found that when these *laeta* hybrids are used as female in crosses, they behave as does *lamarckiana* itself, i.e., the next generation includes both laetas and velutinas; on the other hand, if they are used as the male parent in crosses to the same races, all the offspring are laetas, indicating to him that the *laeta* pangene transmitted through the pollen is now stable and not capable of mutation under the influence of the other parent, though in the eggs it is still labile.*

The situation in the velutinas, according to de Vries, is different. Since the *laeta* pangene of *lamarckiana* has become inactive in the velutinas, both in egg and sperm, the progeny of a *velutina* twin will be wholly *velutina* when it is outcrossed in either direction.

When selfed, laetas and velutinas ordinarily breed true. This is to be expected according to de Vries' formulation. Since labile and active pangenes give identical phenotypic results, and since laetas have labile *laeta* pangenes in the egg and active pangenes in the pollen, he would expect all individuals of the next generation following selfing of a *laeta* to have both labile and active pangenes and these should be typical *laeta*. Velutinas will breed true because they have only *velutina*, i.e., inactivated *laeta* pangenes (1913, p. 125 *et seq.*).

de Vries found one exception, however, to the rule of true breeding, so far as the laetas were concerned. The *laeta* formed by crossing *lamarckiana* with *hookeri* (in either direction) did not breed true, but gave a

---

* This is not true, however, in all cases. For example, although (*biennis* × *lamarckiana*) *laeta* × *lamarckiana* gave both laetas and velutinas (1913, p. 135), the cross (*lamarckiana* × *biennis*) *laeta* × *lamarckiana* yielded laetas only (1913, p. 162). In the other direction, *biennis* × (*biennis* × *lamarckiana*) *laeta* produced only laetas (1913, p. 138), but if he had made the cross *biennis* × (*lamarckiana* × *biennis*) *laeta*, he would have obtained both laetas and velutinas.

splitting progeny when selfed, composed, according to his interpretation, of laetas and velutinas. The velutinas thus produced bred true, but in the F_2 and later generations *laeta* again produced progenies composed of laetas and velutinas. de Vries explained this situation by assuming that the *laeta* derived by crossing with *hookeri* contained both labile and inactive *laeta* pangenes, which segregated in Mendelian fashion, a very different situation from that found in other laetas, where the eggs possess only labile, the pollen only active *laeta* pangenes. He did not realize that his velutinas in this case were pure extracted *hookeri*.

When de Vries crossed *lamarckiana* as male to certain races, as for example *biennis Chicago* (= *chicaginensis*), the twins were sufficiently different in phenotype from *laeta* and *velutina* so that he gave them different names, *laxa* and *densa* (de Vries and Bartlett, 1912). These twins, however, showed a behavior somewhat comparable to that of laetas and velutinas. *Densa* bred true as did the velutinas. *Laxa* used as female in crosses produced twins, but these were not *laxa* and *densa*, but rather *laxa* and a different, true-breeding form which he called *atra*. de Vries originally considered *laxa* and *densa* to be based upon a different set of pangenes from the ones responsible for the *laeta-velutina* split. Later (1918a), when he came to accept the idea that *lamarckiana* regularly produces two kinds of gamete, he concluded that *densa* and *laxa* differ from *laeta* and *velutina* because of the characters supplied by the other parent and not because of differences in the contributions made by *lamarckiana*.

In concluding this discussion of de Vries' work on isogamous forms, it should be added that the work of Renner has later shown that there is no foundation for the conclusion that a *laeta* pangene is active in the pollen, but labile in the eggs of laetas, and that it is inactive in both egg and pollen in the velutinas. de Vries thought that all broad-leaved hybrids resulting from a given cross of *lamarckiana* belonged in a single class (the laetas), the same being true of the narrow-leaved hybrids (the velutinas). Renner's work, however, showed that two classes of broad-and/or narrow-leaved plants were present following certain crosses, their occurrence being readily explained on the basis of his theory of complexes (see Chapter 3). Table 2.1 lists a number of crosses upon which de Vries based his conclusions with regard to the nature of laetas and velutinas (see de Vries, 1913, pp. 133–144). In each case de Vries' interpretation is followed in the table by one based on Renner's analysis. After the reader has become familiar with Renner's findings, he will understand the nature of the results that de Vries obtained, and will realize that these results have another basis than the presence of active, inactive or labile pangenes.

B. Heterogamous Races

These were defined as races which tended to produce uniform off-spring in a given cross, but very different types of offspring in reciprocal crosses. They included certain of the "older" species, as de Vries called them, such as *muricata* and *biennis*. (The "younger species" included the various mutant forms derived from *lamarckiana*.) Other races with which he worked should also have been placed in this group, such as *cockerelli* and *strigosa* which he considered to be isogamous.

de Vries found, however, that the distinction between isogamous and heterogamous races was not always clear-cut. Some races, such as *biennis Chicago* (1913, p. 32) behaved in some crosses as though they were isogamous, in others as though they were heterogamous. In other cases, hybrid progenies were not entirely uniform, but contained exceptional individuals which he called metaclines and which he ascribed to mutated germ cells. These phenomena are readily understood on the basis of Renner's theory of complexes.

de Vries made crosses in all possible combinations between certain heterogamous races and found that the offspring followed quite definite rules (1911). These can be illustrated by the case of *biennis* (B) and *muricata* (M), as shown by the following formulae:

reciprocal crosses
   biennis × *muricata* differs phenotypically from *muricata* × *biennis*,
   i.e., BM derived from *biennis* × *muricata* is different from MB derived
   from the reciprocal.
double reciprocal crosses
   $(M \times B) \times (B \times M) = M$
   $(B \times M) \times (M \times B) = B$
   (progeny resembles species in the peripheral position in the formula)
sesquireciprocal crosses
   $B \times (M \times B) = B$
   $(M \times B) \times M = M$
   (progeny resembles species in peripheral position)
iterative crosses
   $(M \times B) \times B = MB$
   $(B \times M) \times M = BM$
   (progeny combines characters of species in peripheral position, but
   reciprocal combinations are different)

From these studies de Vries pointed out (1913, p. 100) that such hybrids receive from the mother only those hereditary determiners or pangenes that the mother received from its mother, i.e., via the egg; from the father they receive only the inheritance that the father received through the sperm. Hybrids between heterogamous races do not receive inheritance from their mother's father or their father's mother.

de Vries early adopted an hypothesis (1903b) based on the conclusions of Van Beneden (1883) to account for the separation into different gametes of paternal and maternal hereditary factors, a phenomenon that characterizes isogamous as well as heterogamous races. He assumed that the pronuclei at the time of fertilization do not fuse completely but retain their identity throughout the life of the individual, finally separating in germ cell formation into different gametes. At meiosis, the two become very closely associated, so that it is possible, where Mendelian inheritance occurs, for exchanges of pangenes to take place between the two sets, resulting in the assortment required by Mendel's second law. This exchange, however, was visualized as a process similar to what we now know as crossing over, not related to the segregation of independent structural units such as chromosomes. In *Oenothera*, he thought that such exchanges occur only rarely, so that maternal and paternal sets of determiners are as a rule segregated from each other intact. To explain the absence of the paternal set of pangenes in the eggs and the maternal set in the sperm of heterogamous races, de Vries (1911) made two suggestions, but did not consider the experimental evidence sufficient to distinguish between them. (a) The maternal and paternal sets of pangenes separate into different germ cells, but the pollen grains that receive the maternal set degenerate, as do the eggs that acquire the paternal set. This is essentially the concept of balanced gametophytic lethals, though de Vries seems not to have visualized the situation in such concrete terms at this time. (b) The maternal and paternal sets of pangenes are both received into all reproductive cells, but the maternal set remains inactive in the pollen, as does the paternal set in the eggs. de Vries did not choose one of these alternatives rather than the other at this time. Later (1913, p. 32, footnote) he suggested that the pangenes transmitted by egg and sperm were identical, but were bound in such a way that they were free to function in only one sex—in other words these plants were isogamous so far as the actual pangenes were concerned, but hetero-gamous in their ability to transmit these characters. Still later, he modified this viewpoint, in the light of Renner's work, and postulated the presence of lethal factors (1924a) which caused degeneration of the maternal set in the pollen, resulting in empty pollen grains, and the failure of the paternal set to survive in the eggs. Although it might seem that the presence of two distinct sets of hereditary determiners, one derived from the female, the other from the male, would suggest heterozygosity, de Vries insisted that this situation arose, not as the result of hybridization, but by the process of mutation of pangenes in pure lines. It should be remembered that de Vries did not visualize a high degree of heterozygosity in the various races of *Oenothera*. Associated genomes differed in respect to only a few pangenes. Except for these, his races were considered to be pure lines.

de Vries further pointed out that the various heterogamous races differ among themselves in regard to the degree to which the male or female gametes transmit the visible characters of the plant. In the case of *biennis*, the pangenes determining the characteristics of the race are transmitted almost exclusively through the sperm ("patroclinous inheritance"); the eggs transmit characters that are scarcely discernible in the race. In *muricata*, on the other hand, racial characteristics seem to be divided between the paternal and maternal sets—some are transmitted through the egg, some through the sperm. de Vries found the various races to vary in this respect, running the gamut from cases where all visibly active pangenes are in one set, male or female, to cases where they are shared more or less equally by the paternal and maternal sets.

de Vries had little to say in regard to the reason why some races are heterogamous, some isogamous, and some show varying degrees of intermediacy. He seems, however, to have entertained the hypothesis that *lamarckiana*, an isogamous form, may have descended from an originally heterogamous stock, since some of its mutants behaved as heterogams: *lamarckiana*, he suggested, is carrying the trait for heterogamy in the latent or inactive condition (1913, p. 346).

2. Second System of Classification

Let us turn now to the second type of classification of races (and mutants) used by de Vries (1913, p. 283). He grouped oenotheras not only on the basis of isogamy *vs.* heterogamy, but also on the basis of whether hybridization was followed by splitting or not, and whether splitting occurred in the F_1 or F_2. We will consider these types of behavior and show how de Vries attempted to explain them in terms of his theory of Intracellular Pangenesis. First let us list these categories and then take up each type of behavior in turn.

A. Splitting Occurs Following Hybridization

(1) In the second generation. This occurs when inactive and active pangenes are brought into association.

(2) In the first generation. This occurs when one parent contributes an inactive, the other a labile pangene.

B. No splitting occurs following hybridization, i.e., the F_1 is uniform and breeds true

(1) When the pangenes that are combined are in the same condition (both are active, or inactive or labile).

(2) When a pangene from one parent finds no antagonist from the other—one parent contributes to the offspring a pangene or cluster of pangenes that is not represented in the other parent.

(3) When an active pangene has a labile pangene as its antagonist.

A.1. Splitting occurs in the second generation, the first being uniform

This is typical Mendelian behavior. As we have seen, de Vries called a hybrid containing active and inactive pangenes a "true" hybrid in contrast to other classes of hybrids to be mentioned below, which he classified as "false" hybrids. He found in the earlier years of his *Oenothera* work only one character that showed typical Mendelian behavior, namely the recessive trait *brevistylis*, in which the styles are much shortened and the stigmas more or less crippled. This was actually one of the first two aberrants found in 1887 at Hilversum (Mutationstheorie I, p. 224) and the only one that involved a floral character. de Vries interpreted typical Mendelian behavior as resulting when two lines are crossed, one of which has at one time suffered a retrogressive mutation whereby an active pangene (in this case, for development of the pistil) has become latent or inactive. A true hybrid, therefore, has alternative pangenes, one inactive, the other active (i.e., capable of becoming active in the cytoplasm of appropriate cells).

A.2. Splitting occurs in the first generation

This takes place when one parent contributes an inactive pangene, the other a labile pangene. Two groups of plants show this behavior: (a) an isogamous race such as *lamarckiana* produces twin hybrids when crossed with other species or with certain of its own mutants. In this case, one of its pangenes is in the labile condition, whereas the corresponding pangene in the plant to which it is crossed is in the inactive condition. (b) Certain mutants of *lamarckiana*, including *lata* and *scintillans*, give twins when crossed back to *lamarckiana*, not because they have inactive pangenes corresponding to a labile one in the latter, but because the reverse is true. These mutants have arisen according to de Vries by premutation of an inactive pangene to a labile condition. They are therefore degressive mutants. When crossed, not only to *lamarckiana*, but to other species, such as *biennis, hookeri* and *cockerelli*, they give splitting progenies in the F_1. In either case, the labile pangene in one parent is induced by the other parent to mutate in a proportion of the zygotes thus resulting in a splitting F_1.

B.1. No splitting occurs when hybridization brings together antagonistic pangenes that are in the same condition (both are active, or inactive, or labile)

One would naturally expect, of course, that active × active, or inactive × inactive, would result in uniform progenies, but it may seem surprising that labile × labile does not result in splitting. According to de Vries, however, labile + labile plants should not be regarded as hybrids but as genetically pure. A labile pangene behaves like an active gene except that it is in a condition where it is easily changed into the active or inactive state by the external or internal environment. In selfed line it will probably mutate more often than an active or an inactive pangene, but it requires an environmental stimulus to bring about such a mutation. According to de Vries, these stimuli are furnished by association with foreign pangenes which are in the inactive condition.

B.2. Splitting fails to occur when a pangene has no antagonist

This happens when one parent has pangenes that the other lacks, or when each parent lacks certain pangenes that the other possesses. In the resultant hybrid a given pangene will find nothing to oppose it. As a result, when germ cells are formed, all gametes will, according to de Vries' reasoning, receive this pangene and the plant will breed true when selfed. In reasoning thus de Vries was not thinking, of course, in Mendelian terms. On the latter basis, an unopposed pangene should go into half the gametes, the other half lacking it. de Vries, however, postulated (1903d, pp. 50, 51) that a pangene which has no antagonist will divide into two at the stage when antagonists become closely associated prior to germ cell formation, so that all gametes will receive this pangene, which will therefore be passed on to all of the progeny .de Vries explained all hybrids that are true-breeding and intermediate between their parents in terms of pangenes that have no antagonists (1913, pp. 283, 285).

A comparable situation was considered to be present in the case of the mutant *gigas* from *lamarckiana*, where de Vries believed that a pangene had been newly created over and above the normal complement, and hence lacked an antagonist. It represented a "progressive"mutation. one in which the total number of different kinds of pangene had been increased, thus producing a new elementary species. de Vries considered this of special importance in relation to the process of evolution. At first it was not known that *gigas* was a tetraploid. When this was discovered, de Vries rejected the idea that doubling of the chromosome number was the cause of the changed phenotype and behavior. Rather, he claimed that the increase in chromosome number was one of the results of the progressive mutation that produced *gigas* (1918c, p. 217).

B.3. It is possible for a hybrid to receive an active pangene from one parent and a labile pangene from the other

Since both would produce the same phenotypic effect, the hybrid would appear to breed true when selfed, although it would actually be transmitting two types of pangenes.

We can see from this brief résumé (see Table 2.II) that de Vries attempted to explain all types of genetical behavior, Mendelian or non-Mendelian, in terms of active, inactive and labile pangenes, the key to the genetical peculiarities and to the origin of his so-called mutants being found in the behavior of the labile pangenes.

In concluding this brief review of de Vries' early genetical studies, a few points should be emphasized or re-emphasized since they were

TABLE 2.II

de Vries' classification of oenotheras

On the basis of comparison of reciprocal hybrids following interspecific crossing	On the basis of breeding behavior of interspecific hybrids
A. Isogamous (reciprocal hybrids alike) 1. Hybrid progenies uniform 2. Hybrid progenies split into two types B. Heterogamous (reciprocals differ)	A. Splitting occurs following hybridization 1. In the second generation (Mendelian segregation) (active and inactive pangenes have become associated) 2. In the first generation (inactive and labile pangenes are contributed by the parents) B. No splitting follows hybridization (hybrids are uniform and breed true) 1. Associated pangenes are in the same condition (i.e., both active, or inactive or labile) 2. A pangene has no antagonist 3. Active and labile pangenes are associated

central to his total concept. In the first place, de Vries emphasized most strongly (1915) that *lamarckiana* and the other *Oenothera* species with which he worked were "pure species", and not, as Davis (1911, 1915, 1917), Heribert-Nilsson (1912, 1920b), Renner (1914) and others argued, of hybrid origin. To have admitted impurity would have been to throw doubt upon the evolutionary significance of his mutants. Even at a later date, when the work of Renner and his own experiments forced him to admit the formation of two genetically different classes of sperms or eggs in most of the races, he still clung to a belief in the essential purity of these forms. All genetical heterogeneity that he found within a race he ascribed to mutations. He repudiated the idea that it was the result of hybridization. When contrasting pangenes are present in a plant, e.g., *laeta* and *velutina* pangenes, one of these has arisen from the other by mutation. Furthermore, as we have pointed out (p. 5), de Vries considered that it was only in respect to one or at most a few pangenes that heterogeneity existed in a race. The pangenes for most characters were uniform; the races, therefore, apart from these few, were pure.

Secondly, de Vries supposed that species may occasionally pass through a mutable period, more or less transitory, during which labile pangenes appear as the result of "premutation". *O. lamarckiana* is passing through such a period at the present time. The other oenotheras with which de Vries dealt, however, were not thought to be in a mutable period, unless it might be that *biennis* and perhaps *suaveolens* were entering or leaving such a period, since these species showed a minor tendency to produce mutants, and therefore probably had a few labile pangenes. de Vries emphasized, however, that active or inactive pangenes are also capable of mutating, although such happenings will be relatively few and far between. *Lamarckiana*, then, differs from other races in being in a mutable period, and therefore in experiencing mutations at a relatively high rate, totaling as high as 1–2% or higher.* Hence de Vries' term "Gruppenweise Artbildung"—the appearance of mutants or new elementary species in groups or clusters.

Finally, it should be stressed that the findings and interpretations just reviewed belonged to the earlier period in de Vries' *Oenothera* studies. During this period, extending over more than 25 years, the concept of Intracellular Pangenesis so dominated his thinking that he made little attempt to relate his findings to the rapidly developing field of cytogenetics. In time, however, his ideas began to change in certain respects. The last paper to mention pangenes as such appeared in 1915, although

---

* Later (1929) de Vries found that under certain conditions mutations could occur with a frequency up to 8–10%. They occur much more frequently early in the season than later, and are more numerous under favorable growing conditions.

he still expressed support of his pangenesis theory as late as 1925 (1925b). After 1915, de Vries spoke only of factors and characters instead of pangenes. By 1917 he had accepted the concept that most oenotheras produce two kinds of gamete (two kinds of egg and sperm in isogamous forms, one kind of functional egg but a different kind of functional sperm in heterogamous races). He ascribed the presence of two kinds of gamete in each race to mutation that had occurred at some prior time in that race. In accepting the presence of two classes of gamete in forms such as *lamarckiana*, he removed the inconsistency in his earlier reasoning according to which twinning results when a genome derived from one parent causes a labile pangene from the other parent to mutate in the hybrid zygote, a process out of line with his general idea that mutations of labile pangenes occur prior to germ cell formation. According to his original hypothesis, each *velutina* individual in a *lamarckiana* cross represents a new and separate mutation of the labile *laeta* pangene, induced by the plant with which *lamarckiana* has been crossed. According to his modified view, *lamarckiana* has suffered at some time in the past a mutation of a *laeta* pangene into the *velutina* condition and so has carried both *laeta* and *velutina* pangenes ever since.

de Vries also adopted (1918a, b) the idea of lethal factors and even spoke of *lamarckiana* as being heterozygous for the lethals, although he held throughout his life to the belief that this heterozygosity had come about through mutation within the race rather than by crossing between races. He continued to think in terms of active, inactive and labile genes, even though he often substituted the terms dominant, recessive and mutable for the earlier adjectives. We will refer later (chapters 3, 15, 16) to de Vries' activities and ideas during the last twenty years of his life, when he had the benefit not only of Renner's work, but also of the increasing knowledge of chromosome behavior, in *Oenothera* as well as in other organisms.

CHAPTER 3

Renner and the Theory of Complexes

Let us now turn to Renner and his work on *Oenothera*. In one respect, the careers of Renner and de Vries were alike: de Vries had established an outstanding reputation as a physiologist before he turned to evolution and genetics; Renner was by training more a plant physiologist than a geneticist, and throughout his entire life carried on an active program and trained many students in this area, especially in the field of water relations in plants (Fig. 3.1).

FIG. 3.1. Otto Renner (1883–1960).

Renner's work on *Oenothera* began in 1913 while he was assistant professor of Plant Physiology and Pharmacognosy at the University of Munich. Richard Goldschmidt, then in the zoological faculty at Munich, published a paper in 1912 in which he attempted to explain certain cases of apparent patroclinous inheritance in *Oenothera* in terms of merogony, a situation in which the sperm nucleus replaces the egg nucleus in the zygote. Goldschmidt thought that he had found cytological evidence to support this hypothesis. This led Renner to make a careful check, not only of Goldschmidt's material, but also of freshly prepared material of his own. He found Goldschmidt's conclusion to be erroneous, and in 1913 published his first *Oenothera* paper, indicating that the cytological behavior in the zygotes of the material in question was entirely normal. This meant that the apparent patroclinous inheritance must have some other explanation, and Renner became interested in discovering the causes of this behavior and of other peculiarities of *Oenothera* genetics.

Almost at once, Renner uncovered a fact that gave him a clue to the true situation (1914, 1917a). He found that if *lamarckiana* is selfed, half the seeds are sterile. *Lamarckiana* had been shown by de Vries to produce two kinds of offspring (*laeta* and *velutina*) when outcrossed: on the other hand, except for an occasional "mutant", it bred true when selfed. Renner postulated that the sterile seeds represent *laeta·laeta* and *velutina·velutina* combinations that are apparently inviable. In contrast, *biennis* and *muricata* produce almost no sterile seeds. These are heterogamous races according to de Vries, which produce one kind of egg, but a different kind of sperm. These races also breed true when selfed, but not because homozygotes are eliminated. With but one kind of egg and another kind of sperm, the only progeny possible after selfing are heterozygotes—resulting from re-union of the two kinds of gamete. All zygotes in heterogamous races, therefore, should be viable and there should be no sterile seeds, as was shown by Renner to be the case.

The following tests carried out by Renner supported the interpretation that sterile seeds represent inviable genome-combinations (1917a).

(1) If *lamarckiana* gives *laeta* and *velutina* when crossed as female with another species, all its seeds resulting from this cross should be good. Renner found this to be true in the case of *lamarckiana* × *cockerelli*, *lamarckiana* × *hookeri* and others.

(2) The corollary should also be true: when *lamarckiana* as female produces only good seeds, the progeny must be two-formed. de Vries described what seemed to be a negation of this prediction in the case of *lamarckiana* × *muricata*, which gave good seeds but appeared to yield only a single type. Renner found, however, that there are two classes in this F_1, one class composed of yellow seedlings which die early.

(3) When *lamarckiana* as female produces only one type of hybrid progeny, it should produce a class of inviable seeds. This was found to be true in *lamarckiana* × *biennis*.

(4) When another species is pollinated by *lamarckiana*, the percentage of good seeds should be the same as when the female parent is selfed. This proved to be the case when *biennis*, *muricata*, and *biennis Chicago* were pollinated by *lamarckiana*.

(5) When the twins *laeta* and *velutina* are crossed, they should yield *lamarckiana*. This had been found to be true by de Vries.

(6) When heterogamous forms are crossed with each other, e.g., *biennis* × *muricata* and *muricata* × *biennis*, all seeds should be good. Renner showed this to be the case.

On the basis of these findings, which were supported by a wealth of data presented in later papers, Renner formulated his theory of complexes which will now be briefly outlined (see Renner, 1914; 1917a, b, c; 1918a, b; 1919a, b).

1. In most of the oenotheras with which he worked (in large part derived from de Vries,* and all belonging to the subgenus *Oenothera*) all genes seem to belong to a single linkage group. It is as though there was present only a single pair of chromosomes, although *Oenothera* has, in actuality, a diploid number of 14. If a plant had but a single pair of chromosomes, the members of the pair would synapse in meiosis and later separate into different germ cells: except for crossing over, the maternal chromosome of the pair, with all its genes, would go into one cell, the paternal chromosome and genes into the other. Thus, only two kinds of germ cells would be formed, genetically speaking, and these would be identical with the two that united to form the plant. This is the situation in most oenotheras. Only two kinds of germ cell are formed. One kind is genically identical with the egg, the other with the sperm, that united to form the plant.

2. Races of *Oenothera* that show this behavior are highly heterozygous. The genomes derived from father and mother differ with respect to many genes. Renner was inclined to believe that the high degree of heterozygosity is an indication of hybrid origin of the races involved (1917b).

3. Most, though not all, races (*lamarckiana* is an exception) which show this type of behavior are normally self-pollinating, the anthers bursting over the receptive stigma as much as 24 h or more before the

---

* de Vries was exceedingly generous with his material. He was willing to provide other workers with seed, even when they were critics of his work.

flower opens.* Races that are open pollinated are, in the subgenus *Oenothera*, also self-compatible and can be selfed.

4. One would expect a plant heterozygous for one or more genes to give, on selfing, a splitting progeny, as shown by Mendel. In *Oenothera*, however, these races breed true. Renner ascribed this behavior to the presence of balanced lethals in the single linkage group,† i.e., both genomes have lethals, though at different loci. Lethals prevent the development of individuals in which the same set of genes is present in double dose. He found two types of lethal in *Oenothera* (Fig. 3.2). In

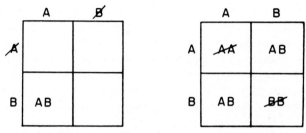

Fig. 3.2. Diagram to show the effect of balanced lethals. At the left, one complex is inactivated by one gametophytic lethal, the other complex by the other lethal. At the right, zygotes that receive the same zygotic lethal from both parents fail to develop. Reprinted from Proceedings of the American Philosophical Society by permission.

some lethal-bearing races (possibly all), zygotic lethals are present. A zygotic lethal causes, usually at a very early stage, the death of any individual in which it is present twice. Renner and certain of his students (Renner, 1914, p. 139; 1946, p. 216; Gerhard, 1929; Langendorf, 1930; Krumholz, 1930) studied the early development of the embryos in such plants and found that in some races zygotic lethals kill the fertilized egg, in others the embryo succumbs at a few-celled stage, in still others, rather well developed embryos develop. In all cases, however, they usually die within the seed, the resulting seeds being inviable. Zygotic lethals may not be genes, but merely the absence of essential loci: if both genomes in a zygote lack the same essential locus, it will die.

* Hoff (unpublished) found in one experiment that in 57·4% of the buds studied (about 1000 buds from 10 self-pollinating races) dehiscence of the anthers occurred 24 h or more before anthesis. In another experiment, 48% dehisced more than 24 h before flowering. In both cases, dehiscence tended to occur more frequently in the morning than during the rest of the day. Thus, in the first experiment, 64·8% of the buds dehisced between the hours of 8 a.m. and noon on the two mornings prior to anthesis; in the second experiment, 57% dehisced during these hours.

† The term "balanced lethals" was first used by Muller (1917) to describe a somewhat comparable situation in *Drosophila*.

The other kind of lethal is a so-called gametophytic lethal, which prevents or handicaps the male or female gametophyte, as the case may be, from developing or reproducing properly. The nature of gameto-phytic lethals will be discussed later. When gametophytic lethals are balanced, one lethal prevents or handicaps the development of the female gametophyte, the other inhibits the male gametophyte. As a result, one genome is transmitted only or mainly through the sperm, the other through the eggs, and only the heterozygotes are formed when the plant is selfed. This is the situation in heterogamous races.

5. Linkage of all genes into a single group, coupled with self-pollination and the presence of balanced lethals, means that a given set of genes will be transmitted intact through countless generations (Fig. 3.3). It is an entity of indefinite duration. Renner emphasized the fact that such a genome is as permanent an entity as is the race of which it is a part. He therefore called such a genome a "complex" (1917a) and gave to each complex a Latin name.* Thus, the complexes of *O. lamarckiana* are *velans* and *gaudens*. Such indefinitely existent genomes are now known

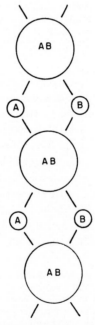

Fig. 3.3. Complexes are transmitted intact from generation to generation. Reprinted from Proceedings of the American Philosophical Society by permission.

* Renner was a profound student of the classics. He owned a complete set of the Greek classics in the original, and read Greek almost daily.

as "Renner complexes". Renner called races that have two such genically diverse complexes "complex-heterozygotes".

6. A complex-heterozygote breeds true, not because it is pure or
homozygous, but in spite of the fact that it is highly heterozygous. Each
generation in a naturally selfed line is identical genetically with the
parental generation, each individual of a given generation is identical
with all the other individuals of that generation. Such a race behaves as
a clone, except that it reproduces sexually instead of vegetatively.

7. Being highly heterozygous, a complex-heterozygote gives results
in outcrosses that are quite different from those produced in selfed line.
Outcrosses between complex-heterozygotes occur only rarely in nature
because of the self-pollinating habit which characterizes most of them,
but when they do occur, fertile offspring are usually produced. In cases
where gametophytic lethals are absent, a plant will produce two kinds
of egg and the same two kinds of sperm. Outcrosses in either direction
will therefore yield two kinds of progeny, provided the other parent
produces only one kind of functional gamete. If both parents produce
two kinds of functional gamete, four classes of progeny are to be expected.

When balanced gametophytic lethals are present and fully functional,
reciprocal hybrids will be unlike, since one complex will come through
the egg only, the other only through the sperm. Thus, if complex A in
an AB race is transmitted through the eggs only, and B only through
the sperm; and if complex C in CD comes only through the eggs, D
through the sperm, the cross AB × CD will result only in AD plants; the
reciprocal cross CD × AB will produce only CB individuals. Reciprocal
hybrids may therefore be very unlike in appearance, as de Vries originally
showed.

Gametophytic lethals are not always fully functional, however, for
reasons that will be explained later. In such cases, a percentage of B
eggs and/or C sperm may be formed and the complex-combinations BC
and/or AC may result. Hybrid combinations that appear in small
percentages as the result of failure of gametophytic lethals to function
completely are the so-called "metacline" hybrids of de Vries (1913,
p. 308).

8. The peculiarities outlined above apply to the group of races designated as complex-heterozygotes. Renner found, however, that some
races, mostly from the Pacific southwest, are exceptional in that they do
not display the peculiarities that characterize most of the group. They
have no lethals, and are open pollinated. They have only one kind of
complex, which is present in double dose, so that they are essentially
homozygous, and their true-breeding is the result of homozygosis. *O.
hookeri* of de Vries is an example of this group. Renner referred to such
races as "complex-homozygotes".

9. So far, the findings that we have summarized refer to naturally occurring races of the subgenus *Oenothera*. Renner's findings with regard to the behavior of hybrids between such races are still to be mentioned. He found that such hybrids, in contrast to their parents, often fail to breed true, but instead may produce splitting progenies when selfed. In some hybrids, the genes are found to belong to two linkage groups, in others there are three, in still others four or more. Some, on the other hand, have only a single linkage group, as did the parents. In a given hybrid combination, the number of linkage groups is always the same, no matter how often the cross is made. The number is often different, however, in different hybrids.

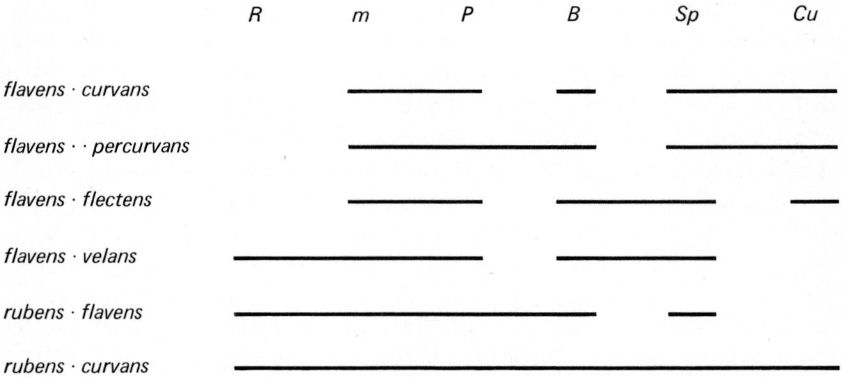

	R	m	P	B	Sp	Cu

flavens · curvans

flavens · · percurvans

flavens · flectens

flavens · velans

rubens · flavens

rubens · curvans

FIG. 3.4. Linkage relations of genes (shown above) in different hybrid combinations (shown at left). All genes were in a single group in the parents of all combinations. After Renner, 1928. Reprinted from Proceedings of the American Philosophical Society by permission.

This extraordinary situation is illustrated by a diagram (Fig. 3.4) adapted from a paper by Renner (1928). The complex-combinations to the left of the diagram are hybrid combinations. Across the top are listed certain genes. The horizontal lines indicate which genes are linked in each hybrid. Thus, in *flavens·curvans* m and P (which may be alleles) are independent of B, which in turn is independent of Sp and Cu, which are linked. In *flavens·percurvans* m, P and B are linked, Sp and Cu are also linked. In *flavens·flectens* B and Sp are linked, m, P and Cu are independent. In all cases, both parents of the listed hybrids had but a single linkage group, but in the hybrids, the genes studied were differently linked in different hybrids.

The breaking up of the single parental linkage group into two or more groups in many hybrids indicated to Renner (1928) that the basis for extensive linkage of genes in the races is not the inclusion of these genes within a single pair of chromosomes as postulated by Shull (1925); there must be some mechanism in most races by which all genes are inherited in a single block, although scattered among the various chromosomes. This mechanism was found in the peculiar behavior of the chromosomes, which will be brought out in the following chapter.

It can be seen from the above outline of Renner's findings that he and de Vries arrived at quite different interpretations of essentially the same facts. There was little disagreement in regard to the empirical data— the differences were primarily interpretive. Renner considered that most of the races upon which both of them worked were of hybrid origin, de Vries considered them to be pure species. Renner found all of the genes by which one race is distinguished from another, and presumably all genes in a complement, linked in inheritance, there being but one linkage group in a race, for which (except for *hookeri* and a few others) it is heterozygous. de Vries visualized the difference that produced the two kinds of gametes as involving, not all pangenes, but only a single pangene or an associated group of pangenes. Presumably, even a mutable race such as *lamarckiana* was pure for most of its pangenes, and this was also the case for the heterogamous races in which not the whole of the pangenic complement was involved in the differences between sperm and egg. With respect to *hookeri*, de Vries failed to realize that this was indeed a homozygous race and as such was very different from the other races with which he dealt. He considered *hookeri* to be an isogamous form along with *cockerelli* and *strigosa* and could not understand why its hybrids split in F_2, throwing plants that were indistinguishable from pure *hookeri*, when hybrids of the other two races bred true. Renner recognized *hookeri* as an alethal complex-homozygote in contrast with the complex-heterozygotes with balanced lethals.

A summarization of his findings and interpretations was published by de Vries in 1913 in his "Gruppenweise Artbildung". It was only a year later that Renner's paper on fertilization and embryology in *lamarckiana* and related species appeared, throwing doubt upon these interpretations. In this paper Renner first showed the relation between the presence of sterile seeds and the production of splitting F_1 hybrids: sterile seeds represent genetic combinations that, because of the presence of zygotic lethals, are inviable. In selfed line they represent zygotes that have received the same complex in double dose. In *lamarckiana* × *biennis* they represent the *gaudens* × *rubens* combination, demonstrating the presence of the same lethal in these two closely related complexes.

de Vries at first (1915, 1916b) resisted the suggestion that sterile seeds represent specific genetic combinations rendered inviable by lethals. He ascribed their presence to mutation, and invented the concept of "semi-lethals" (1916b) which arise through mutation in races that give splitting progeny, and which kill half the zygotes. By semi-lethals, de Vries originally meant pangenes that cause partial sterility, that reduce seed set by roughly 50%, but without reference to the genetic composition of the zygote being sterilized. He considered that these semi-lethals behave as recessives in hybrids involving races that have empty seeds (*lamarckiana* or *suaveolens*), since crossed seeds of these races are mostly viable. To de Vries, empty seeds were caused by the mutation of a pangene that was different from, and independent of, the *laeta* pangene whose mutation resulted in *velutina* gametes.

Later (1917b), de Vries revised his ideas somewhat, and adopted the concept of balanced lethals, but without admitting a hybrid origin for his races. What led him to this position was a study of certain forms which he called "half-mutants". These were derived directly or indirectly from *lamarckiana*. When selfed, they reproduced themselves in about two-thirds of the progeny and threw a new true-breeding type as the other third. He concluded that such forms had an unbalanced lethal situation. Half the gametes had a lethal, the other half were lethal-free. The latter could combine with themselves to produce lethal-free, true-breeding segregants. The reunion of lethal-bearing and lethal-free gametes gave a continuation of the half-mutant. From this he derived the concept of *lamarckiana* itself being a kind of half-mutant, i.e., a sex cell had mutated to *velutina* which had combined with a normal gamete to give the present *lamarckiana*. Such a plant should give a 1:2:1 ratio when selfed, but both homozygotes are prevented from appearing because of lethals and only the heterozygotes (he avoided this term and used instead "mixed types") were able to survive. In outcrosses, both types of gamete could function and thus give rise to twin hybrids. de Vries suggested that *grandiflora* is also a half-mutant, since it throws a true-breeding segregant in each generation: *ochracea*. Also, the races that Bartlett (1915b, e) had studied in which "mass mutation" occurred were considered to be "half-mutants".

This concept was in general agreement with that of Renner so far as the lethal situation was concerned. However, de Vries still refused to think of his races as truly heterozygous or of hybrid origin. He considered that they were essentially pure, and the differences that resulted in twin hybrids, or produced unlike reciprocals or "mass mutation", involved at most only a few pangenes. Renner, however, not only thought in terms of entire genomes, or complexes as he called them, but set out to analyze the various complexes from the standpoint of their genic content, and was able to show (1925) that different complexes, and even the complexes associated in a single species, differ in respect to many, not just a few genes. Renner's analysis of the genetic structure of the oenotheras was, therefore, much more penetrating than that of de Vries, who never got down to the level of the individual gene, though he thought that he was dealing with individual pangenes or pangene clusters when explaining the appearance of twin hybrids, or of most of the mutants.

The work of Renner just described was carried on during World War I, and communication between Germany and the United States was almost completely cut off for some years. During this period, two persons in the United States, working independently, came to conclusions that paralleled in some respects the findings of Renner, and these will be mentioned at this point.

One of these persons was H. H. Bartlett of the U.S. Department of Agriculture, later at the University of Michigan (Fig. 3.5). His early papers dealt with the systematics of *Oenothera*, but beginning in 1915 (1915a, b, c, d, e; 1916; Cobb and Bartlett, 1919) he turned his attention increasingly to evolutionary problems. He considered himself a follower of de Vries and he attempted to explain the appearance of aberrants in terms of mutation rather than segregation. Nevertheless, it has been claimed that in one respect he approached Renner, namely in his concept of "heterogametism" as he called it (Cobb and Bartlett, 1919), that is, the presence of different sets of genes in egg and sperm. This concept was developed as a result of a study of *O. pratincola* from Lexington, Kentucky, and *O. reynoldsii* from Knoxville, Tennessee. He found aberrants in the progenies of these races, appearing in the F_2 or later generations of selfed lines, the most striking feature of which was the relatively large numbers in which they appeared in certain lines—in some cases making up more than 50% of a culture. Bartlett called the process by which such large numbers of aberrants appeared "mass mutation". He ascribed their appearance to mutation of unstable factors present in the female gametes that have no counterpart in the male gametes. Thus he arrived at the idea that male and female gametes carry different potentialities, a concept that seemed to be borne out by his subsequent experiments. In at least three respects, however, his concept differed from that of Renner:

(1) He considered that the difference between female (α) and male (β) gametes involved, not the whole of the female or male set of genes, but only a portion of the female set that had no counterpart in the male (and possibly of the male that had no counterpart in the female, although there was no direct evidence of the latter). This portion was thought to consist of a chromosome, or more likely a group of chromosomes that

remain in association at meiosis, that carry factors not duplicated in the complementary gamete. He called this the "characteristic portion" of the gamete. The rest of the chromosomes of the α and β gametes were considered to be freely segregating and to carry Mendelizing factors. "Mass mutants" result from mutations in the "characteristic portion"

FIG. 3.5. Harley Harris Bartlett (1886–1960).

which occur with very high frequency. Renner, on the other hand, did not envisage one complex as having loci not present in its associated complex, nor did he find evidence of Mendelizing factors in his complexes. Heterogametism was based, in his view, on the presence of many heterozygous genes in associated complexes whose loci were homologous throughout.

(2) Heterogametism is maintained, according to Renner, by the

presence of balanced lethals. Bartlett did not visualize the presence of lethals but ascribed the maintenance of heterogametism rather to the frequent occurrence of mutations in the "characteristic portion" of the female gametes.

(3) Heterogametism has arisen as the result of hybridization in Renner's view; according to Bartlett, it is due to mutation.

In view of these rather fundamental differences in point of view, it can scarcely be argued that Bartlett was a co-discoverer of the phenomenon of complex-heterozygosity.

The second person whose work related to that of Renner was H. J. Muller, who discovered in *Drosophila* (1917, 1918) a case where balanced lethals in a pair of chromosomes maintains the heterozygosity of the pair. Muller found that the character "beaded wings" is lethal when homozygous, but behaves as a morphological dominant when hetero-zygous. He developed a stock, however, which bred true for beaded even though this is lethal when homozygous. The reason for this is that the chromosome homologous to the one carrying beaded in this stock also had a lethal, at a different locus, which prevented that chromosome from existing in homozygous condition. Since both homologues had lethals, only the heterozygote could be formed, and this showed the beaded character, so that the strain bred true for beaded. In this case there was also an inversion in the beaded chromosome that prevented crossing over of the lethals, thus excluding even the exceptional appearance of homozygous individuals. Muller invented the term "balanced lethals" to describe this situation.

At the time that he analyzed this situation Muller was unaware of Renner's work, but he was acquainted with de Vries' early work on *Oenothera* in which twin hybrids and true-breeding hybrids were de-scribed; and he pointed out that de Vries (1911) had arrived at a kind of balanced lethal concept to explain his double reciprocal crosses. Muller drew attention to one difference between *Oenothera* and *Drosophila*: in *Oenothera* reciprocal crosses are often unlike, whereas in the beaded situation they are essentially alike. He concluded that in *Oenothera* this probably meant that lethals often act on the germ cells rather than on the zygotes. Muller therefore proposed that a balanced lethal situation was responsible for many of the peculiarities in *Oenothera* heredity.

As an addendum to his 1918 paper, Muller referred to more recent papers by de Vries (1916b, 1918a) in which a lethal situation, not only in *lamarckiana* but also in *grandiflora* and other forms, was definitely advocated. From one of these papers, Muller first learned of Renner's work, and of his scheme of balanced lethals. It is interesting that Muller was led, on the basis of finding a somewhat comparable condition in

Drosphila, to postulate a situation in *Oenothera* that Renner had already demonstrated empirically. Renner, who only learned of Muller's work after the war, was not inclined at first to use the term "balanced lethals" for *Oenothera*, since he felt that there was no way at that time to determine whether lethals in *Oenothera* were individual loci or not, as had been so clearly established in several cases in *Drosophila*.

We have seen that Renner's work began with a study of *Oenothera* embryos and seed structure. The early degeneration of many embryos, leading to the presence of empty seeds, led him to the concept of balanced zygotic lethals. The fact that in many oenotheras only one complex is transmitted through the egg, the other through the sperm, suggested the presence of lethals that affect the gametes rather than the zygotes. This led Renner to examine pollen and embryo sac development to see if he could find visible evidence of the work of gametophytic lethals. He found that it was possible in many cases to distinguish the pollen grains carrying the two complexes of a plant (1919a, b). In some heterogamous races, the two classes of pollen are very distinct, one class (the "actives") plump, with spindle-shaped starch grains, and viable; the other class (the "inactives") shrunken, with spherical starch and unable to germinate (Fig. 3.6). For example, in *muricata*, the grains carrying *rigens* are inactive, the grains carrying *curvans* are active. In addition there are very small completely empty grains of uncertain origin. Isogamous forms lack the inactive class; in some isogamous forms, however, two classes of actives can be recognized based on size, measurements of pollen grains yielding a bi-modal curve (Hoeppener and Renner, 1929; Rudloff, 1929). Renner thus found good morphological evidence of the

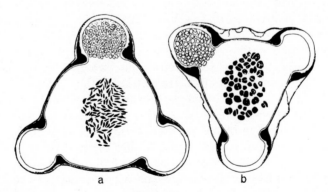

FIG. 3.6. Pollen grains of *O. muricata* Venedig (from Renner, 1919b). (a) *curvans* pollen, active, with spindle-shaped starch grains; (b) *rigens* pollen, shrunken, with spherical starch grains.

death of a class of pollen grains in some heterogamous forms, and the presence of two classes of viable pollen in certain isogamous races.

It was possible, therefore, for Renner to demonstrate the effects not only of zygotic lethals, by the presence of inviable seed, but also of pollen lethals, by the sterility of a class of pollen.

On the female side, Renner found that complexes differ among themselves in respect to their ability to further, or their tendency to inhibit, embryo sac development (1921a). In heterogamous forms the megaspores containing the pollen complex rarely develop into embryo sacs whereas those carrying the other complex rarely succeed. The results of Renner's study of the behavior of complexes in relation to embryo sac development will be considered later in another connection (Chapter 9).

PART II
PHYSICAL BASIS OF THE GENETICAL
PECULIARITIES OF OENOTHERA

CHAPTER 4

Chromosome Catenation

The early work on the genetics of *Oenothera* was done without reference to chromosomes. The chromosomes of *Oenothera*, however, had been examined, and the fact that they behave peculiarly in meiosis was realized by early workers in this field. The nature of these irregularities, however, and their genetical implications, were not at first understood.

The first papers dealing with *Oenothera* cytology appeared in 1907, and for a period of 5 years thereafter contributions to this subject were published at frequent intervals, some eighteen papers appearing during this time. Interest in the field then died down and during the period 1912–1921 inclusive only five cytological papers appeared, one of which dealt exclusively with somatic mitosis. The pioneer investigators were Gates (1907a; 1908a, b; 1909a, b, c; 1911c); Lutz (1907, 1908, 1909); Geerts (1907, 1909); and Davis (1909, 1910) (Figs 4.1, 4.2). Lutz's papers dealt primarily with chromosome numbers, but the other authors also attempted to describe meiotic behavior. Although Davis found normal behavior during meiosis in a strain of *O. grandiflora* (1909) he and Gates found in other material a large amount of seeming irregularity in meiotic chromosome behavior, the chromosomes failing to pair properly or to line up in pairs on the first meiotic spindle. Neither worker discovered regularity in the behavior of those forms in which some or all of the chromosomes do not form pairs. In some of the material that Gates studied there is no regularity, including the trisomic *lata* and the tetraploid *gigas*, but other materials which he and Davis studied do display a regularity of behavior which was later discovered.

Geerts (1909) studied meiosis in *O. lamarckiana* and found a type of regularity which does not in fact exist; he thought that the chromosomes form pairs at metaphase I in this race, whereas most of them are unpaired.

It was not until 1922 that regularity was discovered in forms that had some unpaired chromosomes. Cleland in that year published an account of meiosis in *O. franciscana*, a California race in which four of the chromosomes fail to pair. These chromosomes, it was found, form regularly a closed circle of four chromosomes in diakinesis, the other chromosomes being paired. The fact that the unpaired chromosomes form a closed circle with a high degree of regularity suggested that other oenotheras in which behavior seemed to be abnormal might also show regularity in behavior. Investigation of other materials soon showed that this was indeed true: a form known as *O. franciscana sulfurea* (a hybrid between Davis' *biennis* and his *franciscana* B) was found to have a circle of 12 chromosomes and one pair; *biennis* of de Vries had ⊙6, ⊙8,* and *muricata* had ⊙14 (Fig. 4.3) (Cleland 1923, 1924). *Oblonga*, a "mutant" of *lamarckiana*, was described as having a chromosome configuration which later

Fig. 4.1. Reginald Ruggles Gates (1882–1962).

proved to be incorrect. *O. lamarckiana* was soon shown to have ⊙12, 1 pair, and certain derivatives of *lamarckiana* showed the following configurations: *O. rubrinervis* ⊙6, 4 pairs; *O. rubricalyx* ⊙8, 3 pairs; *O. blandina* and *O. deserens* 7 pairs (Cleland, 1925). Thus, it seemed that unpaired chromosomes behaved as regularly as did paired ones.

In addition, however, a further peculiarity of behavior was found (Cleland, 1923, 1924). It was discovered that the circles remain intact throughout metaphase I and become arranged across the spindle in such a way that adjacent chromosomes are directed toward opposite poles.

* ⊙ = circle.

Since all chromosomes in the circle are essentially equal in size and all have median centromeres, the chromosomes show a regular zigzag appearance as seen from the side (Fig. 4.4). As a result, adjacent chromosomes move to opposite poles in anaphase I, a process that occurs with a

FIG. 4.2. Bradley Moore Davis (1871–1957).

high percentage of regularity in *Oenothera*. In both *muricata* and *lamarcki-ana*, for example, about 80% of the cells were found to have perfect regularity in this respect. It was soon realized that the linking of all 14 chromosomes into a single closed circle is characteristic of many forms of *Oenothera* found in nature. We now know that ⊙14 is the all but universal configuration among the members of the subgenus *Oenothera* from the Rocky Mountains eastward, an amazing fact, when one stops to think about it.

The genetical implications of this type of behavior were early recognized. Shull (1923a, b) had found that most of the genes in *lamarckiana* with which he worked were linked, and Renner, as we have seen,
had found that all genes were in a single linkage group in most races of
Oenothera. It was natural, therefore, to raise the question as to whether
linking of the chromosomes into a chain was correlated with the extensive

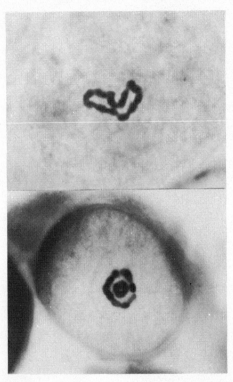

Fɪɢ. 4.3. (above) Circle of 14 chromosomes, prometaphase, first meiotic division.
Proceedings of the American Philosophical Society by permission. (below) Circles of 4
and 10, polar view, metaphase I. Magnification at microscope × 1350.

linkage of genes found by Shull and Renner. A single assumption was
all that was necessary to tie chromosome linkage and gene linkage
together. This assumption was that chromosomes of paternal and
maternal origin alternate in the circle. If they alternate in the circle, then
all paternal chromosomes of the circle will go to one pole, all maternal
chromosomes to the other in any cells with the regular zigzag arrangement. If a circle of 14 is present, all paternal genes will enter one germ
cell, all maternal genes another. Half the resultant gametes will have the

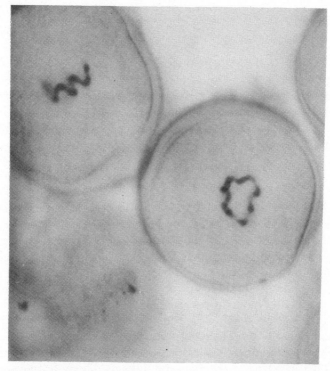

FIG. 4.4. Upper left, first meiotic metaphase, side view, showing beginning separation of adjacent chromosomes to opposite poles. Lower right, same stage, polar view. Magnification at microscope × 1350.

intact paternal genome, half will have the maternal genome. Only two kinds of germ cells will be formed and these will be identical genetically with the gametes that united to form the plant (see Fig. 4.5). If, on the other hand, smaller circles are present, or more than one circle, then one may expect separation of all paternal from all maternal chromosomes within a given circle, but different circles, or circles and pairs will segregate independently. Instead of a single linkage group, there will

FIG. 4.5. Diagram to show that alternation of paternal and maternal chromosomes in the circle, coupled with disjunction of adjacent chromosomes, results in the formation of only two kinds of gamete, which are identical with those that united to form the plant.

then be as many groups as there are chromosome groups. Thus, chromosome behavior will account both for the linkage of all genes into a single
group in most races, and for the presence of two or more linkage groups
in many hybrids. A hybrid will be expected on this basis to have as many
linkage groups as it has chromosome groups. This suggestion was made
independently by Cleland (1923) and Oehlkers (1924) and all subsequent
work has confirmed its validity.

The assumption that paternal and maternal chromosomes alternate
in the circle was not at first accepted by all. The mechanism by which
chromosomes are attached end to end was not at that time understood
and it was assumed by all *Oenothera* cytologists that a circle of 14 consisted of seven pairs of homologues, linked together into a common
spireme according to the telosynaptic interpretation of meiosis. Since
it was believed, therefore, that every chromosome in a circle has its
homologue in the circle, one could assume either that homologues are
scattered at random through the circle, or that they are attached together end to end, and if the latter is the case one could further assume
that it will make no difference whether the paternal or the maternal
member of a pair of homologues occupies the left-hand or the right-hand
position in the circle. There was, therefore, at that time no good reason
for assuming that paternal and maternal chromosomes alternate, except
that such an assumption would make it possible to explain extensive
gene linkage in terms of chromosome linkage. There developed, therefore, two alternative hypotheses, one which assumed regular alternation
of paternal and maternal chromosomes, and consequent relation of
chromosome linkage to gene linkage; the other which took the standpoint that it is a mere matter of chance which one of a pair of homologues
occupies the right or the left position. If the latter alternative were
correct, then either member of a pair could go to either pole, and consequently no relationship would exist between chromosome linkage and
gene linkage. This would mean that genes to be linked must lie in the
same chromosome, and the fact that most genes seem to belong to a
single linkage group must imply that one chromosome of the set contains
most of the genes that distinguish one race from another—the other
chromosomes are either empty or carry only those genes that are
common to all races.

The chief advocate of the latter hypothesis was Shull (Fig. 4.6), who
stated (1928a, p. 104): "the orientation of the members of a pair is a
mere matter of chance, so that in following a series of spiremes from left
to right, any chromosome of maternal origin will lie to the left of its
paternal homologue in 50 percent of the cases, on the average, and to the
right in the remaining 50 percent." Shull presented as his chief evidence

against the "chromosome cohesion theory" (as Gates called it) the supposed fact that *lamarckiana*, with ⊙12 and 1 pair, had three linkage groups. One of these groups was large, the others small; in fact one group was represented by only one character, *brevistylis*, or short style.

Fig. 4.6. George Harrison Shull (1874–1954).

We now know, however, that *brevistylis* is dependent upon a gene that lies at the extreme end of a circle chromosome and hence exhibits 50% crossing over with respect to the two complexes (Emerson and Sturtevant, 1932; Cleland and Brittingham, 1934). It belongs, therefore, in the large linkage group, and there are only two linkage groups present in *lamarckiana*.

It was, of course, not difficult to test these two hypotheses. According to Shull's point of view, linkage relations should in no way be affected by the type of chromosome configuration present, since the integrity of individual chromosomes would not be affected by the way in which they were arranged. On the other hand the Cleland-Oehlkers hypothesis would call for a close correlation between chromosome and gene linkage— extensive genetic linkage would be maintained only so long as the chromosomes involved remained a part of a large circle. The obvious test, therefore, was to look for plants with different chromosome configurations and test them genetically to determine whether a correlation existed between the number of chromosome groups and the number of linkage groups.

The first evidence bearing upon this question came from the work of Oehlkers (1926) (Fig. 4.7). He crossed *suaveolens* ($\odot$12) by *strigosa* ($\odot$14) and in the F_1 obtained two kinds of hybrids: *albicans·stringens*

FIG. 4.7. Friedrich Oehlkers (1890–1971).

(⊙12) and *flavens·stringens* (7 pairs—should have had ⊙4). The former showed no splitting in the following generation, the latter showed independent segregation of three gene pairs, thus supporting the chromosome cohesion hypothesis.

It became possible to carry out a more extensive test of this hypothesis in 1927 and 1928. A Guggenheim Fellowship enabled Cleland to spend two summers and the intervening winter in Europe, partly with Oehlkers and partly with Renner. Several weeks were also spent in Amsterdam and Lunteren, collecting material in de Vries' garden. During this period, Oehlkers and Renner between them grew a large number of F_1 hybrid combinations. Cleland studied their cytology, and Oehlkers and Renner analyzed their genetical behavior.

Prior to this time, a total of eight different chromosome configurations had been found by Cleland in the races, mutants and hybrids he had examined, so it was known that different configurations are possible. Assuming that a circle will contain an equal number of paternal and maternal chromosomes, there are a total of 15 configurations theoretically possible in forms with 14 chromosomes (Table 4.1). It was hoped that the hybrid combinations prepared by Oehlkers and Renner would show a variety of configurations and this proved to be the case. Ten different configurations were found, ranging from ⊙14 to ⊙4.

The results of this experiment are summarized in Table 4.II. It will be noted that, except for a factor governing flower size, which showed exchange between complexes even in the presence of ⊙14, there was a

TABLE 4.I

Possible chromosome configurations in a plant with 2n = 14

⊙14
⊙10, ○4
⊙8, ○6
⊙6, ○4, ○4
⊙12, 1 pair
⊙8, ○4, 1 pair
⊙6, ○6, 1 pair
○4, ○4, ○4, 1 pair
⊙10, 2 pairs
⊙6, ○4, 2 pairs
⊙8, 3 pairs
○4, ○4, 3 pairs
⊙6, 4 pairs
⊙4, 5 pairs
7 pairs

TABLE 4.II

Correlation between chromosome configuration, as determined by Cleland, and genetical behavior, as analyzed by Oehlkers (O) and Renner (R), of certain F_1 hybrids of *Oenothera*

Form	Configuration	Expectation as to maximum number of linkage groups	Extent of factor exchange between complexes of F_1	
(*grandiflora* × *hookeri*) truncans · ʰhookeri	⊙14	1 group	No exchange (except flower size)	(O)
(*grandiflora* × *lamarckiana*) acuens · gaudens	⊙14	1 group	No exchange (except flower size)	(O)
(*suaveolens sulf.* × *lamarckiana*) albicans · velans	⊙14	1 group	No exchange (except flower size)	(R)
(*muricata* × *chicaginensis*) rigens · punctulans	⊙14	1 group	No exchange (except flower size)	(O)
(*grandiflora* × *lamarck.* + recip.) truncans · velans	⊙4, ⊙10	1 larger, 1 smaller group	No exchange (except flower size)	(O)
truncans · gaudens	⊙4, ⊙10	1 larger, 1 smaller group	Exchange of flower size, crippling of main stem	(O)
(*suaveolens sulf.* × *lamarckiana*) albicans · gaudens	⊙6, ⊙8	2 groups	Probably no exchange except flower size	(O)
(*R-biennis* × *chicaginensis*) rubens · punctulans	⊙6, ⊙8	2 groups	No exchange (except flower size)	(R)
(*r-lamarckiana* × *chicaginensis*) gaudens · punctulans	⊙6, ⊙8	2 groups	No exchange (except flower size)	(R)
(*strigosa* × *suaveolens sulf.*) deprimens · flavens	⊙12	1 large and 1 small group	No exchange	(O)
(*lamarckiana* × *suav. sulf.* + recip.) gaudens · flavens	⊙12	1 large and 1 small group	Occasional exchange of flavens lethal, exchange in flower size	(O)
(*suaveolens sulf.* × *strigosa*) albicans · stringens	⊙12	1 large and 1 small group	Exchange of pollen sterility factor	(O)
(*suaveolens* × *cockerelli*) albicans · elongans	⊙12	1 large and 1 small group	No exchange (according to de Vries)	(O)
(*biennis sulf.* Hanover × *cockerelli*) albicans · elongans	⊙12	1 large and 1 small group	Exchange of flower size, leaf breadth and crinkling of leaves (linked)	(O)
(*cockerelli* × *suaveolens*) curtans · flavens	⊙12	1 large and 1 small group	No exchange	(O)
(*r-lamarckiana* × *chicaginensis*) velans · punctulans	⊙12	1 large and 1 small group	Exchange of flower size only	(R)
(*strigosa* × *lamarckiana cruciata*) deprimens · velans	⊙10	1 large, 2 small groups	Occasional split in leaf breadth	(O)
(*chicaginensis* × *R-biennis*) excellens · rubens	⊙10	1 large, 2 small groups	R-r exchanged	(R)
(*hookeri* × *chicaginensis*) ʰhookeri · punctulans	⊙10	1 large, 2 small groups	Exchange in flower size	(R)
(*chicaginensis* × *r-lamarckiana*) excellens · gaudens	⊙10	1 large, 2 small groups	Exchange of R-r and flower size	(R)
(*R-biennis* × *chicaginensis*) albicans · punctulans	⊙10	1 large, 2 small groups	Pus-pus exchanged	(R)
(*suaveolens* × *chicaginensis*) albicans · punctulans	⊙10	1 large, 2 small groups	Pus-pus exchanged	(R)
(*suaveolens* × *cockerelli*) flavens · elongans	⊙8	1 large, 3 small groups	Independent exchange of flower size, hypanthium length, leaf width and leaf color	(O)
(*suaveolens* × *chicaginensis*) flavens · punctulans	⊙8	1 large, 3 small groups	Independent exchange of Pus-pus B-b, (leaf width) and a hair-producing factor	(R)
(*grandiflora* × *lamarckiana* and reciprocal) acuens · velans	⊙4, ⊙6	4 linkage groups	Rich splitting, in flower size, habit, red flecking of leaves, leaf breadth, crinking	(O)

Table 4.II—*continued*

Form	Configu-ration	Expectation as to maximum number of linkage groups	Extent of factor exchange between complexes of F_1	
(*chicaginensis* × *r-lamarckiana*) excellens · velans	⊙6	1 large, 4 small groups	Independent exchange of *R-r*, *P-p*, flower size and factor for stature	(R)
(*grandiflora* × *hookeri* and reciprocal) acuens ·ᵸhookeri	⊙4, ⊙4	5 linkage groups	At least 4 independent factors	(O)
(*lamarckiana* × *suav. sulf.*) velans · flavens	⊙4, ⊙4	5 linkage groups	At least 3 independent factors	(O)
(*chicaginensis* × *suaveolens*) excellens · flavens	⊙4, ⊙4	5 linkage groups	*R-r* exchange, probably linked with *p* *let* probably exchanged	(R)
(*suaveolens sulf.* × *strigosa*) flavens · stringens	⊙4	6 linkage groups	At least 4 independent factors	(O)
(*chicaginensis* × *hookeri*) excellens ·ᵸhookeri	⊙4	6 linkage groups	Complicated splitting not fully analyzed, including flower size and probably *R-r*	(R)

clear-cut correlation between chromosome behavior and genetic behavior: the degree of factor exchange between complexes in the F_1 was related to the number of chromosome groups present (Cleland and Oehlkers, 1929, 1930; Renner and Cleland, 1933). Flower size (Co) has since been shown (Emerson and Sturtevant, 1932; Cleland and Brittingham, 1934) to behave as does *brevistylis* (*br*). It is situated at the extreme end of its chromosome and therefore shows 50% crossing over between the complexes. Apart from this character, however, no exchange between complexes was observed when ⊙14 was present, and conversely, the frequency of exchange increased with increase in the number of chromosome groups. This series of tests was extensive enough to demonstrate clearly the correctness of the chromosome cohesion hypothesis: paternal and maternal chromosomes alternate in a circle and the separation of adjacent chromosomes to opposite poles results in all the paternal chromosomes and genes in the circle disjoining from the entire maternal set. The larger the circle, the larger the number of genes that behave as though they were linked. A circle of 14 results in the apparent linkage of all genes and produces the kind of genetical behavior that characterizes most races in nature. The differing numbers of linkage groups that are present in different hybrid combination are in agreement with the number of independent chromosome groups in these hybrids. Thus the genetical phenomena uncovered by Renner are explicable in terms of chromosome behavior.

All subsequent work has fully confirmed these findings. Every one of the 15 possible configurations has been found over and over again, and wherever the genetical and cytological behavior of complex-combinations has been tested simultaneously the former has conformed to the latter in every case.

In the early days of *Oenothera* cytology all workers were agreed on a telosynaptic interpretation of meiosis in *Oenothera*. The tendency for chromosomes to be attached end to end in diakinesis seemed to constitute prime evidence for the existence of a continuous spireme. The apparent singleness of the threads in earlier meiotic prophase stages and the presence of what was called a strepsitene stage also argued for the telosynaptic theory, according to which pairing of homologous chromosomes in the univalent spireme did not take place until the strepsitene stage, when loops were formed which radiated out from the center, each loop consisting of a pair of homologues. This theory seemed to receive additional support when it was found that in most wild races of *Oenothera* all 14 of their chromosomes are arranged in a single circle—a seemingly perfect example of a continuous and univalent spireme. The evidence for this interpretation seemed cogent enough to cause E. B. Wilson to state, in the third edition of his "The Cell in Development and Heredity" (1925, p. 565): "In the case of *Oenothera*, as already indicated, a 'telosynaptic' association of the chromosomes in early diakinesis seems indubitable."

Some *Oenothera* cytologists accepted the telosynaptic theory as applying to all organisms, though Gates as early as 1908 admitted the possibility of parasynapsis in organisms other than *Oenothera*. However, Gates changed his mind later, claiming that telosynapsis is present in most plants and probably in some animals (1922).*

As long as the telosynaptic interpretation held sway for *Oenothera*, it was assumed that each chromosome in meiosis had a homologue, and the two homologues were assumed to lie end to end in the spireme. This led to certain misconceptions that were only removed when it was realized (see below) that no chromosome in a circle is wholly homologous with any other. One of these misconceptions was the belief that irregularities in the zigzag arrangement could result in seven-chromosome gametes which had exchanged one or more chromosomes of one complex for the corresponding chromosomes of the other complex. This was thought to be a possible mechanism for the production of mutations of various sorts. Among the workers who entertained this possibility were Cleland (1926a, b), Sheffield (1927, p. 803) and Gates (1928a, p. 754). We now know that this is not possible. Any irregularity in the zigzag arrangement, which does not alter the number of chromosomes going to each

---

* In 1931, Gates (with Goodwin) again modified his stand by accepting the parasynaptic interpretation for all-pairing forms of *Oenothera*, though not for those with circles, only to reverse his stand once more in 1935 (with Nandi) in favor of the concept of "acrosyndesis," according to which *Oenothera* chromosomes pair only at their ends, and are held together by electrostatic attraction, chiasmata not being necessary.

FIG. 4.8. *Oenothera* workers at the Fifth International Congress of Genetics in Berlin, 1928. Seated, left to right: J. A. Honing, R. R. Gates, G. H. Shull, O. Renner, N. Heribert-Nilsson. Standing, left to right: Fr. Oehlkers, T. Stomps, E. Lehmann, R. E. Cleland. J. Schwemmle.

pole, will result in duplications and deficiencies of certain segments, and consequent inviability of the products formed. Mutations cannot arise in this way.

A second misconception (Cleland, 1926a) was the idea that a circle consists of *pairs of homologues* which are relatively highly heterozygous, in contrast with paired chromosomes, which are relatively homozygous. This opinion was based on the fact that forms with large circles are highly heterozygous, whereas forms in which all or most of their chromosomes are paired are usually homozygous. It is true that the presence of a large circle is evidence of a high degree of heterozygosity, the two complexes forming the circles are very different in their genic content. However, since no chromosome in a circle is wholly homologous with any other, one cannot speak of heterozygous *pairs* of chromosomes in the circle.

In the case of paired chromosomes, on the other hand, it is true that in many cases these are highly or wholly homozygous. For instance, such forms as *deserens* or *decipiens*, which are all-pairing derivatives from so-called half-mutants of *lamarckiana*, are completely homozygous. Wild races with all-paired chromosomes, such as *hookeri*, are highly, though not necessarily wholly homozygous. In many hybrids that result from outcrosses, however, pairs may be present in which the homologues differ with respect to one or more genes, or they may even be quite heterozygous.

It is still generally true, therefore, that complex-combinations that have large circles are highly heterozygous, whereas combinations in which the chromosomes are all or mostly paired are wholly or largely homozygous. What is not true is that heterozygous *pairs of homologues* exist within circles of chromosomes.

Fortunately the belief in the presence of wholly homologous pairs of chromosomes within a circle did not necessarily affect the chromosome cohesion hypothesis of Cleland and Oehlkers. Irrespective of its homology, each chromosome is either paternal or maternal in origin, and the hypothesis of alternation of paternal and maternal chromosomes is tenable. The concept of total homology of pairs within the circle did, however, affect the *acceptance* of the chromosome cohesion hypothesis, since this concept made the idea of regular alternation of paternal and maternal chromosomes a mere assumption, supported only by the fact that, on such an assumption, chromosome linkage could be shown to be responsible for extensive gene linkage.

CHAPTER 5

The Cause of Circle Formation in *Oenothera*

Before 1927 the situation in *Oenothera* was thought to be as follows: (1) *Oenothera* was the most outstanding example of telosynapsis; (2) two opposing hypotheses regarding the arrangement and distribution of the genes were held, one that all genes for characteristics that distinguish one form from another are linked in a single chromosome, the other that such genes are scattered through the chromosome set but are effectively linked together because paternal and maternal chromosomes alternate in the circle and hence are separated to opposite poles and into different germ cells.

Beginning in 1927, however, this picture changed drastically. In the first place, the chromosome cohesion hypothesis was extensively tested and found to be correct, as we have seen in the preceding chapter. In the second place the concepts of chromosome structure and the basis for circle formation underwent a profound change, the result largely of new hypotheses by Belling and Darlington followed by tests of Belling's ideas by Emerson and Sturtevant, and by Cleland and Blakeslee.

In 1926 Belling and Blakeslee described a trisomic *Datura* in which some cells showed a trivalent in the form of an open chain of three, the chromosome at one end being smaller and evidently not a full homologue of the other two, though homologous with the others at one end. This plant was in the progeny of a hybrid between two 12-paired races, and Belling developed a concept of "segmental interchange" to explain the situation. His hypothesis was to the effect that an interchange of segments between non-homologous chromosomes had taken place in the ancestry of one of the parents of the hybrid, so that the hybrid had received two interchanged chromosomes from one parent and two unmodified chromosomes from the other (Fig. 5.1). The trisomic appearing in the progeny of a hybrid plant received unmodified chromosomes through both gametes plus an interchange chromosome, one of whose ends was homologous with one end of each of two unmodified chromosomes. Thus, if the original hybrid had received chromosomes 1·2 and 3·4* from one parent and 1·3 4·2 from the other, one of its progeny derived by selfing might have received 1·2 3·4 through both sperm and egg, plus 1·3. In the sporocytes of this trisomic individual, a trivalent would

---

* Two digits connected by a dot stand for a chromosome, the digits representing the terminal segments of the chromosome.

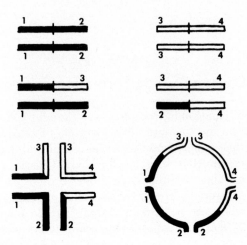

FIG. 5.1. Diagram showing Belling's hypothesis of segmental interchange. Non-homologous chromosomes 1·2 and 3·4 exchange segments to produce 1·3 and 2·4. In an individual that receives 1·2 and 3·4 from one parent and 1·3 and 2·4 from the other, pairing of corresponding segments in synapsis will produce a circle of four chromosomes.

often be formed, with the 1·3 chromosome occupying one end of the chain.

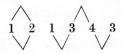

Since the exchanged chromosome was smaller than the others, its position in the chain would be easy to observe. Other configurations would also be possible, and these were found.

If this hypothesis were correct, then one should find a circle of four chromosomes in the F_1 hybrid between the two races used, and this was soon found. In his 1927 paper Belling suggested that circles in *Oenothera* might also have arisen via segmental interchange, a proposal that at first sight seemed rather bizarre. So-called non-reciprocal translocations had been observed genetically in *Drosophila* as far back as 1919 (Bridges), but the idea of reciprocal translocations, or segmental interchanges, as Belling called them, was new. To suggest that a phenomenon hitherto undetected in Nature had been going on in wholesale fashion in *Oenothera* seemed at first sight rather improbable. To build up a circle of fourteen it would be necessary to postulate a sequence of at least six interchanges. Despite the seeming improbability of such a process, however, the suggestion was taken seriously by several workers. Darlington (1929) and Håkansson (1928, 1930b) attempted to apply the hypothesis to *Oenothera*, mainly from a theoretical point of view. Emerson and Sturtevant (1931) and Cleland and Blakeslee (1930, 1931) tried to devise experimental tests of the concept.

Darlington accepted Belling's suggestion with enthusiasm. He did so because it made it possible to do away with the telosynaptic hypothesis and bring *Oenothera* into line with the parasynaptic interpretation of meiosis of which Darlington, originally a telosynaptist, had become a strong advocate. Darlington developed and urged the idea that a given chromosome in a circle is homologous at one end with one chromosome, and at the other end with a different chromosome—a necessary consequence of segmental interchange. He advanced the hypothesis that *Oenothera* chromosomes pair parasynaptically in zygophase, as they do in other organisms, but the pairing extends only a short distance back from their ends. The chromosomes are thus held together end to end by the chiasmata that develop in the pairing ends. The proximal portions of the chromosomes do not synapse; most of the length of the chromosome, therefore, remains unpaired throughout diplophase,* a fact originally considered to be evidence for telosynapsis, but now shown to be compatible with parasynapsis occurring only in the end segments.

Håkansson also was strongly inclined toward Belling's hypothesis but in his first paper (1928) felt that a difficulty in the way of acceptance arose from the behavior of the chromosomes in de Vries' half-mutants, which he thought should show open chains instead of closed circles. Håkansson, however, was visualizing the origin of half-mutants as resulting from irregularities in the zigzag arrangement at metaphase I. He was still thinking in terms of the telosynaptic theory and the presence of fully homologous pairs of chromosomes in the circle. Darlington, however, attempted to explain the origin of half-mutants through a process of interchange between chromosomes in the circle of twelve in *lamarckiana*. In a second paper (1930b), Håkansson accepted this interpretation, which removed his doubts and made it possible for him to accept Belling's segmental interchange hypothesis as valid for *Oenothera*. It may be remarked parenthetically, however, that it is not possible for

---

* It is not difficult to observe in pachyphase or diplophase cases where chromosomes, still thin, are paired side by side for a short distance back from the tip, the remainder of the chromosomes remaining unpaired (Fig. 8.1). No convincing evidence of synapsis farther back from the ends has been presented, even in all-paired forms, although Weier (1930) has claimed to have observed pairing throughout the length of the chromosome in *O. hookeri*. It can be safely assumed that synapsis is side by side in *Oenothera*, but only in the terminal segments, and that the mechanism that holds the chromosomes end to end is the presence of chiasmata resulting from crossing over in the pairing segments.

half-mutants to arise from *lamarckiana* by a single interchange of segments. The situation is more complicated than this, as will be mentioned later.

Neither Darlington nor Håkansson attempted to demonstrate experimentally that Belling's hypothesis would explain the origin of circles in *Oenothera*. This remained for Emerson and Sturtevant (1931) and Cleland and Blakeslee (1930, 1931), who independently set out to test the validity of this concept as applied to *Oenothera*.

The reasoning behind these tests was as follows: If circles have arisen through segmental interchange, there must have occurred in the course of evolution a considerable shuffling of pairing ends through interchange. Each complex now existing must have acquired in this process one, and only one, arrangement of end segments out of the many arrangements theoretically possible. Actually, 135 different arrangements are possible in *Oenothera*, i.e., 14 ends can be arranged into seven groups of two in $13 \times 11 \times 9 \times 7 \times 5 \times 3$ ways. Each complex has one of these. If two complexes are combined, each with one specific arrangement of ends, then the resultant hybrid can have but one chromosome configuration out of the 15 theoretical possibilities. If one could then determine the segmental arrangement in two complexes, one could predict which one of the 15 possible chromosome configurations would be found when these were combined. If this prediction were verified, it would demonstrate that each complex does indeed have a specific arrangement of ends. This could only have been acquired through a history of interchange in the course of evolution—the only alternative would be to deny that the various complexes have come from a common source, that each was specially created.

Both pairs of investigators, therefore, set out to analyze the arrangement of segments in a variety of complexes in terms of a selected standard, in the hope that predictions could be made and tested.

Emerson and Sturtevant, and also Cleland and Blakeslee, chose [h]*hookeri** as their standard complex. This was done because it was thought that this complex might be primitive and close to the ancestral line of the subgenus *Oenothera*, since it belongs to a lethal-free, open-pollinated, complex-homozygote. Actually it turned out later that the ancestral arrangement for the subgenus is probably one interchange removed from the [h]*hookeri* arrangement.

Both pairs of authors adopted the convention of using numbers to represent the pairing ends, the two ends of a single chromosome being

---

* = haplo-hookeri. Superscript [h] is used for complexes belonging to complex-homozygotes.

connected by a dot. The formula assigned to the standard complex was therefore:

$$^{h}hookeri = 1·2 \quad 3·4 \quad 5·6 \quad 7·8 \quad 9·10 \quad 11·12 \quad 13·14$$

While the methods of determining segmental arrangements of the various complexes differed in detail, depending upon what crosses the two groups of investigators happened to make first, in essence the methods were the same and produced the same results. At first, arbitrary selections among multiple possibilities played a part, but it was not long before all ends had been defined in terms of the standard and the other complexes with which it had been combined. The first few formulae were derived using only cytological data (i.e., chromosome configurations), but the use of available genetical data helped in a number of cases to reduce the possibilities for a complex. After the number of complexes that had been analyzed completely became large enough, it became possible once more to utilize only chromosome configurations in the analysis of additional complexes. Morphological differences among chromosomes could not be used, inasmuch as all circle chromosomes in all complex-combinations where circles are present are so similar in morphology that they cannot be distinguished in meiosis.

To illustrate the way in which the analysis started, we may use the work of Emerson and Sturtevant (1931).

Let $^{h}hookeri = 1·2 \ 3·4 \ 5·6 \ 7·8 \ 9·10 \ 11·12 \ 13·14$. *Flavens* gives ⊙4 and 5 pairs with $^{h}hookeri$ and differs from it, therefore, in respect to two chromosomes. Let *flavens* have 1·4 3·2.

$$flavens = 1·4 \quad 3·2 \quad 5·6 \quad 7·8 \quad 9·10 \quad 11·12 \quad 13·14.$$

Velans also gives ⊙4 and 5 pairs with $^{h}hookeri$, but gives ⊙4, ⊙4, 3 pairs, with *flavens*. It differs from $^{h}hookeri$, therefore, in respect to two different chromosomes from the two involved in the *flavens* exchange. Let *velans* have 5·8 7·6.

$$velans = 1·2 \quad 3·4 \quad 5·8 \quad 7·6 \quad 9·10 \quad 11·12 \quad 13·14.$$

These selections leave 1·2 and 3·4 equivalent and 5·6 and 7·8 also equivalent. They may be distinguished by the use of *gaudens*, which gives ⊙10, 2 pairs, with $^{h}hookeri$ and ⊙12, 1 pair, with *velans* and *flavens*. *Gaudens* therefore has a chromosome that is present in $^{h}hookeri$ but not in *flavens*, i.e., 1·2 or 3·4; it has another chromosome that is present in $^{h}hookeri$ but not in *velans*, i.e., 5·6 or 7·8. The authors arbitrarily selected 1·2 (not 3·4) and 5·6 (not 7·8). *Gaudens*, therefore, has 1·2 and 5·6.

The strain of *franciscana* used by Emerson and Sturtevant (its complex here called ^{h}*franciscana E and S*) gave $\odot$4, 5 pairs, with ^{h}*hookeri** and $\odot$4, $\odot$4, 3 pairs, with *flavens*, but gave $\odot$6, 4 pairs, with *velans*. It also gave $\odot$10, 2 pairs, with *gaudens*. It has two chromosomes with the same arrangements of ends as ^{h}*hookeri* but which differ from *flavens*. It must, therefore, have 1·2 3·4. It has one chromosome like ^{h}*hookeri* which is not in *velans* (5.6 or 7·8). If it has 5·6, then the second pair with *gaudens* is accounted for and the circle of four with ^{h}*hookeri* must involve an exchange between 7·8 and one of the last three chromosomes of the ^{h}*hookeri* formula. Two of the last three ^{h}*hookeri* chromosomes will also be present in ^{h}*franciscana E and S*. If on the other hand it has 7·8, then it cannot have 5·6, which must be in the $\odot$4 with ^{h}*hookeri*, and the second pair with *gaudens* is still to be sought. The second pair with *gaudens* would then have to involve one of the remaining ^{h}*franciscana E and S* chromosomes which are identical with ^{h}*hookeri* chromosomes. *Gaudens* would then give too many pairs with ^{h}*hookeri*. The second alternative is therefore incorrect. ^{h}*franciscana E and S* has 5·6. The exchange with ^{h}*hookeri* involves 7·8. The authors arbitrarily selected for ^{h}*franciscana E and S* 7·10 9·8.

$$\text{^{h}}\textit{franciscana E and } S = 1\text{·}2 \quad 3\text{·}4 \quad 5\text{·}6 \quad 7\text{·}10 \quad 9\text{·}8 \quad 11\text{·}12 \quad 13\text{·}14.$$

This is perhaps enough to show the way in which, one by one, chromosomes and chromosome ends were originally defined in terms of the standard complex and of other complexes previously determined. Only a few more experiments of this sort sufficed to define all the ends. After this point was reached, arbitrary selections could no longer be made. It was possible, however, to use genetic data in some cases to assist in the determination of the proper segmental arrangements. Genetical studies had shown that certain pairs of genes segregated independently of the complexes and of each other in some hybrids; in other hybrids they were independent of the complexes but linked together. In still other hybrids they were linked to each other and to the rest of the complex. Certain possibilities for the segmental arrangement of a given complex could be ruled out because they would not permit the kinds of linkage relations known to exist in combinations of this complex.

The result of analyses of segmental arrangements made it possible to predict the chromosome configurations in certain hybrid combinations, and several predictions were made in the early papers by Emerson and Sturtevant, Cleland, and Blakeslee and Cleland (1932, 1933). All

---

* The ^{h}*franciscana* of de Vries' strain gives seven pairs with ^{h}*hookeri*.

but one of these* predictions proved to be correct, thus demonstrating beyond question that each complex has its own particular arrangement of pairing segments, and that different complexes have different arrangements. Assuming an evolutionary development from a common ancestor, such differences can only be accounted for by a process of segmental interchange. Various sequences of interchange have occurred in the ancestry of the various complexes found in nature, thus leading to diverse segmental arrangements.

The diversity in segmental arrangement that now exists in the population is no doubt very great. Although the materials that have been examined thus far represent but an infinitesimal fraction of the total population,† we have already found in our laboratory 162 different segmental arrangements in North American races. Additional arrangements have been found in European material derived from North America. The total number in existence must be up in the thousands. Another indication of the diversity that exists is seen when we realize that every possible combination of ends into groups of two has already been found. There are 91 possible associations of 14 ends by two's—1·2, 1·3, 1·4—2·3, 2·4, etc. As will be seen in Fig. 5.2 and Table 4.III every one

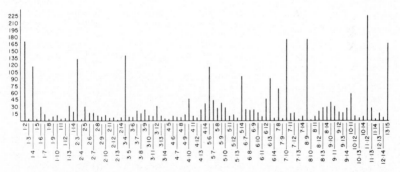

FIG. 5.2. Diagram to show the frequency with which each of the 91 possible chromosomes from the standpoint of association of end segments has been found among the North American strains studied by Cleland and associates.

* In one case, incorrect determination of a chromosome configuration led to an incorrect prediction (*velans·maculans* has ⊙12, not ⊙14; the prediction of ⊙12 for ʰ*hookeri·maculans* was therefore incorrect. It should have been ⊙10, 2 pairs.).

† The map in Fig. 18.4 shows that large areas within the range of the subgenus are as yet unexplored. Furthermore, those areas where collections have been made are represented, as a rule, by material derived from a single individual, no attempt having been made to analyze populations as a whole.

of these has been found. The shuffling of ends in the course of evolution has, therefore, been extensive. It has involved all ends. So far as one can judge, exchange can occur equally well between any two ends (but see Chapter 18 for certain limitations).

TABLE 5.I

Frequency with which individual chromosomes have been found among the members of the subgenus *Oenothera* (all 91 possible chromosomes have been found)

Chromosome	Frequency	Chromosome	Frequency	Chromosome	Frequency
1·2	171	3·12	31	7·8	68
1·3	4	3·13	10	7·9	3
1 4	117	3·14	4	7·10	174
1·5	4			7·11	16
1·6	31	4·5	4	7·12	18
1·7	15	4·6	10	7·13	2
1·8	2	4·7	8	7·14	11
1·9	9	4·8	6		
1·10	12	4·9	12	8·9	175
1·11	4	4·10	47	8·10	11
1·12	5	4·11	11	8·11	10
1·13	31	4·12	4	8·12	20
1·14	19	4·13	24	8·13	28
		4·14	37	8·14	34
2·3	134				
2·4	5	5·6	116	9·10	39
2·5	31	5·7	45	9·11	31
2·6	16	5·8	25	9·12	18
2·7	16	5·9	38	9·13	17
2·8	11	5·10	27	9·14	27
2·9	8	5·11	10		
2·10	11	5·12	14	10·11	58
2·11	6	5·13	6	10·12	10
2·12	7	5·14	95	10·13	6
2·13	1			10·14	10
2·14	7	6·7	25		
		6·8	23	11·12	225
3·4	139	6·9	23	11·13	27
3·5	9	6·10	18	11·14	4
3·6	8	6·11	11		
3·7	23	6·12	47	12·13	16
3·8	17	6·13	91	12·14	7
3·9	24	6·14	5		
3·10	11			13·14	165
3·11	10				

CHAPTER 6

Method of Analysis of Segmental Arrangements of Complexes

The analysis of segmental arrangements, which was begun to test the interchange hypothesis, has continued to be carried on in several laboratories. A number of complexes were determined by the writer during the 1930's. Catcheside contributed several new determinations in 1940. In 1950, Cleland and Hammond, Preer, and Geckler published between them the formulae for 80 additional complexes. Since the first determinations were made, Renner and his students have routinely analyzed all complexes with which they have worked, and other laboratories have also contributed significantly (see Appendix I for a list of complexes so far determined).

In all later work in our laboratory, which has been done in connection with studies of the evolution of the group, reliance has been placed solely on cytological data. This is because the complexes analyzed have belonged to races brought in from the wild, about whose genetical behavior little or nothing was known at the time when segmental analyses were made. The data used were entirely the chromosome configurations found in combinations of the complex being tested with other complexes whose segmental arrangements had been previously determined. Since all circle chromosomes in all races are essentially identical in morphology, there are no cytological markers that can be used in such determinations.*

In order to illustrate the methods of reasoning used in the determination of segmental arrangements, some illustrations will be given at this point. These have not been previously published. It should be mentioned that we no longer give Latin names to complexes. There are too many of them. Instead, the complex in a race that tends to be transmitted exclusively or predominantly through the egg is called the alpha complex; the one that comes mostly through the sperm is the beta complex. Races are ordinarily named after the locality where the original seed was obtained. One additional comment: the reasoning followed in each case was based on the data available when that particular analysis was made. The process in some cases would have been simpler and less

---

* It has recently been found in a number of forms in our laboratory that the larger nucleolus is apparently associated with end number 3 (unpublished). We have not attempted, however, to use this as a marker in determining the segmental arrangements of previously unknown complexes.

roundabout if the analysis had been made at a later date when more cytological data were available.

Alpha Warsaw

(Egg complex of a race from Warsaw, Indiana)

Complexes used as testers

ʰ*blandina*	1·2	3·4	5·6	7·10	9·14	11·12	13·8
ʰ*franciscana de Vries*	1·2	3·4	5·6	7·8	9·10	11·12	13·14
β*Nebraska*	1·4	3·2	5·10	7·6	9·8	11·12	13·14
neo-acuens	1·13	3·2	5·6	7·10	9·8	11·12	4·14
punctulans	1·4	3·9	5·2	7·8	6·12	11·10	13·14
rigens	1·2	3·4	5·6	7·11	9·10	8·14	13·12
velans	1·2	3·4	5·8	7·6	9·10	11·12	13·14

Gives ⊙4, 5 pairs, with ʰ*blandina* and ⊙6, 4 pairs, with ʰ*franciscana* de V.; has, therefore, one ʰ*blandina* chromosome* not present in ʰ*franciscana*, namely, one of 7·10 9·14 13·8. Gives ⊙8, 3 pairs, with *neo-acuens*, so has a ʰ*franciscana* chromosome not present in *neo-acuens*, namely, one of 1·2 3·4 7·8 9·10 13·14. It cannot have 7·8 or 13·14, however, since these are in *punctulans* with which it gives ⊙14. Gives ⊙8, 3 pairs, with *velans*, so has a ʰ*franciscana* chromosome not present in *velans*, namely, 5·6 or 7·8. We have seen, however, that it cannot have 7·8, so it must have 5·6. Gives ⊙4, ⊙4, ⊙4, 1 pair, with β *Nebraska*. In order to complete a circle with β *Nebraska* when 5·6 is present, it must have 7·10. It cannot, therefore, have 9·10 in common with ʰ*franciscana*. Since it does not have 7·8, 9·10 or 13·14, these three must be in the ⊙6 with ʰ*franciscana*. The four pairs with the latter must result, therefore, from the presence of 1·2 3·4 5·6 11·12.

It is now possible to write out all formulae that will give ⊙6, 4 pairs, with ʰ*franciscana*. There are eight such formulae:

1·2	3·4	5·6	11·12	8·9	10·13	14·7
					10·14	13·7
				8·10	9·13	14·7
					9·14	13·7
				8·13	14·9	10·7
					14·10	9·7
				8·14	13·9	10·7
					13·10	9·7

When each of these is tested against ʰ*blandina* and *rigens* (⊙6), all are eliminated but one:

---

* Reference is made here only to the matter of end arrangement. Nothing is implied with respect to genic content.

$\alpha Warsaw = 1 \cdot 2 \quad 3 \cdot 4 \quad 5 \cdot 6 \quad 7 \cdot 10 \quad 9 \cdot 13 \quad 11 \cdot 12 \quad 8 \cdot 14^{*}$

This arrangement gives the correct configuration with all complexes with which α *Warsaw* has been combined. In addition to those mentioned above, these include β *Bestwater II* (⊙14), α *Bestwater II* (⊙10), β *Birch Tree II* (⊙12), β *Iowa II* (⊙4, ⊙4, ⊙6), *truncans* (⊙14), β *Paducah* (⊙14) and β *Warsaw* (⊙14). (See appendix I for segmental arrangements of these complexes.)

Beta Delaware

(Sperm complex of a race from Delaware, Ohio)
Complexes used as testers

ʰ*blandina*	1·2	3·4	5·6	7·10	9·14	11·12	13·8
excellens	1·2	3·4	5·6	7·10	9·8	11·12	13·14
ʰ*franciscana* de V.	1·2	3·4	5·6	7·8	9·10	11·12	13·14
maculans	1·2	3·8	5·6	7·10	9·12	11·13	4·14
punctulans	1·4	3·9	5·2	7·8	6·12	11·10	13·14
rigens	1·2	3·4	5·6	7·11	9·10	8·14	13·12
velans	1·2	3·4	5·8	7·6	9·10	11·12	13·14
truncans	1·13	3·7	5·2	4·6	9·14	11·10	8·12
α*Delaware*	1·2	3·4	5·14	7·10	9·8	11·12	13·6

Gives ⊙6, ⊙6, 1 pair, with *rigens*. The chromosome in common with *rigens* cannot be 1·2, 3·4, 5·6 or 9·10 since these are all in ʰ*franciscana* de V. with which β *Delaware* gives ⊙14. It must possess one of 7·11 8·14 13·12.

Gives ⊙4, ⊙8, 1 pair, with *maculans*. The pair cannot involve 1·2 or 5·6 or 7·10 since these are all in *excellens* with which it gives ⊙14. The pair must involve one of 3·8 9·12 11·13 4·14. If it has 7·11 of *rigens*, it cannot have 11·13 of *maculans*. If it has 8.14 it cannot have 3·8 or 4·14. If it has 13·12 it cannot have 9·12 or 11·13. The possible combinations of chromosomes forming pairs with *rigens* and *maculans* respectively are:

7·11	3·8	8·14	9·12	13·12	3·8
7·11	9·12	8·14	11·13	13·12	4·14
7·11	4·14				

---

* In order to make the reading of formulae easier, and to facilitate comparison of formulae, the following convention has been adopted by the author: so far as possible, individual chromosomes are written with the odd-numbered end to the left, and the chromosomes are so arranged in the formula that the left-hand digits are in ascending order (1, 3, 5, etc.), reading from left to right. When two odd- or even-numbered ends are associated in a single chromosome, the lower number is to the left. Chromosomes with two even-numbered ends are placed in spots not occupied by those which have odd-numbered ends.

Gives ⊙10, 2 pairs, with *punctulans*. The chromosomes common to *punctulans* cannot be 7·8 or 13·14 since these are in ᴴ*franciscana* of de Vries. It has one of the following combinations of two *punctulans* chromosomes:

1·4	3·9	3·9	5·2	5·2	6·12
1·4	5·2	3·9	6·12	5·2	11·10
1·4	6·12	3·9	11·10	6·12	11·10
1·4	11·10				

We can combine these two sets of two chromosomes in all possible ways and obtain the following possibilities:

7·11	3·8	1·4	5·2	8·14	11·13	1·4	3·9
		1·4	6·12			1·4	5·2
		5·2	6·12			1·4	6·12
7·11	9·12	1·4	5·2			3·9	5·2
7·11	4·14	3·9	5·2			3·9	6·12
		3·9	6·12			5·2	6·12
		5·2	6·12	13·12	3·8	1·4	5·2
8·14	9·12	1·4	5·2			1·4	11·10
		1·4	11·10			5·2	11·10
		5·2	11·10	13·12	4·14	3·9	5·2
						3·9	11·10
						5·2	11·10

We may now take each of these in turn and arrange all possible combinations that will yield ⊙6, ⊙6, 1 pair, with *rigens*. When duplications are eliminated, there are 24 such combinations. These may be tested against *velans* (⊙14), *excellens* (⊙14), ᴴ*blandina* (⊙14), *punctulans* (⊙10), ᴴ*franciscana* of de Vries (⊙14), α *Delaware* (⊙14), *truncans* (⊙12). As a result, only one formula survives:

$$\beta \ Delaware = 1·4 \quad 3·8 \quad 5·7 \quad 2·14 \quad 9·6 \quad 11·10 \quad 13·12$$

This arrangement gives the correct configuration with each of the complexes mentioned above and also with the other complexes with which it has been combined; namely, *neo-acuens* (⊙14) and *jugens* (⊙14).

Beta La Salle I

(Sperm complex of a race from La Salle, New York)

When cytological data are scanty, or when a complex gives mostly large circles with testers, it is still possible sometimes to determine the segmental arrangement of a complex, though the work is rather laborious. An example involves the analysis of β *La Salle I*, which gives large circles with most of the complexes with which it has been combined.

Complexes used as testers

[h]*franciscana* de V.	1·2	3·4	5·6	7·8	9·10	11·12	13·14
curtans	1·7	3·4	5·8	2·10	9·11	6·12	13·14
neo-acuens	1·13	3·2	5·6	7·10	9·8	11·12	4·14
punctulans	1·4	3·9	5·2	7·8	6·12	11·10	13·14
α *La Salle I*	1·6	3·2	5·7	4·10	9·8	11·12	13·14
excellens	1·2	3·4	5·6	7·10	9·8	11·12	13·14
jugens	1·6	3·2	5·14	7·13	9·8	11·12	4·10
velans	1·2	3·4	5·8	7·6	9·10	11·12	13·14
rigens	1·2	3·4	5·6	7·11	9·10	8·14	13·12
gaudens	1·2	3·12	5·6	7·11	9·4	8·14	13·10
truncans	1·13	3·7	5·2	4·6	9·14	11·10	8·12

Gives $\odot$4, $\odot$8, 1 pair, with [h]*franciscana* de V. and *velans*. The paired chromosome with [h]*franciscana* cannot be 3·4 (gives $\odot$14, with *curtans*), 5·6 ($\odot$6, $\odot$8 with *neo-acuens*), 7·8 or 13·14 ($\odot$4, $\odot$10 with *punctulans*), 11·12 ($\odot$14 with α *La Salle I*). It must, therefore, have 1·2 or 9·10 in common with [h]*franciscana* de V. This will account for the pair with *velans*. The pair with *excellens* ($\odot$12) cannot be 3·4 or 5·6 or 11·12 or 13·14, nor can it be 9·8 ($\odot$14 with *jugens*). It must, therefore, have 1·2 or 7·10 (not both) in common with *excellens*. If it has 7·10, however, it cannot have 9·10 and vice versa. It cannot, therefore, have either 7·10 or 9·10 and must have 1·2.

Since it gives $\odot$4, $\odot$8 with [h]*franciscana* and $\odot$12 with *excellens*, chromosomes 7·8 and 9·10 of [h]*franciscana* must be in different circles with β *La Salle I*. We may first assume that 7·8 of [h]*franciscana* de V. is in the circle of four, and 9·10 in the circle of eight, and list all the possible [h]*franciscana* chromosomes that could exchange with 7·8 to form a circle of four. Of the various possibilities, 3·4, 11·12 and 13·14 cannot be in a circle of four with 7·8, for then it would not be possible for β *La Salle I* to give $\odot$4, $\odot$8 with both [h]*franciscana* and *velans*. If 7·8 is in the circle of four with [h]*franciscana*, the other chromosome must be 5·6. β *La Salle I* will then have 7·5 6·8 (if it had 7·6 5·8, it would give too many pairs with *velans*).

If we assume that 9·10 is in the circle of four with [h]*franciscana* and 7·8 in the circle of eight, it will not be possible to write a formula that will give $\odot$4, $\odot$8 with both [h]*franciscana* and *velans* if 5·6 is in the circle of four with 9·10. If 9·10 is in the circle of four, the other [h]*franciscana* chromosome must be 3·4 or 11·12 or 13·14.

The possibilities for β *La Salle I* so far are:

1·2	7·5	6·8		1·2	9·11	12·10
1·2	9·3	4·10			9·12	11·10
	9·4	3·10		1·2	9·13	14·10
					9·14	13·10

We may now write out all formulae that will give $\odot 4$, $\odot 8$ with ^h*franciscana*. There are 336 such formulae, many of which, however, do not have to be written down since they contain chromosomes that are present in the various complexes with which β *La Salle I* gives no pairs. The formulae that are written down may be tested by the use of *neo-acuens* ($\odot 6$, $\odot 8$), *rigens* ($\odot 12$), *gaudens* ($\odot 12$), *truncans* ($\odot 14$), α *La Salle I* ($\odot 14$) and *punctulans* ($\odot 4$, $\odot 10$). As a result, all are eliminated except one.

$$\beta\ La\ Salle\ I = 1 \cdot 2 \quad 3 \cdot 5 \quad 6 \cdot 8 \quad 7 \cdot 12 \quad 9 \cdot 13 \quad 11 \cdot 4 \quad 10 \cdot 14$$

This formula gives the correct configuration with all complexes with which β *La Salle I* has been combined. In addition to those mentioned above, these include β *Magnolia* ($\odot 6$, $\odot 8$), and *albicans* ($\odot 6$, $\odot 8$).

By the use of methods such as those just illustrated, we have fully determined in our laboratory the segmental arrangements of 438 complexes and have found 162 different segmental arrangements. A number of other workers have added to the list (see Appendix I). Most of our work has been done in connection with a study of the evolution of the group, which will be discussed in a later chapter. It should be emphasized that this makes a completely consistent system. Every complex in the list has given the chromosome configuration called for by its formula with every other complex with which it has been combined, and the configurations of all combinations that have not been made can be predicted.

In conclusion it is clear that, through a series of segmental interchanges, or reciprocal translocations, each of the various complexes existing in Nature has acquired its own arrangement of end segments, the number of different arrangements in the population being exceedingly large. When complexes with different arrangements of end segments are combined, the pairing of homologous ends will often result in the formation of circles, the size and number depending upon the degree of similarity or dissimilarity in segmental arrangement between the associated complexes. The all but universal presence of $\odot 14$ in most of the oenotheras from the Rocky Mountains eastward shows that naturally associating complexes are as a rule very different from each other segmentally. The reason for this will become clear when we discuss the evolution of the group.

PART III
SPECIAL ASPECTS OF OENOTHERA CYTOGENETICS

CHAPTER 7

Chromosome Structure

In order to understand the process by which the subgenus *Oenothera* has evolved, a matter to be discussed later, it is necessary to become acquainted with certain additional features of its cytology and genetics. We shall first consider some aspects of its cytology, then review certain matters of genetical interest.

As we shall see later, when discussing the evolutionary history of the subgenus *Oenothera*, the story as we know it today would not have been possible had it not been for the fact that the common ancestor possessed chromosomes with the following features: they were all of the same size, had median centromeres, and possessed areas of heterochromatin on either side of the centromere. The importance of these facts will be pointed out later, but it is interesting to inquire at this point regarding the situation today. Are all *Oenothera* chromosomes still equal in size and isobrachial, or have there been modifications in chromosome morphology in the course of time?

Not all workers have been fully agreed on this point. Some have asserted that the chromosomes cannot be distinguished with certainty from one another. Others claim that recognizable differences exist, at least in somatic cells. In meiosis any differences that might exist become obscured as the chromosomes condense in diakinesis and the first metaphase.

Among those who have claimed that recognizable differences exist in the soma are van Overeem (1922), Darlington (1931), Wisniewska (1935), Marquardt (1937) and Bhaduri (1940). van Overeem referred briefly to the situation in *lamarckiana*, stating that it was possible to distinguish slight differences in the chromosomes. He recognized two long pairs, strongly bent in the middle, two long pairs only slightly bent, one middle-sized pair and two short pairs. Lewitsky (1931, pp. 128–130), on the other hand, was unable to distinguish size classes among the chromosomes of *lamarckiana*; they formed a graded series, the shortest being three-quarters as long as the longest. He did not think that these differences were significant. He did find some difference in the position of the centromere in the various chromosomes.

Darlington (1931) examined somatic mitosis in four oenotheras. One of these was a tetraploid, one a triploid, one was said to show inferior fixation, the fourth was a *Raimannia* (*O. berteriana*). In all of these, differences in size and morphology were described, no two chromosomes being exactly alike, in line with the fact that in complex-heterozygotes no chromosome is wholly homologous with any other (although there should be identical chromosomes in tetraploids and triploids). The differences were not of such magnitude, however, that they would be easily detectable in meiosis (range in somatic cells 1·5–2·5 μm). Darlington was unable to observe differences in size as between *berteriana* and the other strains, although Schwemmle (1927) stated that *berteriana* chromosomes are smaller than those of *muricata* and we have found in extensive studies that the chromosomes of the subgenus *Raimannia* (including *berteriana*) are definitely smaller than those of the subgenus *Oenothera*.

Bhaduri (1940) also described differences in somatic chromosome structure in several oenotheras, including *lamarckiana*. He found the shortest chromosomes to be about two-thirds the length of the longest, the range in *lamarckiana* being 2·5–3·8 μm. In *lamarckiana* he found a heteromorphic pair. He classified chromosomes according to size into long (L), medium (M) and short (S), and according to position of the centromere into median (M) or submedian (S). Thus, in *lamarckiana* he found 3LM, 1LS, 6MM, 2MS, and 2SM chromosomes, where the first letter refers to length, the second to centromere location. The LS and one of the LM chromosomes constituted the heteromorphic pair.

Measurements such as have been presented by Bhaduri and others must be taken with reservation. As has just been noted, *Oenothera* chromosomes are very small. Since the limit of resolution of the light microscope in white light is not better than 0·25 μm, too much credence cannot be put on measurement in terms of fractions of a micron, especially when these have been obtained with the use of a camera lucida, which introduces an additional source of inaccuracy. Darlington's figures, for example, were drawn at a magnification of 6400 diameters.

Wisniewska (1935) and Marquardt (1937) both studied *O. hookeri* and both concluded that the chromosomes showed distinct size differences, but they disagreed in their classification of these differences. Wisniewska found that one chromosome of the haploid set was larger, the others smaller and equal to each other. The large chromosome had a knob at both ends, five chromosomes had a knob at one end (one of these with a satellite at the other end), and one chromosome had no knobs. The centromeres were subterminal in one chromosome, submedian in another and median in the remainder. Marquardt, on the other hand, classified the haploid chromosomes of this species into two large, one doubtfully large, two median, one doubtfully small and one small chromosome. He considered knobs to be inconstant in occurrence. The centromere was submedian in one chromosome, median in four others, and apparently median in the other two.

These two highly competent workers, therefore, studied presumably identical material and yet were unable to agree on the structure of the chromosomes. My own observations of *hookeri* have been confined to meiotic stages. In some diakinesis cells one can detect a recognizable difference in size between the largest and smallest pairs; in other cells, it is difficult or impossible to choose the largest from among the larger chromosomes. At the other extreme, however, the smallest pair can usually be distinguished, and this distinction can often be made as late as metaphase I, thus suggesting that one chromosome of the set is indeed smaller than the others. The differences, therefore, are on the whole slight and of minor significance. When one considers the small size of *Oenothera* chromosomes, it is not surprising that differences of opinion might arise regarding size differences. The effects of fixing, staining and dehydration, and the fact that the living chromosomes probably condense unevenly during prophase, introduce a degree of variation that is at least equal to the observed variations in length. Before one can accept the conclusion that such small chromosomes are significantly different in size, one must be sure that these differences are at a level greater than what one would expect as the result of unequal condensation or of handling. It would seem that significant size differences in chromosome length have not been satisfactorily demonstrated in the literature, at least in forms with large circles.

In our own laboratory, a careful study of chromosome morphology has been carried out by Gardella (1953), who examined root tip and pollen tube mitoses in four races, two of them all-pairing, the other two with a circle of 14. A certain amount of variation in chromosome length and centromere position was seen in all forms, but these differences were so slight that individual chromosomes, except for the satellite chromosome, could not be identified. Differences were not greater than would be expected to result from inequalities of fixation, differential contraction and inaccuracies of measurement in such small chromosomes. Variations within a single anaphase cell were no greater than those between sister chromatids of single separating chromosomes. The variation among all of the chromosomes of a set within any one race was of the same order of magnitude as that between the satellite chromosomes alone. Variation between races was no greater than between the chromosomes of a single cell (Fig. 7.1).

In races where the chromosomes are paired in meiosis, it makes no difference, so far as their ability to disjoin is concerned, whether nonhomologous chromosomes are equal or unequal in size. In the case of

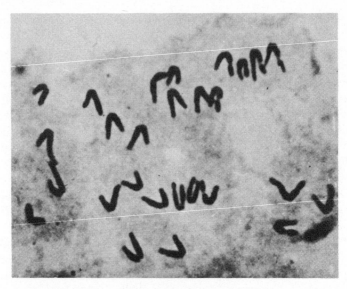

FIG. 7.1. Anaphase in root tip cell of *O. suaveolens*, showing uniformity in size and structure among the chromosomes (Gardella, 1953).

circle chromosomes, however, the case is different. In circles where the chromosomes are unequal in size or in which the arms are of unequal length, as in *Rhoeo*, the centromeres are unequally spaced around the circle and the forces responsible for the movement of the centromeres to the poles are unevenly distributed, and this may result in a certain amount of non-disjunction and consequent sterility. In *Oenothera*, however, where the circle chromosomes are equal in size and isobrachial, centromeres are evenly spaced, and regular disjunction occurs in a high proportion of the cells (Fig. 7.2). In the course of evolution the translocations that have occurred, or at least those that have survived, have involved equal exchanges. Consequently, in spite of, in

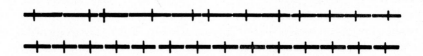

FIG. 7.2. Where chromosomes are unequal in size and have unequal arms (above), centromeres are scattered at irregular intervals around the circle resulting in frequent nondisjunction. When chromosomes are equal and have equal arms (below), centromeres are evenly distributed, and non-disjunction is relatively infrequent. Reprinted from Cytologia by permission.

many cases, long histories of interchange, the chromosomes in large circles have retained essentially unaltered structure; in fact, our experience has shown that the chromosomes in large circles in *Oenothera* are on the average more uniform in size and morphology than are those in all-pairing forms. This conclusion goes counter to the concepts that Darlington developed in regard to chromosome structure in *Oenothera* (1931, 1933, 1936). Darlington has claimed that interchanges do not have to be equal. Breaks may occur anywhere; consequently, some chromosomes may become enlarged, others reduced as the result of interchange. Darlington's conclusions will be discussed in more detail in a later section (p. 121).

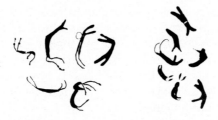

Fig. 7.3. Late prophase chromosomes from the generative cells of *O. suaveolens* (Gardella, 1953).

When examined during somatic prophase, individual *Oenothera* chromosomes tend to show differentiation in structure (Fig. 7.3). In the proximal region, the chromatids are closely appressed, possibly coiled about each other. In the distal areas, however, the chromatids are often separated, are quite straight and frequently knobbed. It is tempting to equate the distal region with the pairing ends and the proximal portion with the region which does not synapse in meiosis. There is, of course, no pairing in somatic cells. Some authors have claimed that the central region of the *Oenothera* chromosomes is largely heterochromatic in nature, the distal segments euchromatic (Seitz, 1935, Marquardt, 1937, Japha, 1939). Japha states that the euchromatic portion of the chromosome constitutes about three-quarters of its length in pachyphase but that this portion condenses more rapidly than the heterochromatic portion in later stages so that by the first metaphase the euchromatic segment is reduced to a small region at either end of the heterochromatin. These workers claim that the distinction between euchromatin and heterochromatin is clear at all stages of meiosis, the heterochromatin being more condensed than the euchromatin.

There are two facts regarding chromosome morphology in *Oenothera* that are beyond question. First, metabolic nuclei in meristematic regions show prochromosomes. This suggests that a portion of each chromosome is heterochromatic. Secondly, the mature chromosome shows a distinction between the proximal and distal regions (Fig. 7.3). The central region appears thicker than the distal segments.

The development and structure of somatic nuclei has been studied in our laboratory by Gardella (1953). She finds that the chromosomes spin out in telophase beginning at the distal end. More and more restricted regions remain condensed until in the early metabolic stage only the region contiguous with the centromere remains pycnotic. During interphase, the prochromosomes become progressively more spun out and less prominent until prophase approaches, at which time they begin to condense again and the process leading up to, and during, prophase is the reverse of what is observed during telophase and early interphase. In nuclei of root tips that have reached the region where mitosis will no longer take place, the prochromosomes continue to spin out and finally tend to disappear. This raises the question as to how much of the chromosome is truly heterochromatic, since there is little evidence of heteropycnosis in nuclei that no longer divide.

If heterochromatin is to be defined in terms of differential condensation, i.e., the tendency to remain pycnotic when the euchromatin has spun out, one may argue that regions approaching the centromere in *Oenothera* are progressively more heterochromatic, with the greatest concentration of heterochromatin closest to the centromeres. Although a given prophase chromosome shows a sudden transition from the apparently more condensed to the less condensed region, this should probably not be interpreted as showing the point at which heterochromatin ceases and euchromatin begins. The probability is that the transition point gradually shifts from nearer the centromere to nearer the ends during prophase as the region of condensation grows.

The genes that comprise the Renner complexes can hardly be situated in the pairing ends if pairing is due to chiasmata which in turn represent crossovers, for the genes would then be exchanged from one complex to the other through crossing over, something that rarely occurs in complex-heterozygous races. Most of the genes of the complexes, therefore, must reside in the proximal region of the chromosomes. It is therefore not likely that the proximal region is wholly heterochromatic in the genetic sense. It probably contains both euchromatic and heterochromatic sectors with the proportion of heterochromatin increasing as one approaches the centromere.

1. EXTRA DIMINUTIVE CHROMOSOMES

In a number of organisms extra diminutive chromosomes are known, generally designated as "B" chromosomes. In many cases these are largely heterochromatic in nature, and are either genetically inert or have rather vague effects, especially when present in large numbers.

A few cases of supernumerary chromosomes are known in the subgenus *Oenothera* (Cleland, 1951). These have been found only in races of *O. hookeri* (Fig. 7.4). The two races that have been investigated in detail

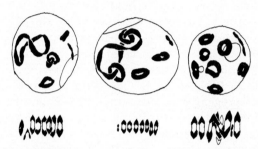

FIG. 7.4. *O. hookeri* Mataguey, diakinesis above, first metaphase below. Diakinesis figures show, left and center, ⊙6, 4 pairs, plus two diminutives; to right, 7 pairs plus three diminutives (Cleland, 1951). Reprinted from *Evolution* by permission.

(*Mono* and *Mataguey*) were derived from localities in California about 500 miles apart. Most plants of these strains carry two diminutives, which, except for size, do not differ appreciably from normal chromosomes, either in morphology or in behavior. They are not heteropycnotic. In meiosis, they pair and disjoin normally. They never associate with the other chromosomes, showing that their pairing ends are not homologous with any of the 14 pairing ends present in ordinary *Oenothera* chromosomes. They have no noticeable effect on the phenotype. They are transmitted through pollen as well as through the eggs.

When races that have diminutives are outcrossed, these chromosomes fail to associate with any of the other chromosomes no matter what the chromosome configuration of the hybrid. There is a tendency, however, for the diminutives to multiply, as shown by the fact that, whereas one would expect to find only a single extra in hybrid F_1 plants, a majority of the plants examined have shown two or even three extras.

Crosses were made between *Mono* and *Mataguey* (Cleland and Hyde, 1963), and while the diminutives never associated with normal chromosomes in the hybrids, the extras from the two races were able to synapse with each other, showing that they were at least partially homologous, and have come from a common source, even though they were resident in regions 500 miles apart. The extras in *Mono* were larger than those of *Mataguey*, so they are not wholly homologous.

It was suggested in the two papers cited that the extras could have arisen in one of two ways: (1) They might have arisen by loss of pairing ends through deletion, the extras in the two races having descended from the same chromosome; (2) They might have been derived from another subgenus as a result of intersubgeneric crossing followed by back crossing to *hookeri*. The latter hypothesis was favored.

In a third paper (Cleland 1967a) evidence was presented showing that the second hypothesis was at least within the realm of possibility. The hybrid (*Mono* × *Mataguey*) was crossed as female with a number of races of *Raimannia* (the reverse cross usually resulted in plastid incompatibility), producing in some cases viable F_1s; these F_1s were then crossed as female back to *hookeri* as male, resulting in typical and vigorous *hookeri* progeny with extra diminutives. This lends plausibility to the suggestion that the extra diminutives in races of *hookeri* have been derived from another subgenus by a similar sequence of events. It should be emphasized that the chromosomes of *Raimannia* and other subgenera are distinctly smaller than those of the subgenus *Oenothera*.

These studies, however, have involved only the subgenus *Raimannia*, which is not found except as an adventive in California. We have not had the opportunity of conducting similar experiments using *Anogra* or other subgenera native to this area and do not know whether viable hybrids can be obtained between them and *hookeri*.

CHAPTER 8

Meiosis

The early stages of meiosis are difficult to study in most material, but they are especially difficult in the case of *Oenothera*. The sporocytes are small; they form, as a rule, only one or two rows of cells in a loculus, enclosed within an anther wall several cells thick, often with a heavily cutinized outer surface. As a result, fixing fluids penetrate the anthers with difficulty. Smearing techniques are also difficult; one cannot prepare squashes in the usual manner—the sporocytes are the softest cells in the anther and the first to succumb to the pressure of the knife. The only way one can obtain smears is to squeeze cells out of the ends of the anthers by a rolling pressure in the direction of the end. This results in a small clump of cells being extruded in a relatively unharmed condition. This technique works quite well for older stages, but the earlier stages of meiosis, where the sporocytes are still firmly attached to one another, will not respond. For early meiosis, one must depend primarily on fixed anthers which can then be sectioned, or subjected to hydrolysis, squashed and stained. The quality of fixation when the latter methods are used leaves much to be desired, especially in view of the fact that it is in these early stages that the chromosomal material is in its most delicate and easily disturbed condition. As a consequence, all accounts of the early stages of meiosis in *Oenothera* must be accepted with considerable reservation. A good deal of constructive imagination has entered into the published descriptions of these stages (see Appendix II for notes on techniques).

In discussing meiosis we shall use the terms leptophase, zygophase, pachyphase and diplophase instead of the more commonly used lepto-tene, zygotene, etc. The latter terms are adjectives and strictly speaking should not be used as nouns; the terms leptonema, etc., are nouns which, however, apply to the threads, not to the stage in which the threads find themselves. The terms we shall use are logical, and unambiguous except for the term "diplophase" which is often used in contrast with "haplo-phase". This should pose no difficulty, however, since the context should always indicate the sense in which the term is used.

1. Leptophase

This stage has been described in *Oenothera* by a very few persons. Weier (1930), Wisniewska (1935) and Marquardt (1937) have all at-tempted descriptions. The drawings of Weier and Wisniewska bear little resemblance to the appearance of the thread system as I have observed it. The photographs of Marquardt (his Plate 1, Figs 1-3) give a better picture. The leptophase stage is characterized by a rather extensive and very delicate thread system which, in even the best fixed material that I have been able to secure, seems to form an anastomosing network. It is possible that perfect fixation would show fewer anastomoses and more independence among the threads. One striking feature of this stage is the presence of what look like prochromosomes, which are illustrated in Marquardt's photograph (1937, Plate 1, Fig. 1). These are very irregular in size, shape and structure. Their number is greater than the haploid chromosome number, and not greater than the diploid number. Marquardt's interpretation of these structures is as follows: the chromocenters, as he calls them, are heterochromatic in nature. They become split during the interphase just prior to meiosis. Homologous chromocenters then pair in early leptophase, but the euchromatic segments do not pair until zygophase. At the end of leptophase, therefore, the central heterochromatic regions of the chromosomes have already paired, the distal portions are as yet unpaired.

Marquardt did not mean to imply by the pairing of the chromocenters that chiasmata are formed or that crossing over takes place in this region. On the contrary, he explained the permanence of the Renner complexes by assuming that the genes comprising these complexes are scattered throughout the largely heterochromatic proximal region of the chromosomes, and are prevented from crossing over because the condensed condition of these segments does not permit chiasma formation. He considered the portion of the chromosome that pairs in early leptophase to be the region that does not experience true synapsis, i.e., association of homologues is not so intimate in this region that crossing over can occur. True synapsis takes place only in what he called the euchromatic (distal) segments in *Oenothera*.

To postulate pairing of homologous segments in early leptophase of course goes counter to Darlington's "precocity theory", as Marquardt admits (p. 168); it is also opposed to the generally accepted view that leptophase is the period after the chromosomes have lost most of the interchromosomal connections which bind them together in the metabolic reticulum, and before they have begun to synapse.

A careful study of leptophase stages has led me to a different interpretation from that proposed by Marquardt. The prochromosomes in leptophase are so irregular in shape and structure that it is not possible to interpret them either as split in the earlier stages or paired in the later stages of leptophase. Some could be interpreted as split or doubled, but others show many diverse shapes. It is more probable that these bodies simply represent the regions contiguous to the centromeres, which have remained pycnotic during the last premeiotic interphase and are still pycnotic in leptophase. They have not paired—pairing will not begin until the chromosomes have entered zygophase, and true synaptic pairing will then involve the pairing ends only, in line with the concept which has been proposed by Darlington.

It will not be amiss to mention again the small size of the nucleus (average 10 μm) and the fineness of the threads with which we are dealing. Wisniewska's drawings of the early stages of meiosis were examined under a 30× ocular and drawn at 4800 magnification. In view of the fact that oculars above 15× begin to lose resolution, and considering the loss in accuracy involved in using a camera lucida or similar equipment, one cannot expect a nucleus so greatly enlarged to show a very precise picture, especially when it is filled with threads so delicate that they approach the limit of the resolving power of the light microscope. In evaluating interpretations given to these stages, one must bear in mind the limitations under which the investigator works in examining material so finely dispersed.

2. Zygophase

Several workers have suggested that synapsis in *Oenothera* is of the polarized type, i.e., synapsis begins at the ends of chromosomes, which are oriented toward one side of the nucleus. Catcheside (1932) and Japha (1939) have both suggested this, as has Darlington (1931, p. 431). There can be little doubt that synapsis begins at the ends of the chromosomes, in fact that it proceeds only a short distance back from the tip. It may be open to question whether it is prevented from proceeding farther toward the center because of the condensed or coiled condition of this region, or because it meets a region in which the pairing ends are no longer homologous. Darlington (1931) postulates that a history of unequal exchanges has resulted in some or perhaps all chromosomes possessing "differential segments" which are not homologous in chromosomes whose adjacent end segments are homologous. If synapsis between homologous segments begins at the end and proceeds inward it will cease when a region is reached where the arms are no longer homologous. The other alternative is the one that I favor. We have seen that the regions near the centromeres remain condensed, forming prochromosomes in interphase, that they remain more or less condensed throughout leptophase, and they are still no doubt condensed in zygophase; in fact the process of progressive condensation has already begun by this time, so that we have two processes going on at the same time, a synaptic pairing from the ends inward and a progressive coiling or condensation proceeding from the center outward. When the two processes meet, further synapsis becomes impossible, since the necessary intimacy of association is not possible when the chromosomes are condensed.

Whether the pairing ends are all attracted to one side of the nucleus, resulting in a polarized synapsis, is uncertain, but it is less open to doubt that synapsis occurs only at the ends of the chromosomes. This is probably true even where the chromosomes form pairs rather than circles. The central portions of two completely homologous chromosomes will lie parallel because synapsis has occurred at both ends, but true synapsis will not have occurred in this region. Catcheside (1931) and Weier (1930) claim on the contrary that paired chromosomes synapse throughout their entire length.

It should be added that synizesis, i.e., the clumping of the threads into a more or less tight ball at one side of the nucleus, is characteristic of most fixed cells in this stage. Originally considered a natural phenomenon, this has long been recognized as an artifact. The contraction takes place to the side of the nucleus nearest the outer wall of the cell, which strongly suggests that it occurs on the side where the fixing fluid first reaches the nucleus.

3. PACHYPHASE

In earlier papers on *Oenothera* cytology (Cleland, 1922, etc.) a stage was recognized that was called the "hollow spireme" stage. This term, which was commonly used by telosynaptists, stood for what is now recognized as the pachyphase stage. These early studies made a point of the fact that the threads at this stage were univalent, not bivalent— that they represented unpaired chromosomes, not synapsed ones. It is an interesting fact that, although the telosynaptic hypothesis has been given up, most of the thread system at this stage is still considered to be univalent because of the fact that synapsis only involves the ends of the chromosomes.

This again is contrary to the interpretation of Marquardt who finds the chromosomes during this period paired from end to end (in material with mostly independent bivalents). Some of his drawings show doubleness in some of the chromosomes. This, however, does not mean bivalence, since each chromosome has two chromatids. The double chromosome shown in his text-figure 10 (1937) has probably passed beyond pachyphase and has reached the diplophase. Weier also (1930) concluded that homologues in the all-paired *hookeri* are associated throughout their entire length, but those in *lamarckiana* are paired only at their ends. Håkansson (1926), on the other hand, interpreted doubleness in pachyphase as indication of the double nature of univalent chromosomes.

The threads in pachyphase are considerably thicker than they are in leptophase. In most organisms this is because the pachyphase threads are bivalent whereas in leptophase they are univalent. In *Oenothera*, however, the thickness is apparently due to gradual condensation of the chromosome, beginning at the centromere and progressing distally. This thickening has probably occurred during zygophase and is responsible for the restriction of synapsis to the extreme distal portions of the chromosomes.

4. DIPLOPHASE

This stage is ill-defined in *Oenothera* and seems not to be of long duration. The transition from pachyphase to diakinesis is a rapid one. It commonly involves a tangling of threads at one region of the nucleus resulting in what used to be called "second contraction" or the "strepsitene" stage. This is equally characteristic of forms with paired or ring chromosomes. At this stage it is often possible to observe side by side pairing of chromosomes extending only a relatively short distance back from the tip (Fig. 8.1). The remainder of the chromosome is apparently unpaired. Because pairing occurs only at the ends, it is not possible to tell exactly where pachyphase ends and diplophase begins. In other organisms the beginning of diplophase is signalled by the beginning separation of synapsed homologues, but since most of each chromosome is unpaired in *Oenothera*, this phenomenon is not evident.

When thickening and shortening of the chromosomes begin, they proceed rapidly. It is often possible to find in the same clump of smeared cells those in which very little thickening has occurred, alongside others that are in full diakinesis. When thickening begins, it seems to take place in a distal direction, the end segments being the last regions to coil. The chromosomes are commonly still somewhat tangled in the middle of the nucleus when thickening begins. They gradually untangle as they

FIG. 8.1. Early diplophase chromosomes paired at the tips.

shorten and thicken and by diakinesis are free of each other, except that, where two or more chromosome groups are present they are often linked or interlocked. The thickening that occurs during diplophase is no doubt related to the onset of major spiralling.

5. DIAKINESIS

Interlocking of chromosome groups is a very frequent occurrence in *Oenothera*. In all-paired forms, one occasionally finds in diakinesis several ring bivalents linked in tandem to form a chain, or they may be linked in a branching formation. Where two circles are present, they may be interlocked, and where the configuration is made up of circles and pairs, the pairs may be linked to the circles (Fig. 8.2).

FIG. 8.2. Interlocked chromosomes in diakinesis of *O. franciscana B* (Cleland, 1922).

Such interlockings have completely disappeared by metaphase I. It is of interest therefore, to ask whether the separation of interlocked bivalents (or circles) results in the formation of open bivalents (or chains), or whether the ends that are separated are able to rejoin. The members of a bivalent or a chain are presumably held together by terminal chiasmata. When synapsed chromosomes separate, as must occur when interlocked rings or pairs become unlinked, does a chiasma simply slip off the end, as can happen if there is no telomere, so that no chromatid break occurs; or are the two chromatids of a chromosome attached at the end by a common telomere, in which case the process must necessarily involve the breaking of chromatids? In the former case, there would seem to be no mechanism that would tend to further the re-association of the separated chromatids, so that a circle or ring bivalent would normally become transformed into an open chain. In the latter case, however, raw ends of chromatids would be produced as a result of breakage, which could reunite, thus possibly re-establishing a closed ring or pair, or they could heal without fusion, resulting in an open chain (cf. McClintock, 1941).

It is *a priori* more probable that the ends of sister chromatids are attached by a telomere than that they are not, and for the following reason: in the case of circle chromosomes, the chain is undoubtedly subject to many stresses and strains as the chromosomes condense and shorten. If chiasmata could slip off the ends and disappear, chromosomes in the chain would easily part company under these stresses and the chain would shatter. The fact that they hold together quite tenaciously in spite of the stresses to which they are subjected seems to indicate that sister chromatids are firmly attached at the ends, and separation can only occur as the result of application of considerable force. The same reasoning no doubt applies to paired as well as circle chromosomes.

It is probably safe to assume, therefore, that separation of interlocked bivalents or circles causes the breaking of one chromatid in each of the synapsed chromosomes in one bivalent or circle at the point where separation takes place (Fig. 8.3b). In this case, the broken ends may fuse again, or they may heal without fusion, resulting in minute terminal deletions, and an open bivalent (Fig. 8.3c). If they fuse, two possibilities exist. The original threads can reunite (Fig. 8.3d), in which case the members of the bivalent or circle will remain attached; or fusion may occur in the reverse manner (Fig. 8.3e), in which case a chiasma will be eliminated and an open bivalent produced.

What, then, are the facts that bear on this question? I have analyzed diakinesis and metaphase I nuclei in two of de Vries' races—*hookeri* and *franciscana* (both with seven pairs) with the result shown in Table 8.I.

If separation of interlocked chromosomes usually results in the formation of an open bivalent, one might expect to find in metaphase I a maximum frequency of open bivalents equal to the frequency of open bivalents in diakinesis plus the frequency of interlockings whose separation could result in the opening of one of the separating bivalents. If this expectation were realized, one might expect to find, in metaphase I of *hookeri*, 13% of the bivalents open, instead of 2·7%; and in the case of *franciscana*, one would expect 12·1%, instead of 4·3%. When one subtracts from the observed percentage of open bivalents in metaphase the percentage of open bivalents found in diakinesis and presumably still present in metaphase, one discovers that only about 10% of interlockings in *hookeri*, and 11·3% in *franciscana* actually result in open bivalents.

In arriving at this conclusion certain qualifications should be mentioned.

One qualification arises from the probability that some open bivalents observed in diakinesis were the result of prior separation of interlocked bivalents. The results, therefore, fail to indicate the total extent of interlocking present in these races.

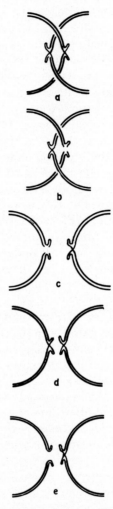

FIG. 8.3. Diagrams to illustrate separation of interlocked chromosomes. (a) interlocked chromosomes; (b) separation causes breakage of a chromatid in each of the synapsed chromosomes in one bivalent or circle; (c) ends may heal without fusion resulting in an open bivalent or chain, two chromatids having suffered minute terminal deletions; (d) broken chromatids may reunite in their original position, thus preserving the chiasma; (e) they may fuse in the reverse way, resulting in an open bivalent or chain.

TABLE 8.I

Frequency of interlocking and of open bivalents observed in diakinesis and metaphase I of *O. hookeri* and *O. franciscana* of de Vries

O. hookeri

Diakinesis

No. of cells scored	136	
No. of bivalents scored	952	
No. of interlockings	110 =	11·5% of number of bivalents
No. of open bivalents	14 =	1·5% of bivalents
		13·0% of bivalents expected to be open in metaphase I if every interlocking results in one open bivalent

Metaphase I

No. of cells scored	122	
No. of bivalents scored	854	
No. of interlockings	0	
No. of open bivalents	23 =	2·7% of bivalents (*vs.* 13% expected if every interlocking in diakinesis produces one open bivalent)

1·5% of bivalents already open in diakinesis; therefore 1·2% of bivalents have opened as a result of separation of interlocked bivalents (*vs.* 11·5% maximum expectation).
Conclusion: about 10% of interlockings result in an open bivalent

O. franciscana Sh.

Diakinesis

No. of cells scored	143	
No. of bivalents scored	1001	
No. of interlockings	89 =	8·8% of number of bivalents
No. of open bivalents	33 =	3·3% of bivalents
		12·1% of bivalents expected to be open in metaphase I if every interlocking results in one open bivalent

Metaphase I

No. of cells scored	110	
No. of bivalents scored	770	
No. of interlockings	0	
No. of open bivalents	33 =	4·3% of bivalents (*vs.* 12·1% expected if every interlocking in diakinesis produces one open bivalent)

3·3% of bivalents already open in diakinesis; therefore 1·0% of bivalents have opened as a result of separation of interlocked bivalents (8·8% maximum expectation)
Conclusion: about 11·3% of interlocking results in an open bivalent

A second qualification should be mentioned—in early diakinesis it is not always easy to distinguish between true interlocking and cases where bivalents have not yet become fully detached from the tangle into which they have been thrown as a result of the rapid shortening and thickening of the chromosomes in late diplophase. A given bivalent often finds itself lying wholly between the members of another bivalent. This is not true interlocking, and the two bivalents can slip apart readily. Because it is difficult to distinguish in early diakinesis between interlocked bivalents and those that are merely closely associated, our survey has omitted these stages. In so doing, however, we no doubt have missed many true interlockings which had separated before the stage when the cells were examined.

It should be added that all observations were made on sectioned material rather than smears, in order to avoid disturbance in the way the chromosomes lay in diakinesis. In sectioned material, however, many nuclei are cut across and not all chromosomes are visible in a section. We have included in our analysis, therefore, only whole uncut nuclei, in which all chromosomes were clearly visible.

All in all, the number of open bivalents in metaphase I is so much smaller than that to be expected on the assumption that separation of interlocked bivalents results in the opening of one bivalent, that it seems reasonable to conclude that reunion of the broken parts of single chromatids takes place in a considerable proportion of the cases so that the members of a bivalent become re-attached. This no doubt means that the two chromatids of a chromosome are united at the ends in a common telomere, a conclusion consistent with the fact that chromosomes in circles are held together with considerable tenacity in spite of the stresses to which they are subjected.

A. The Mechanism of Translocation

It has been suggested by Renner (1943a, p. 81) that interlocking may furnish a mechanism by which translocations can occur; separation of interlocked chromosomes might result in breakage of chromatids of *both* of the interlocked bivalents or rings followed by reunion in the wrong way. While this is not the only conceivable method by which translocations might occur (e.g., they might take place at the time when replication occurs), it is quite possible that translocations could occur as the result of separation of linked rings or pairs, and occur in such a way as to give equal or essentially equal exchanges. Such exchanges might involve the terminal chiasmata, or they might take place at the

region of the heterochromatin that impinges upon the centromeres.*
If interlocked chromosomes pull apart at the terminal chiasmata,
breakage could involve both crossover chromatids of both chromosomes,
resulting in eight raw ends (Fig. 8.4a). This of course would happen only
rarely. If, however, all four crisscross strands break, healing can occur
in several ways; if it occurs in certain ways, as shown in Figs 8.4b, c,

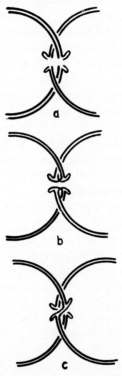

FIG. 8.4. Separation of interlocked chromosomes may bring about reciprocal translocations. (a) All four chromatids may break during separation; (b, c) two ways in which they
may heal are illustrated, both of which will result in a translocation.

viable cells will result with altered configurations—reciprocal translocations will have occurred. Translocations brought about by this
mechanism will occur late in meiotic prophase since it is in late diplophase
and diakinesis that interlocking disappears.

* They might also be interlocked in such a way that the ends of a pair of synapsed
chromosomes in one unit might impinge on the condensed mid-region of a chromosome
of the unit with which it is interlocked. Separation of interlocked units in this case could
not result in breakage of chromatids in both of these units, and so could not give rise to a
translocation.

The other possible way by which the separation of interlocked chromosome groups can bring about translocation is if the heterochromatic regions next to the centromeres in interlocked chromosomes come into contact. It is possible that sister chromatids have not fully separated in the heterochromatin at the time of diakinesis, so that only one morphological entity is present in each chromosome in this region. If the heterochromatic regions of interlocked chromosomes should become appressed to each other, breakage might occur and reunion might take place in such a way as to produce a translocation. I am inclined to guess that it is easier for mechanical reasons for interlocked chromosomes to pull apart at the ends than in the middle so that translocation resulting from elimination of interlocking is more likely to involve the terminal chiasmata. Exchanges in the central heterochromatic region are more likely to occur at earlier stages—they could occur, of course, before or during replication.

Either of these methods would result in translocated chromosomes having essentially unaltered morphology, such as are characteristic of *Oenothera*.

Translocations can theoretically occur either in the soma or during meiosis. In the former case, which for logistical reasons seems unlikely, there should result clusters of sporocytes with an altered chromosome configuration, or perhaps whole sectors of a plant. In the latter case, an altered configuration would be found only in the single sporocyte where the translocation had occurred. Until recently, no direct evidence for either alternative had been found. We have recently, however, discovered two cases of altered configuration in single cells—not in the subgenus *Oenothera* but in *Raimannia*. In *longiflora* Villa Ortuzor strain B × *stricta* Santa Barbara the configuration is ⊙4, ⊙4, 3 pairs. One cell was found, however, with a ⊙8 (Fig. 8.5). In another hybrid involving the same mother (*longiflora* Villa Ortuzor strain B × *selowii*), with a normal configuration of ⊙4, ⊙6, 2 pairs, one cell was found with ⊙10 (Fig. 8.6). Each of these cells had undoubtedly suffered a translocation by a mechanism such as described above, probably during the course of separation of interlocked circles (two circles of four becoming a circle of eight in one case; ⊙4, ⊙6 becoming ⊙10 in the other). Thus a small amount of direct evidence points to the possibility that translocation may occur during meiotic prophase, and lends plausibility to Renner's suggestion that the separation of interlocked chromosomes may occasionally result in interchange.

B. Major Spirals

Although *Oenothera* chromosomes are small, they contain major spirals, as do plants with larger chromosomes. Not uncommonly, one

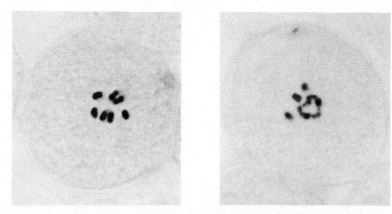

FIG. 8.5. *O. longiflora* Villa Ortuzor strain B × *O. stricta* Santa Barbara. Left, normal cell with ⊙4, ⊙4, 3 pairs; right, exceptional cell in which reciprocal translocation has resulted in ⊙8, 3 pairs (metaphase I, microsporocyte). Magnification at microscope × 1350.

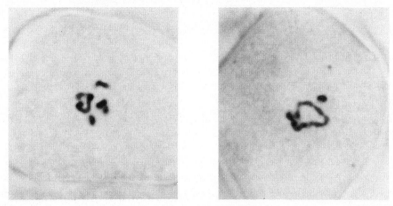

FIG. 8.6. *O. longiflora* Villa Ortuzor strain B × *O. selowii*. Left, cells with ⊙4, ⊙6, 2 pairs; right, exceptional cell with ⊙10, 2 pairs, resulting from reciprocal translocation. Magnification at microscope × 1350.

finds cells in which they show beautifully (Fig. 8.7). There are two or three gyres per arm.

During diakinesis the chromosomes condense greatly. While it is not easy to obtain exact measurements in early diakinesis because of the fact that the chromosomes are in the process of shortening and the rate is not always the same from chromosome to chromosome, the average length of chromosome obtained in our laboratory from several races in this stage was 3·5–4 μm. In premetaphase, the average was 1·5 μm. Measurements were made in premetaphase rather than in metaphase because the chromosomes become V- or U-shaped after alignment on the spindle and length cannot be measured as accurately.

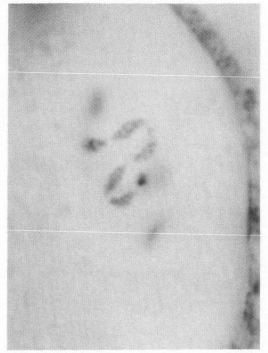

FIG. 8.7. Major spirals in prometaphase I, microsporocyte of *O. biennis* I (Hendricksville) × *lamarckiana* (Ft. Lewis) α Hendricksville·velans. Magnification at microscope × 1350.

6. LATER STAGES

Alignment of circle chromosomes on the spindle is a gradual process and seems to involve some shifting of spindle attachments. The percentage of cases where adjacent chromosomes, or chromosomes separated by two, four or six chromosomes, are being attracted to the same pole is much higher at the beginning of the process than at full metaphase. This must mean that some chromosomes change their alignment during the orientation process, losing attraction for one pole and substituting an attraction to the other pole. When the metaphase alignment is complete, adjacent centromeres have as a rule formed attachments to opposite halves of the spindle.

The percentage of irregularity in the zigzag arrangement is surprisingly low. In arriving at an estimate of such percentages, it is better to use sectioned material than smears. The smearing process is apt to upset the alignment in some cases, but sections that contain uncut spindles will include metaphase figures in which the chromosomes are showing their true alignment.

Table 8.II gives data on a number of races and hybrids with respect to irregularity in the zigzag arrangement. It will be noted that the larger the circle, the more likely it is that irregularity will occur. Even in those cases where the rate of irregularity is highest, however, most cells are regular, and the loss in fertility resulting from irregularity is not serious.

Irregularities ordinarily conform to one of three basic types, as shown by Emerson (1935) and Catcheside (1936):

1. Two adjacent chromosomes go to the same pole and somewhere else in the circle, two other adjacent chromosomes go to the same pole as the first pair (Fig. 15.2). This will result in eight chromosomes going to one pole, six to the other. Gametes receiving six chromosomes will be lost, but an eight-chromosome egg will probably survive, although an eight-chromosome sperm will ordinarily not be formed, or will not be functional if formed. Such an eight-chromosome gamete will contain a mixture of chromosomes from the associated complexes and the extra ends will belong to different chromosomes. Such an irregularity is one mechanism by which trisomic plants can be produced.

2. Two adjacent chromosomes may go to the same pole, and somewhere else in the circle two other adjacent chromosomes may go to the *opposite* pole (Fig. 15.3). Such a separation will result in gametes with the normal number of chromosomes, but these gametes will all have duplications and deficiencies and consequently will be inviable.

3. Three adjacent chromosomes may go to the same pole together (Fig. 15.1). When this happens, eight chromosomes will go to one pole,

TABLE 8. II

Percentage of irregularity in zigzag arrangement observed in certain *Oenothera* races and hybrids

Race or hybrid	Chromosome configuration	Number of cells in I metaphase	Percentage irregularity	Citation
chicaginensis	⊙12	159	18	Renner and Cleland, 1933
muricata	⊙14	58	24	Cleland, 1926a
biennis de V.	⊙6, ⊙8	151	5·3[a]	Cleland, 1926b
lamarckiana	⊙12	46	26	Cleland, 1929b
suaveolens	⊙12	63	17·4	Cleland and Oehlkers, 1930
suaveolens sulfurea	⊙12	60	8	Cleland and Oehlkers, 1930
cockerelli	⊙14	122	20	Cleland and Oehlkers, 1930
grandiflora de V.	⊙14	65	35·3	Cleland and Oehlkers, 1930
(*grandiflora* × *hookeri*) truncans ·[h]hookeri	⊙14	143	53·8	Cleland and Oehlkers, 1930
acuens ·[h]hookeri	⊙4, ⊙4	358	0·28	Cleland and Oehlkers, 1930
(*hookeri* × *grandiflora*) [h]hookeri × acuens	⊙4, ⊙4	353	1·98	Cleland and Oehlkers, 1930
(*grandiflora* × *lamarckiana*) acuens ·gaudens	⊙14	85	22·3	Cleland and Oehlkers, 1930
truncans ·gaudens	⊙4, ⊙10	27	3·7[b]	Cleland and Oehlkers, 1930
acuens ·velans	⊙4, ⊙6	157	0·64[c]	Cleland and Oehlkers, 1930
truncans ·velans	⊙4, ⊙10	133	10·5[d]	Cleland and Oehlkers, 1930

TABLE 8.II—*continued*

Race or hybrid	Chromosome configuration	Number of cells in I metaphase	Percentage irregularity	Citation
(*lamarckiana × grandiflora*)				
gaudens·acuens	⊙14	141	9·2	Cleland and Oehlkers, 1930
velans·acuens	⊙4, ⊙6	195	2·5[e]	Cleland and Oehlkers, 1930
velans·truncans	⊙4, ⊙10	120	14·1[f]	Cleland and Oehlkers, 1930
(*grandiflora × strigosa*)				
acuens·stringens	⊙4, ⊙4	123	0·0	Cleland and Oehlkers, 1930
truncans·stringens	⊙4, ⊙10	83	9·6[g]	Cleland and Oehlkers, 1930
(*suaveolens sulfurea × lamarckiana*)				
flavens·gaudens	⊙12	131	23·7	Cleland and Oehlkers, 1930
flavens·velans	⊙4, ⊙4	724	1·1[a]	Cleland and Oehlkers, 1930
albicans·gaudens	⊙6, ⊙8	320	10·0[a]	Cleland and Oehlkers, 1930
albicans·velans	⊙14	161	31·0	Cleland and Oehlkers, 1930
(*lamarckiana × suaveolens sulfurea*)				
gaudens·flavens	⊙12	184	13·5	Cleland and Oehlkers, 1930
velans·flavens	⊙4, ⊙4	658	8·5	Cleland and Oehlkers, 1930
(*suaveolens × cockerelli*)				
albicans·elongans	⊙12	30	20·0	Cleland and Oehlkers, 1930
(*cockerelli × suaveolens*)				
curtans·flavens	⊙12	66	20·0	Cleland and Oehlkers, 1930
(*suaveolens sulfurea × strigosa*)				
flavens·stringens	⊙4	109	1·83	Cleland and Oehlkers, 1930
albicans·stringens	⊙12	183	14·2	Cleland and Oehlkers, 1930
(*strigosa × suaveolens sulfurea*)				
deprimens·flavens	⊙12	197	14·2	Cleland and Oehlkers, 1930
(*strigosa × lamarckiana cruciata*)				
deprimens·gaudens	⊙10	200	13·0	Cleland and Oehlkers, 1930
deprimens·velans	⊙10	177	11·3	Cleland and Oehlkers, 1930
(*lamarckiana cruciata × strigosa*)				
gaudens·stringens	⊙14	187	30·5	Cleland and Oehlkers, 1930
velans·stringens	⊙4, ⊙6	188	3·2[a]	Cleland and Oehlkers, 1930
(*chicaginensis × hookeri*)				
excellens·[h]hookeri	⊙4	321	0·0	Renner and Cleland, 1933
(*hookeri × chicaginensis*)				
[h]hookeri·punctulans	⊙10	111	20·0	Renner and Cleland, 1933
(*chicaginensis × suaveolens*)				
excellens·flavens	⊙4, ⊙4	188	5·3[a]	Renner and Cleland, 1933
(*suaveolens × chicaginensis*)				
albicans·punctulans	⊙10	32	9·4	Renner and Cleland, 1933
flavens·punctulans	⊙8	96	5·2	Renner and Cleland, 1933
(*chicaginensis × lamarckiana*)				
excellens·velans	⊙6	87	2·3	Renner and Cleland, 1933
excellens·gaudens	⊙10	61	10·0	Renner and Cleland, 1933
(*lamarckiana × chicaginensis*)				
velans·punctulans	⊙12	12	0·0	Renner and Cleland, 1933
gaudens·punctulans	⊙6, ⊙8	83	3·61[a]	Renner and Cleland, 1933
(*chicaginensis × biennis*)				
excellens·rubens	⊙10	90	10·0	Renner and Cleland, 1933
punctulans·rubens	⊙6, ⊙8	58	1·72[a]	Renner and Cleland, 1933
(*biennis × chicaginensis*)				
rubens·punctulans	⊙6, ⊙8	99	7·0[h]	Renner and Cleland, 1933
albicans·punctulans	⊙10	21	5·0	Renner and Cleland, 1933

[a] Irregularity in one circle only.
[b] ⊙10 irregular in one cell
[c] ⊙6 irregular in one cell
[d] ⊙10 irregular in 13 cells, ⊙4 in one
[e] ⊙4 irregular in two cells, ⊙6 in three cells
[f] ⊙10 irregular in 17 cells
[g] ⊙10 irregular in eight cells
[h] ⊙8 irregular in six cells, both circles irregular in one cell.

six to the other. The set containing eight chromosomes will include one full Renner complex plus a chromosome from the other complex. An egg carrying this eight-chromosome complement can give rise to a trisomic.

One often sees in metaphase an irregularity in which one chromosome is suspended without obvious spindle attachment between two chromosomes, which are directed toward opposite poles. This will be compensated

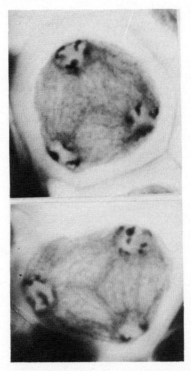

FIG. 8.8. Microsporocytes just prior to cleavage. Spindle-shaped areas develop between nuclei, but cell plates are not formed. Cleavage takes place from outside inward, cutting across the middle of these areas. Magnification at microscope × 1800.

at some other part of the circle by another suspended chromosome or by two adjacent chromosomes going to one pole or the other. These will give results, however, similar to those outlined above: either eight-chromosome gametes capable of producing trisomics will be formed, or the gametes will be deficient for part of the genome and will be unable to function, depending upon which direction the irregularly arranged chromosomes go.

It is a striking fact that oenotheras with large circles have among their pollen grains many that are small, shrivelled and empty. These are quite different from "inactives." It is probable that many of the empties are the result of deficiencies arising from irregularities in the zigzag. Their absence in forms, most or all of whose chromosomes are paired, bears out this hypothesis.

In forms with large circles one also finds a certain proportion of ovules in which all four spores of the linear quartet disintegrate. This may also be due to the presence of deficiencies resulting from irregularity in distribution of the chromosomes, in this case in the megaspore mother cell.

As the diads reach the poles in the anaphase I, the chromatids separate widely except at the centromeres, which remain undivided. Daughter nuclei are constituted but the chromosomes do not return to a metabolic stage. They uncoil to some degree, forming Maltese crosses, which lie widely scattered, usually against the nuclear membrane. As the end of interphase approaches, the diads condense into small X-shaped bodies with short arms. These become oriented on the spindle, the centromeres divide and the monads proceed to the poles in the second anaphase

There is no cleavage of the cytoplasm following the first meiotic division in pollen mother cells. After the second nuclear division, the four nuclei take positions, as a rule, in tetrahedral fashion and the protoplast is cut into four portions synchronously. Cleavage begins at the outside. There develop in front of the four advancing clefts, spindle-shaped areas which run from nucleus to nucleus, but fail to develop cell plates (Fig. 8.8). By the time cleavage is complete, the nuclei have returned to a metabolic condition.

CHAPTER 9

Genetic Analysis of Renner Complexes

Most of the work on the genetic analysis of complexes has been done by Renner and his students, though others have contributed to a considerable extent, including Oehlkers and his students, as well as Emerson and Sturtevant. This has not been easy of accomplishment. The peculiarities of cytological and reproductive behavior in *Oenothera* set up barriers which have impeded progress in this direction. The more important of these difficulties may be mentioned at this point.

First and foremost, of course, is the effect of the presence of large circles of chromosomes and their balanced lethals. This situation prevents recessive genes from segregating out in homozygous condition and thus showing up. Even in the case of a dominant trait in a race with ⊙14, one cannot easily tell whether the gene responsible for this character is in one or both complexes. Selfing does not help to determine this, since Mendelian segregation is suppressed. The only way one can discover whether a dominant is carried by one or both complexes of a plant is to combine both complexes with a third complex in which the character is absent. If one hybrid shows the trait and the other does not, one can assume that only one complex in the plant being tested is carrying the dominant. If neither hybrid shows the trait, then the tester complex is probably carrying a dominant inhibitor.

Genetic analysis becomes more complicated when one realizes that hybrids frequently have some other configuration than ⊙14 so that the progeny of hybrids often reveal independent assortment in regard to genes that were linked in the parental races. Genes often appear to be differently linked in different hybrids, depending upon the chromosome configurations present. Obviously this is a situation that cannot be understood or analyzed without reference to chromosome behavior in all plants involved.

A second complication arises from the fact that, when segregations do occur, the ratios obtained often or usually depart widely from typical Mendelian ratios, making interpretation difficult. There are two major causes of these aberrances.

1. *Megaspore competition and the development of the female gametophyte*

Embryo sac development in most of the *Onagraceae*, including *Oenothera*, is unusual in two respects: it tends to develop from the micropylar

spore of the linear quartet, rather than from the chalazal end as in most plants; and it contains, when mature, only four nuclei—an egg, two synergids and one polar body. There are no antipodals and the polar body unites with one of the sperms to form a diploid endosperm (Geerts, 1909, p. 139 *et seq.*) which fails to survive in the mature seed.

The way in which the embryo sac develops from the linear quartet of spores may vary from race to race and from hybrid to hybrid. Renner was the first to study this matter in detail (1921a). He stated that the embryo sac regularly develops from the micropylar spore in *hookeri* and in the isogamous *lamarckiana*, but that in the heterogamous *muricata* the chalazal spore frequently develops, often in competition with the micropylar spore, which also tries to develop. He concluded that the *rigens* complex permits development of the micropylar spore when it finds itself in the favored (micropylar) position. When, however, it finds itself in the chalazal spore, it causes this spore to develop even though it is not in the more favorable position. In some cases, the *curvans* complex, having the benefit of position, although less effective in promoting development of the embryo sac, will succeed in causing the micropylar spore to develop in competition with the chalazal spore. In a small percentage of cases, the *curvans* spore will actually win out in the competition. The apparent egg lethal in *curvans* is not a true lethal, therefore, but this complex is less capable, in comparison with *rigens*, of promoting the development of the embryo sac. If the *curvans* complex does succeed in developing the embryo sac, the *curvans*-containing egg is perfectly healthy and functional. The term "Renner effect" has been applied to the results of such competition (Fig. 9.1). Although the sperm complex is occasionally transmitted through the egg in *muricata*, other forms are strictly heterogamous, the egg complex winning out in all cases.

That relative ability to promote the rapid development of the embryo sac is an important factor in determining which complex will be present in the egg is shown by the different degrees of success that a given complex may have when in competition with a variety of other antagonists. Thus, Renner pointed out (1921a) that *albicans* in de Vries' *biennis* completely suppresses the *rubens* complex in the ovules (in other strains of *biennis*, *rubens* succeeds in a small percentage of the cases), but in *albicans* · [h]*hookeri*, it is wholly suppressed by [h]*hookeri*, even though it is in its own plasma and [h]*hookeri* in a foreign plasma.

Later on, Renner and his students found cases which failed to conform to the pattern that competition between megaspores is characteristic of heterogamous, as opposed to isogamous forms. Rudloff (1929) found in homozygous *O. purpurata* (seven pairs) that the micropylar spore developed in about half the ovules, and there was competition between chalazal and micropylar spores in the other half, often resulting in two fully developed embryo sacs in an ovule. On the other hand, he found no competition in *O. rubricaulis* (a heterogamous race with ⊙6, ⊙8), the micropylar spore developing into the embryo sac in all functional ovules, all of them carrying *tingens*, the egg complex, rather than *rubens*, the pollen complex. Rudloff interpreted the latter case in terms of a polarization of the spindle in the first meiotic division, by which the *tingens* chromosomes always go to the pole nearer the micropyle.

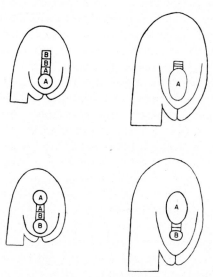

Fig. 9.1. The "Renner effect". Complex A is more potent in promoting development of the embryo sac than complex B. If it is in the favored (micropylar) position (above), this spore develops; if it is in the less favored chalazal position, (below), complex B may succeed in inducing growth of the micropylar spore. If A is sufficiently strong, it may always win out, in which case complex B appears to have a lethal. If the differential is less, B may succeed in the competition in a proportion of ovules. Reprinted from Proceedings of the American Philosophical Society by permission.

Hoeppener and Renner (1929, p. 68) found that *lutescens*, the seven-paired homozygotic *flavens·flavens*, showed the same pattern of embryo sac development as its parent, *suaveolens*, which has ⊙12; in both, competition between micropylar and chalazal genes is commonly found.

Renner (1940a) finally concluded that no type of behavior is confined to any one class of *Oenothera*. The eggs of heterogamous forms are not necessarily formed as the result of competition in the sense that two embryo sacs develop in an ovule, or of polarization of the spindle in the megaspore mother cell. Success of a complex depends largely upon its relative ability to further the development of the female gametophyte, i.e., upon its genetic constitution. When associated complexes differ greatly in their ability to further embryo sac development, one complex will tend to win out no matter which spore it is in, and the plant will be strictly heterogamous with respect to the eggs. In such cases, there will be little or no tendency for chalazal and micropylar spores to develop competitively, other things being equal. Whichever spore possesses the strong complex will develop the embryo sac, irrespective of its position.

When associated complexes are more or less equal in strength, the one that is in the favored position will tend to win out, and in most cases there will be little or no spore competition. Such plants will be isogamous.

In situations where associated complexes differ in strength, but to only a moderate degree, competition will be more likely to occur. Whether both spores attempt to develop will depend, other things being equal, on the balance between the strengths of the two competing complexes. Such races will transmit one complex through most of the eggs, but metaclines will be produced in percentages that reflect the degree of difference in strength between the complexes.

Some complexes are so weak with respect to embryo sac development that they find it difficult to function even in the absence of effective competition. For example, diplarrhene hybrids (in which two sperm complexes are present) may show very few mature embryo sacs and will set very few seeds (Renner, 1940a): only the extreme plus-variants among the female gametophytes are able to survive. In these cases, the complexes in question may be thought of as approaching the condition of having female gametophytic lethals.*

Finally, Harte (1958a, b, c) has introduced an additional factor into the situation. She finds great variability in the degree of success which one of a given pair of associated complexes shows in competition with the other, not only as between plants but also within a single plant or on a single spike. She suggests that local environmental conditions of unknown nature are also important in determining which complex will succeed. This is probably more true of situations in which the competing complexes are fairly evenly matched than in cases where one complex greatly surpasses the other in its ability to develop an embryo sac.

The effect of this situation on the ratios obtained in crosses can be

---

* The existence of competition in homozygous *O. purpurata* and *O. lutescens* suggests that relative ability of complexes to further embryo sac development may not be the sole factor that determines which complex will succeed. It is probable that in some forms one end of the linear tetrad may not possess a distinct inherent advantage over the other end, so that, even in a homozygote where the associated complexes are identical, competition between chalazal and micropylar spores may occur.

readily imagined. A given gene can be transmitted through the egg to a certain proportion of the offspring when its associated allele is carried by one complex, but when the latter is carried by some other complex the ratio may be quite different, even to the point where the gene may be totally excluded in the eggs. Ratios are, therefore, often profoundly modified by the events occurring during the development of the female gametophyte, and this adds to the complexity of genetic analysis in *Oenothera*.

2. *Pollen tube competition*

That different classes of pollen tubes may grow through the style at different rates was first shown independently (1917) by Correns and Renner. Correns found that the faster growing pollen tubes in *Melandrium* tend to produce female plants, and vice versa. Renner discovered in *Oenothera* that the tubes containing one complex may tend to outgrow those carrying the other complex. The term "certation" was applied to this phenomenon by Heribert-Nilsson (1920a) who found that pollen tubes which carry *R* (for red midribs) grow faster than do tubes containing *r* (confirmed by Renner, 1921b). Renner (1919b) described several cases of certation. For example, *velans* tubes grow more rapidly than do *gaudens* tubes in *lamarckiana*. The relative speed of growth in other styles, however, depends in part on the kind of style in which they are growing. For instance, in the hybrid *muricata* × (*lamarckiana* × *muricata*) *velans·curvans*, four times as many *velans* sperm fertilize the *rigens* eggs as *curvans*; in the hybrid (*muricata* × *lamarckiana*) *rigens· velans* × (*lamarckiana* × *muricata*) *velans·curvans*, the ratio of *velans* to *curvans* sperm which function is 12:1. In *biennis* × (*lamarckiana* × *muricata*) *velans·curvans*, the ratio is 30:1, and in *suaveolens* or *lamarckiana* × (*lamarckiana* × *muricata*) *velans·curvans*, no *curvans* progeny are produced. The relative speed of growth of *velans* and *curvans* pollen tubes is therefore dependent on the style pollinated. This difference, however, may be influenced to some extent by the length of the style. The average style lengths in these experiments, in the order in which they are mentioned, are (in millimeters) 45, 45, 55, 55 and 87.

Similar results have been obtained in our laboratory by Hoff (unpubl. thesis). For example, when *Hollis* and *Parras* pollen are placed together on *Nobska* stigmas, *Parras* pollen rarely succeeds in fertilizing eggs, whereas when these two pollens are placed on *Johansen* stigmas, *Parras* pollen completely eliminates *Hollis* from competition. The fact that *Parras* and *Johansen* are both hookeris and therefore closely related may be of some significance in this case.

As a result of such differences, richness of pollination is an important factor in determining ratios. If a style is pollinated richly, the faster class of tubes will reach the ovules before the slower class reaches the ovary and will bring about the fertilization of all available eggs. If, however, the same stigma is pollinated sparingly with the same pollen, there will not be enough of the faster class to accomplish fertilization of all the eggs, and the slower class will then find ovules that have not yet been invaded. Thus, when *lamarckiana* is used to pollinate another race, the ratio of laetas to velutinas produced will depend, not only on the relative speeds of *velans* and *gaudens* pollen tubes in the styles of this particular race, but also on the richness of pollination. With rich pollination, velutinas may tend to predominate over, or even to exclude, the laetas, but with increasing sparsity of pollination the proportions of the two twins may more nearly approach equality.

It can be seen, therefore, that the ratios obtained in *Oenothera* crosses are subject to influences which often tend to prevent them from following the Mendelian pattern. Large circles and balanced lethals prevent segregation, megaspore competition and pollen tube competition result in altered, atypical ratios. Genetical analysis in *Oenothera* must take these factors into consideration and one must be prepared to deal with the difficulties which they impose.

In spite of these handicaps Renner and his students, as well as other investigators, have been able to make considerable progress in the analysis of various complexes. At first, this effort was directed toward identifying genes and assigning them to the proper complex within a plant, as shown by the results of outcrosses (Renner's "factor analysis"). Thus, in *lamarckiana*, the punctate factor P and the narrow leaf factor b were found to belong to *velans*, the *nanella* gene n and the broad leaf factor *B* to *gaudens* (Renner, 1925). Later when it became possible to combine cytological determination of chromosome configuration with genetic analysis, and especially when the segmental arrangements of the various complexes had been worked out, it became feasible to locate a number of genes in specific chromosomes or chromosome arms.

In order to locate genes within a complex, one must study this complex in combination with a variety of others, which give different chromosome configurations with it. One can then determine in which hybrid combinations two genes are linked and in which they are independent. In this way, if one knows the segmental arrangements of the complexes involved, one can narrow down the possible locations of these genes to smaller and smaller groups of chromosomes and, in specific cases, can locate a gene in its proper chromosome or even in a given arm of this chromosome.

To be more explicit with regard to the methods of genetic analysis, suppose that a certain complex-combination has $\odot 12$, 1 pair, with balanced lethals in the circle. This hybrid will breed true in the main. If, however, its progeny shows independent segregation of one or more traits, the genes controlling these traits probably lie in the pair. If in another hybrid a small circle and several pairs are present (e.g., $\odot 4$, 5 pairs, or $\odot 6$, 4 pairs) a certain character may express itself only in F_2 individuals which lack the circle and have pairs instead. This would indicate that the gene involved belongs in the circle. If, however, a gene expresses itself in a proportion of the progeny independently of the presence or absence of the circle, then this gene lies in one of the pairs.

An example of how one may locate a gene may be taken from a paper (1931) by Emerson and Sturtevant. *P* (red papillae) and *R* (red midribs) are independent in *lamarckiana* ($\odot 12$, 1 pair). Since *R* segregates independently of the complexes *velans* and *gaudens*, it resides in the pair, which means that *P* must be in the circle. Since *P* is found only in velutinas when *lamarckiana* is outcrossed to forms lacking *P*, it belongs to *velans*, not to *gaudens*. In *flavens·velans* (*suaveolens* × *lamarckiana* or reciprocal) *P* and *R* are completely linked; in this hybrid, the 1·2 chromosome of *lamarckiana* which carries *R* is a member of a circle of four (1·2—2·3—3·4—4·1). Since *P* cannot be in 1·2, and was brought into the cross from *lamarckiana*, it must be in 3·4. In *flavens·stringens* (both complexes containing 1·4 3·2) *P* (from *stringens*) is independent of *fr* (pollen sterility factor in *stringens*). The latter is in 1·4 of *stringens*, since it segregates independently of the lethals in *albicans·stringens* in which 1·4 is the only pair. Therefore, *P* cannot be in 1·4, and must be in 3·2; but since it cannot be in 2, it must be in 3. The only assumption made in this case is that *P* in *velans* represents the same locus as *P* in *stringens*.

In at least one case, it has been possible to learn something about the relative positions of genes that lie on the same arm. The genes *P* and *s* (sulfur colored petals) both belong to arm number 3. By determining relative crossover percentages between these genes and the lethals, in cases where they lie within large circles that contain balanced lethals, it has been shown that *P* is proximal to *s* (Renner, 1933; Oehlkers, 1933).

Two other genes have also been assigned to this end. A pollen lethal *an* in *albicans* (Renner, 1948) lies in between *p* and *S*, as does the mutant gene *hel* (*helix*) discovered by Renner (1953) and studied by Chrometzka (1955, 1956).

In view of the difficulties confronting genetic analysis in *Oenothera* it is not always easy to tell when one is dealing with a single gene or with a group of genes. In a number of cases, the supposed gene appears to have more than one effect. As long as this factor is situated in a circle, one cannot be sure that he is not dealing with two or more genes situated perhaps in different chromosomes but inherited together. Only if the chromosomes in this circle can be brought into combinations where they are separated into different circles or pairs may it be possible to determine whether the multiple effects are based upon genes in more than one chromosome. If in a hybrid with a single pair, two or more phenotypic effects which segregate independently of the lethals are still found to be inherited together, they are probably in the same chromosome. If in this case they cannot be separated by crossing over, we may be dealing with a single pleiotropic gene. This is not necessarily so, however, since, as we shall see, crossing over is largely confined to the pairing ends of chromosomes. It is therefore possible that the effects observed are due to two or more genes present in the proximal region of a chromosome, which, because crossing over rarely if ever occurs in this region, remain tightly linked.

In view of these restrictions, therefore, it is not surprising that many putative genes with multiple effects have not been shown definitely to represent single loci. This includes a number of lethals that appear to have phenotypic effects when heterozygous.

We may now review what is known about the location of genes on the chromosomes.

R is in chromosome 1·2 in *lamarckiana*. In addition to reddening the midribs, this gene is either lethal in the homozygous condition or it is so closely linked to the lethal that they have never been separated. This gene, however, is supplemental to the zygotic lethals that belong to *velans* and *gaudens*, and can be transmitted along with either. It is interesting to note that although R is in 1·2 in *excellens* also, it is not lethal in the latter race when homozygous, which may mean that we are dealing with a different allele from the one in *rubens* and *gaudens*; or alternatively that R of *rubens* or *gaudens* is not itself lethal but is closely linked to a lethal, which in the case of *excellens* has been lost or never attained.

With respect to the lethal of *velans*, Renner (1933) showed that chromosomes 1·2 3·4 9·10 11·12 and 13·14 can exist in homozygous condition, leaving only 5·8 and 7·6 as possible sites of the lethal. Baerecke later claimed (1944, p. 62) that the lethal is in 5·8, on the ground that *laxans·velans*, which has 5·8 as a pair, gives very few viable seeds when selfed or when pollinated by *p-albicans·velans*. The situation with regard to the *gaudens* lethal, however, is not so clear. All that is known is that it is probably not in 5·6 since the trisomic *lata* has in addition to full *gaudens* and *velans* complexes an extra 5·6 derived from *gaudens* (Catcheside, 1936, 1937b).

Another zygotic lethal whose locus has been traced to a single chromosome is that of *flavens*. It resides in 5·6 in this complex, and does not have a recognizable phenotypic effect, although the gene for acute leaves and sepal tips (*Sp*) was earlier considered to have the lethal effect. Later, however, (see Renner, 1941), it was concluded that *Sp* probably resides in the other arm of 5·6 from the lethal.

A third lethal whose locus is known within narrow limits is *Pil* of *flectens*. This factor, for colorless, silky pubescence, belongs on arm 8 of the chromosome 9·8, and is homozygous lethal (Renner, 1943b, p. 395).* Whether it is lethal in other complexes which seem to have the same locus is uncertain.

The only other zygotic lethal for which a locus has been suggested is *Cu*, which produces the bent stem tips in *curvans*. It may lie in arm 4 in *curvans*, but this is not yet certain (Renner 1942a). Whether the lethal effect, and the effect on stem tip growth, are the result of a single gene or of two loci that have not been separated is not known for sure.

These are the only zygotic lethals which have been in any way identified as to location. There are, however, a number of factors which have semi-lethal effects when homozygous, about whose location something is known. One of these is *Fl* of *flectens* (pollen complex of *atrovirens*), situated in 1·4 (Renner 1942a), and sublethal when homozygous.† It produces bent stem tips and spreading sepal tips.

Another such factor is *lor* in *flectens*, sublethal in homozygous condition, producing very narrow strap-shaped leaves (Renner, 1933). It is apparently a doublet, *lor¹* lying in 11·14, *lor²* in 13·12. The combination is semi-lethal in homozygous condition. When *lor lor* is combined with *FlFl*, the combination is fully lethal (Renner, 1943b).

The factor *b* (narrow leaves) is located in 7·8 of *curvans* (Renner, 1942a). Its status with regard to lethality, however, is unclear. It was originally considered to be a homozygous lethal since the cross (*suaveolens* × *muricata*) *B-flavens·b-curvans* never produced *bb* progeny, although it did produce *BB* (homozygous broad-leaved) offspring (Renner, 1933). Since *B* is in 7·8, it was in one of the circles of six in this hybrid (⊙6, ⊙6, 1 pair). Later (Renner and Cleland, 1933), Renner obtained *bb* segregants in the progeny of certain F₂s and backcrosses involving *flavens* and *punctulans* (e.g., *flavens·punctulans* × *s* and (*flavens·punctulans*) × *punctulans*). In these cases, however, *b* was derived from *punctulans* and

---

* In 1933, Renner placed *Pil* in 7·9 of *flectens*. The formula for *flectens* was later revised (Renner 1942a). Instead of having 5·10 6·8 7·9, it has 5·7 6·10 9·8.

† In this and certain other cases, lethality may result not from the action of a single locus, but may involve other parts—perhaps the whole—of the chromosome.

may have been an alethal allele of the b from *curvans* (or a gene that has become separated from a closely linked lethal).

There are a number of characters which appear in several races, but with different intensities. These have been described as belonging to allelic series. For example, bent stem tips is characteristic of races of what are now considered to be members of the *parviflora* assemblage (see Chapter 18). Renner has assigned different symbols to the responsible genes in different races: in descending order of intensity in their effect they are $Perc > Cu > Ce > Fl > Sbc$, found respectively in *percurvans*, *curvans*, ^{h}*argillicola*, *flectens* and *subcurvans* (Renner, 1942a, p. 128). These probably belong to end 4. Another example is the P-series, for red stem papillae and red flower cones (in end 3). In order of strength of pigmentation the best studied alleles (Renner 1942a, p. 126) are (from strongest to weakest) $P^{rbr} > P^{vel} > P^{pur} > P^{ho} > P^{fra} > P^{rig}$, found respectively in *rubricalyx, velans,* ^{h}*purpurata,* ^{h}*hookeri,* ^{h}*franciscana, rigens*. These all have some pigmentation on the buds. In addition, some alleles produce red papillae but no bud color, e.g., P^{str}, P^{ti} and P^{pun}, present respectively in *stringens, tingens* and *punctulans*.

It is possible, however, that we are not dealing in these cases with different alleles, but rather with a single gene whose effect is influenced by modifiers which may differ from race to race.

Renner and his students have not been the only ones who have engaged in "factor analysis" of the various complexes. Harte (1948) has also been active in this regard. She studied complexes in eight races (*hookeri, franciscana* of de Vries, *suaveolens sulfurea, strigosa* of de Vries, *lamarckiana* of de Vries and Renner, and two strains of *brevistylis*). She recognized 59 genes controlling 34 visible characters, and three lethals, of which 37 were thought to be newly recognized genes. For instance, in *flavens*·^{h}*hookeri* ($\odot$4, 5 pairs) she identified 28 genes in four linkage groups.

Table 9.I lists most of the genes that have been assigned tentatively or definitely to particular chromosomes or chromosome arms. Different workers have sometimes used different symbols for what are probably the same loci. For example, Emerson (1931a) has used a terminology based on qualitative differences in phenotypic effect for the P-series. Thus, P^r = punctate stems and *rubricalyx* color, P^s = punctate stems and striped buds, P = punctate stems and green buds, p = nonpunctate stems and green buds. Shull (1923a) has used still a different set of symbols: R^h = *rubricalyx* color, R^c = red cone and green hypanthia, r^c = green buds, R^s = red stems. Some of these at least are meant to represent the factors in Renner's P-series. Shull's R-series has nothing to do with the R (red midribs) of Renner and Heribert-Nilsson.

TABLE 9.I

List of genes in the subgenus *Oenothera* which have been assigned to particular chromosomes or chromosome arms

Chromosome or Chromosome arm	Symbol	Complex	Phenotypic effect	Reference
1·2	*bu*	velans, gaudens	(Bullata) coarsely crinkled leaves	Shull, 1928b
	R. R^{rub}	velans, gaudens, rubens	Red midrib, homozygous lethal	Renner, 1942a
	R^{exc}	excellens	Red midrib, not lethal	Renner, 1942a
	Sp	velans, gaudens	(Supplena) double flowers	Shull, 1925
	v	velans, gaudens	Old gold flower color (vetaurea)	Shull, 1921, Emerson and Sturtevant, 1932
2 (probably)	*Su*	flavens	Suppresses Co	Harte, 1942
1·4	fr^a	stringens	Pollen sterility	Oehlkers, 1926 Emerson and Sturtevant, 1931
1, 2, 3, or 4	*ang*	ʰfranciscana	Narrow leaves	Harte, 1948
	bco	ʰhookeri	Short bracts	Harte, 1948
	de	Several	Flattened tips of spike	Harte, 1948
	den	ʰhookeri, ʰfranciscana	Toothed leaves	Harte, 1948
	lat	ʰfranciscana	Broad bracts	Harte, 1948
	lin	ʰhookeri, ʰfranciscana	Linear leaves	Harte, 1948
	long	velans	Long style	Harte, 1948
	ra	ʰhookeri, stringens	Red stem tips	Harte, 1948
	ser	ʰfranciscana	Soft, silky calyx hairs	Harte, 1948
3	*M*	curvans =	Red leaf margins, short sepal tips, at least when homozygous; allele of *P*-series?	Renner, 1942a
	P-series	Several	Red papillae on stem, red cone color	Renner, 1942a
	an	albicans	Pollen lethal	Renner, 1948
	s	ʰhookeri, flavens, etc.	Sulfur-colored petals	Renner, 1942a
	str	Several	Striped calyx (close top)	Harte, 1948

TABLE 9.I—*continued*

Chromosome or Chromosome arm	Symbol	Complex	Phenotypic effect	Reference
3 or 2	*na*	flavens	Dwarf habit	Harte, 1948
3·4	A^{cro}	flavens, stringens	Pointed sepal tips	Harte, 1948
	rub	^hfranciscana, ^hhookeri	Red stems	Harte, 1948
	tub	Several	Red hypanthium	Harte, 1948
	vit	^hfranciscana, ^hhookeri	Red-striped capsule	Harte, 1948
4	*Ce*	argillicola	Bent stem tips, allele of *Fl*?	Renner, 1942a
	Cu	curvans	Bent stem tips, homozygous lethals	Renner, 1942a
	Fl	flectens	Bent stem tips, subterminal sepal tips, homozygous semilethal	Renner, 1942a
	n	gaudens	Dwarf habit	Renner, 1942a
	ov	velans	Ovate leaves	Harte, 1948
	pus	punctulans	Dwarf (allele of *n*?)	Renner, 1942a
	serp	velans brevistylis	Tortuous stems	Harte, 1948
	ster	velans brevistylis	Pollen sterility	Harte, 1948
	Sbc	subsurvans	Bent stem tips, weak	Renner, 1942a
5·6	let^{fl}	flavens	Zygotic lethal	Renner, 1942a
	Sp	several	Acute leaves	Renner, 1942a
5·8	let^{vel}	velans	Zygotic lethal	Renner, 1942a
7·8	*B*	Several	Broad leaves	Renner, 1942a
	ery	flavens	Red-striped capsules	Harte, 1948
	Kr	stringens	crippled habit	Harte, 1948
	Te	flavens	Narrow-leaved mutant	Renner, 1942a
7 or 8	*Cer*	velans brevistylis	Dark red hypanthium	Harte, 1948
8	*pil*	albicans	Brownish hairs, homozygous lethal	Renner, 1942a
9·10 or 11·12	*acu*	flavens, stringens	Long, thin sepal tips	Harte, 1948
	car	stringens	Red-striped capsules	Harte, 1948
	hir	flavens	Long hairs on capsule	Harte, 1948
	ru	stringens	Red stem tips	Harte, 1948
11·12	*br*	velans	Short, defective styles	Catcheside, 1954
	def	flavens	Defective sepals	Renner, 1942a
11·13[b]	lor^2	flectens	Narrow, thread-shaped leaves, semilethal	Renner, 1942a
12·14[b]	lor^1	flectens	Narrow, thread-shaped leaves, semilethal	Renner, 1942a

TABLE 9.I—*continued*

Chromo-some or Chromo-some arm	Symbol	Complex	Phenotypic effect	Reference
13·14	A^{de}	ʰfranciscana	Glandular hairs on buds	Harte, 1948
	ano	flavens	Crippled stem tips	Harte, 1948
	Co^c	Several	Large flowers	Harte, 1948
	dil	ʰfranciscana	Pale anthocyanin	Harte, 1948
	his	ʰfranciscana	Strong hairs	Harte, 1948
	lob	ʰfranciscana	Narrow bracts	Harte, 1948
	tec	flavens	Long bracts	Harte, 1948
	tri	ʰfranciscana	Long hairs on buds	Harte, 1948
	un	flavens	Crinkled bracts and leaves	Harte, 1948
	uni	ʰhookeri, ʰfranciscana	Solid calyx color	Harte, 1948
	vir	ʰfranciscana	Green stem color	Harte, 1948

[a] Harte (1942) places *fr* in ends 2 or 3.
[b] 11·13 = 12·13, and 12·14 = 11·14 on Cleland system.
[c] Emerson and Sturtevant (1932) found *Co* and *br* linked with 15% crossing over. Catcheside has located *br* in 11·12. *Co* should therefore be in 11·12, not in 13·14 as stated by Harte.

It should not be assumed that all genes listed in Table 9.I are definitely established as separate and valid entities. Probably some of the genes listed by Harte, for instance, are identical, or at least allelic with genes recognized by Renner. Some may in time be discarded. Others, with pleiotropic effects, may prove to be multiple loci. The references in the table are to the most important, but not necessarily the first paper in which the symbol is mentioned.

In addition to the complexes represented in the table, scores of other complexes have been analyzed by Cleland and his students and by others from the standpoint of the characters which are controlled by, and distinguish each complex, without attempting to assign symbols to the genes controlling these characters, or to locate them on specific chromosomes. This work has been done in connection with studies of the evolution of the group and the relationships that exist among the complexes.

Mention may be made in conclusion of a character first described by Harte (1948) and considered by her to represent a case of fluctuating dominance at a gene locus. This character, which she named *falcifolia* (*fal*) or sickle-shaped leaf, has since been ascribed by Stubbe (1970) to the interaction between different plasma types. It appears in hybrids between *lamarckiana* and *suaveolens* and takes the form of crippling of leaves and floral parts as the result of defective development of vascular structures. Reciprocal crosses differ in the extent of crippling: *lamarckiana* × *suaveolens* yields 90–100% *falcifolia* plants, the reciprocal produces only 20–40%. Backcrosses in either direction to *lamarckiana* give less than 20% falcifolia; to *suaveolens* the percentage is over 50%. This is true whether the F_1 plant used was crippled or normal. The character seems to be independent of genome and plastome and is definitely non-Mendelian.

If Stubbe's hypothesis proves to be correct, this will be the first case found in the subgenus *Oenothera* where a phenotypic effect is based on elements of the cytoplasm distinct from the plastome.

CHAPTER 10

Crossing Over

Since most members of the subgenus *Oenothera* are complex-hetero-zygotes and therefore highly heterozygous, one might expect to find many cases of crossing over in the group. It is therefore surprising to discover that detectable crossing over is rarely observed in the various races of *Oenothera*. That crossing over does occur, however, seems obvious from the fact that the chromosomes are held together at their ends, presumably by chiasmata, which are generally regarded as having been formed as the result of crossing over. Two facts, however, seem to be clear in this connection, as first brought out by Darlington (1931). (1) Crossing over is apparently confined almost exclusively to the pairing ends: the central portion of the chromosome is rarely if ever involved in crossing over, presumably because it seldom if ever becomes synapsed with the region homologous to it. This means that most of the genes which distinguish one complex from another remain together from generation to generation in the unsynapsed region and consequently the Renner complexes survive intact and uncontaminated for indefinite periods. (2) Most races must be homozygous for most or all of the genes that lie in the pairing segments. If this were not the case, crossing over would be frequently detected. At first sight this may seem somewhat improbable, but it is what might be expected to develop in the course of time. Suppose that *velans* carries in one of its pairing ends a dominant gene A, *gaudens* carries the recessive a, and suppose that crossing over between the A locus and the lethals occurs in 20% of the cells. The next generation following selfing will consist of 82% Aa, 9% AA and 9% aa plants. The 9% that are homozygous for the dominant will not show crossing over, nor will their descendants. Of the 82% heterozygotes, 9% of their selfed progeny will again be homozygous for the dominant and show no detectable crossing over. The proportion of the total population homozygous for this locus, other things being equal, will slowly increase from generation to generation in selfed line—and complex-hetero-zygotes are as a rule self pollinating. When one realizes that it is ordinarily impossible to distinguish visually between homozygous dominants and heterozygotes, it is clear that, even in the experimental garden, and with controlled selfing, heterozygotes for genes in the pairing ends may in time disappear (see Renner's work, cited below, p. 117).

It seems, therefore, that crossing over is confined to the pairing segments and that these segments have become homozygous for any genes that lie sufficiently distal to be crossed over from one complex to the other. Under these circumstances one would expect to observe crossing over in *Oenothera* only rarely. True crossing over will be observed only where at least one heterozygous allelic pair lies close enough to the end of a chromosome so that crossovers can at least occasionally occur between it and the lethals. Very few loci have been found where this is the case, and the situation is further complicated by the fact that a gene may find itself in a pairing end in one complex-combination, but proximal to the pairing end in another combination. It will show crossing over in the former case but not in the latter. Furthermore, it may lie, in one combination, near the proximal end of the pairing segment, in another combination nearer the distal end. In the former case, it will cross over rarely, in the latter case, more frequently.*

For example, Shull (1923a) found $6 \cdot 1$–$8 \cdot 7\%$ crossing over between the P and s loci situated in end number 3, in hybrids between "*franciscana sulfurea*" and *rubricalyx*; and Emerson (1931a) obtained an average frequency of $8 \cdot 7\%$ in a cross between *franciscana sulfurea* and a hybrid containing $p^r p^s S s$. On the other hand, Oehlkers (1933, Table 32; 1940, Table 8), obtained much lower percentages in a number of complex-combinations (see Table 10.I). It is probable that the genes involved lay farther from the proximal end of the pairing segment in Shull's and Emerson's material than in Oehlkers'. This may have been due to the ability of the number 3 ends to synapse over a longer distance in the former material than in the latter, which would increase the probability of crossing over between P and s. This can be made clear by reference to Fig. 10.1. If the arms synapse such a short distance in from the tip that neither P nor s is in the pairing arm, there will be no crossing over between them. If, however, the s locus is in the pairing end and the P locus is not, there is then some chance of crossing over between the two. Finally, if pairing extends far enough back to include both P and s in the pairing arm then the full amount of crossing over can occur that corresponds to the map distance between these two loci.

There are alternative explanations, however, for these discordant results. It is possible that structural alterations have resulted in P and s being at a greater distance from each other in Shull's and Emerson's

---

* Renner (1948) found a positive correlation between the frequency of crossing over observed in a complex-combination and the incidence of incompletely terminalized chiasmata during the stages diakinesis through anaphase I, suggesting that the chances for crossing over are greater where the pairing ends are longer.

TABLE 10.I

Percentage of crossing over between *s* and *P* in certain complex-combinations
(Oehlkers 1933, 1940)

Complex-combination	*s* and *P* in	Percentage of crossing over
flavens·stringens	Pair	0·25–0·28
flavens·ʰhookeri (anthers)		0·00–1·82
(ovules)	⊙4	0·61–3·06
flavens·velans	⊙4	1·13
albicans·stringens	⊙12	1·09
albicans·ʰhookeri (anthers)		0·00–0·57
(ovules)	⊙14	4·24
albicans·velans	⊙14	0·00
curtans·flavens	⊙14	0·03
deprimens·flavens	?	0·20
gaudens·ʰhookeri (anthers)	⊙10	2·44
(ovules)		2·14
excellens·ʰhookeri	Pair	0·36

material (they used material from a common source) than in Oehlkers'
material.

Another factor, also, may influence the apparent frequency of crossing
over. It is possible that crossover and non-crossover gametophytes may
differ in their ability to compete in embryo sac formation or in pollen
tube growth. That this may be a factor is shown by Oehlkers (1940),
who obtained a crossover frequency of 0·0–1·82% between *P* and *s* in
flavens·ʰhookeri when the heterozygote was used as male, 0·61–3·06%

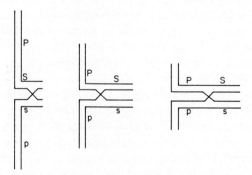

FIG. 10.1. Percentage of crossing-over found to exist between the *P* and *s* loci will
depend upon the distance over which synapsis can occur in the particular complex-
combination being studied.

when it was used as female. Renner (1942a) also stated that crossovers are transmitted through the pollen definitely less frequently than through the eggs.

In most organisms, in order to detect crossing over, one must have two pairs of linked heterozygous alleles between which exchange can occur. In the case of large circles with balanced lethals, however, the lethals themselves can serve in place of one pair of alleles (even though they are not allelic). If a recessive gene in one complex of a complex-heterozygote is able to express itself in an individual, without the chromosome configuration in that individual having been altered, we may take this as an indication that true crossing over has occurred (Emerson, 1931b). The responsible gene, which has apparently been linked to one lethal by reason of its being in the same circle and the same complex, has now been transferred to the other complex and in a particular gamete is now linked to the other lethal. It can now show up in homozygous condition if this gamete functions in fertilization.

One must be careful, however, in drawing conclusions regarding the relationship between apparently linked genes when they are in circle chromosomes. This may be illustrated by reference to the gene n, for dwarf stature. This is situated in end 4 of the 9·4 chromosome of *gaudens* (*lamarckiana*). Shull claimed to have found linkage between r^h (*rubricalyx* = p^r of Emerson) and n, with a crossover value of about 9% (1923a). Renner, however, (1942a) found that n crosses over from *gaudens* into *velans* with a frequency of only 1% or less, and p^r crosses from *velans* to *gaudens* even less frequently. The discrepancy arises from the fact that Shull did this work before large circles were discovered, and hence was unable to consider the question of chromosome configuration in arriving at his conclusion. The gene r^h (= p^r) is situated in the 3-end of the *velans* chromosome 3·4, whereas n is in the 4-end of the *gaudens* chromosome 9·4. In *lamarckiana*, therefore, r^h and n are not linked in the conventional sense. They appear to be in the same linkage group, in repulsion phase, merely because they are in the same circle, with its balanced lethals. It is possible for n to cross over into the 4-end of *velans* and if this should occur in the case of a *velans* chromosome which contained r^h, the two genes would then be truly linked. On the other hand, if r^h should cross over into *gaudens*, it would not be truly linked with n, since it would lie in the *gaudens* chromosome 3·12, whereas n is in 9·4. Such crossing over rarely occurs—at a level of 1% or less, as we have mentioned. The 9% "crossing over" claimed by Shull undoubtedly involved material in which a 3·4 chromosome was present in double dose. The origin of his material was not revealed by Shull other than to say that the crosses which resulted in crossing over were between *rubricalyx* tall and green-budded *nanella*. The *rubricalyx* tall probably had the *latifrons* complex combined with a complex of unknown composition, perhaps including a mixture of *velans* and *gaudens* chromosomes. The *rubricalyx* tall parent may have had 3·4 in double dose. The cytological nature of the green-budded *nanella* is unknown, but it probably had at least one 3·4 chromosome also. The hybrid, with two 3·4 chromosomes, was no doubt heterozygous for both the loci, one in the 4-end, the other in the 3-end. The crossovers found in the F_2 backcross represented the total amount of crossing over between two genes placed at opposite ends of a pair of chromosomes that were presumably homologous throughout.

This example shows the essentiality of knowing the chromosome configurations of all oenotheras that are used in any kind of breeding work.

An interesting fact regarding crossing over in *Oenothera* was brought out by Renner (1942a). He found that the frequency of crossing over is often much greater in hybrids than in the parental races, but tends to decrease in later generations derived from these hybrids Thus, *n* (*nanella*), which crosses over very rarely in *lamarckiana* or *biennis*, appeared in as many as 10–15% of the progeny of a *pingens-rubens* combination through the F_3, but failed to appear in subsequent generations.

The increase in amount of crossing over observed in some fresh hybrids can be attributed, as suggested above, to the fact that in these hybrids certain arms are able to synapse farther back from the tips than is the case in the parental races.* The relatively few genes that occasionally cross over lie apparently in the transition area close to where pairing ceases. In some combinations, pairing fails to reach these loci, in other combinations, these genes are included in the pairing ends and are therefore in a position to cross over.

On the other hand, the loss of ability to show visible crossing over in later generations is probably due to the unconscious selection of dominant homozygotes which cannot be told from heterozygotes, as pointed out above.

In summary, crossing over is confined in *Oenothera* to the pairing ends. In most races the ends are essentially homozygous, and detectable crossing over therefore rarely occurs. In certain hybrid combinations, however, pairing may extend to a point proximal to the locus of a given heterozygous pair of alleles so that visible crossing over between this locus and the lethals may occur, sometimes with rather high frequency.

---

* If this is the case, it might be considered an argument for the presence of the "interstitial segments" of Darlington, but it seems more likely the result of the fact that translocations which arise from the pulling apart of interlocked chromosomes do not always occur at exactly the same position in different lines.

The number of genes in *Oenothera* that have been observed to cross over is small. In order to give a picture of the situation, a representative sample of these cases will be mentioned.

The free pair in *lamarckiana* (1·2) furnishes examples of crossing over. In this chromosome *R* (red midribs), together with certain mutant genes found by Shull, have been located—*vet* or *v* (*vetaurea* or "old-gold"), *sp* (*supplena* or double-flowers) and *bu* (*bullata* or heavily crinkled leaves). Shull (1927) found crossover percentages between *v* and *sp* as follows: 0·057, 0·92, 0·95, and 1·8%. Between *v* and *bu* crossover percentage was found to be 0·87% (1930). The order of the three genes in the chromosome was not determined, nor was the relation of these genes to *R* worked out. Renner, however, obtained a crossover percentage between *v* (*vet*) and *R* of 1·4% (1942a, p. 171). It is not known which end contains any of these genes.

There is one chromosome end, as we have seen, in which more than one gene is known to reside. End number 3 in certain complexes carries the *P*-series of alleles. The gene *s* for sulfur-colored petals also resides in this end as does the pollen lethal in *albicans* (*an*), and the gene for the mutant character *helix* (*hel*) which produces abnormal growth (Chrometzka 1955, 1956). Chrometzka found that *hel* is in the 3-end of *velans* and crosses over with respect to *P* with a frequency of 6%. It lies distal to *P*, but its relation to the *s* locus is undetermined. The exact position of *an* has not been determined (Renner, 1948). As we have seen, *P* and *s* have been reported to show varying percentages of crossing over, probably depending upon the length over which the 3-ends are able to synapse in different complex-combinations. The *P* locus is apparently proximal to *s* and in some combinations synapsis does not reach far enough back from the tip to include this gene.

Other genes which have either been located singly in their respective chromosomes, or whose loci have not been determined, are known to be involved in crossing over on occasion. To cite a few examples (Renner, 1942a): *Cu* is the gene in *curvans* which causes bending of the stem tips and subterminal position of the sepal tips. It is situated in arm number 4. It is lethal in homozygous condition or closely linked to such a lethal. This has been observed to cross over in one case in the complex-combination ᴴ*hookeri·curvans*, resulting in a slender F₂ *hookeri* individual with bent stem tips and subterminal sepal tips. Although ᴴ*hookeri·curvans* has ⊙6, ⊙8 normally, this particular crossover individual had ⊙6, 4 pairs. The pairs were 1·2 3·4 5·6 13·14. Evidently what happened was that this plant received the *hookeri* half of the circle of eight through both sperm and egg, but one of the parental gametes was carrying *Cu* which had been received from *curvans* by crossing over. The new form then had, in addition to ⊙6, four normal *hookeri* chromosome pairs, except that one of the 3·4 chromosomes had received by exchange a small segment of the *curvans* chromosome carrying *Cu*, so that the plant was heterozygous for *Cu* (*Cu cu*). This F_2 plant was selfed and produced in F_3 typical *hookeri*, as well as a repetition of the new form, but in addition a number of weak plants which failed to bloom, and which were undoubtedly *Cu* homozygotes. Apparently, the lethal effect of the *Cu* chromosome is reduced to a semi-lethal effect when only the *Cu* portion of the *Cu* chromosome is present.

Fl is a factor in *flectens* that produces bent stem tips, and when homozygous is semilethal or associated with such a lethal. It is situated in arm 4 of the 1·4 chromosome in *flectens*. A hybrid combining *flavens* and *P-velans* produced a plant with the typical *flavens·velans* composition except that 1·2 3·4 of *velans* was present in duplicate (Renner, 1942a). This in turn was crossed with *atrovirens* and produced a *P-flavens·flectens* in which 1·2 3·4 of *velans* was present instead of the 1·4 3·2 of *flavens*. This *P-flavens·flectens* was pollinated by *biennis* and produced *P-flectens·rubens* which, because the 4-end contributed by *flectens* had come from *velans*, did not have bent stem tips. However, there was one individual that did have bent stem tips, at the same time that it had *P*. Evidently *Fl* had crossed over from the 1·4 of *flectens* into a *velans* 3·4 chromosome (with *P* in its 3-arm) so that in this case the *flectens·rubens* had both *P* and *Fl*. From this plant were obtained progeny homozygous for *P* and *Fl* which were less weak than would have been the case had they possessed the entire 1·4 chromosome of *flectens* in duplicate.

The zygote lethal of *flavens* (*let*) lies in 5·6. This crosses over into the opposing complex *albicans* in *suaveolens* with a frequency of about 2% so that *flavens·flavens* (*lutescens*) offspring are obtained. The segmental arrangement of the *flavens* complexes in *lutescens* is unaltered, so that the transfer of *let* to the opposing complex is the result of crossing over.

The recessive gene *n* for dwarf habit is present in *gaudens* of *lamarckiana* and *rubens* of *biennis*. de Vries obtained in *biennis* 0·1–0·2% *nanella* plants (1915) and in *lamarckiana* he obtained 1% (1913, p. 205). *Nanella* plants have the chromosome configurations characteristic of the species that have given rise to them. Renner and others have also observed crossing over of this gene in various combinations.

The gene *pil* in *albicans* produces olive-green foliage with brownish hairs. A single plant showing these characters (*pil·pil*) appeared in 1928, in Renner's garden, in the F_2 of a hybrid culture (*biennis* × *lamarckiana*) *albicans·velans* (*velans* carries the dominant *Pil*, *albicans* the recessive *pil*). This individual, with the unaltered ⊙14, arose as the result of crossing over and bred true through at least twelve generations for *pil*

(Renner, 1942a). The reverse crossover was detected when the same hybrid, *pil-albicans·Pil-velans*, was crossed to *pil·pil-albicans·flectens*. A single *albicans·flectens* plant among the progeny had green foliage and colorless hairs (*Pil·pil*).

The factor for sulfur flowers (*s*) was shown by de Vries (1918d) to cross over with a frequency of 0·3% in *biennis* and 0·1% in *suaveolens*.

P has been found in one or two cases to cross over in *lamarckiana*. Shull's so-called mutant *pervirens* (1921) was probably *p-velans· p-gaudens*. Renner has found *p-velans* in a couple of trisomics from *lamarckiana*, but not in *lamarckiana* itself, nor has he found *P-gaudens* in *lamarckiana*. Since *P* is situated proximal to *s* in arm 3, it is not surprising that it crosses over to the *gaudens* complex less frequently than does *s*.

Finally, reference must be made to two loci that are situated so close to the distal end of the chromosome that they show 50% crossing over with respect to lethals. These are *Co* and *br*.

Co is a flower size factor, for which a series of alleles is thought to exist in different races (Renner, 1925). *Co* represents small flowers, *co* large flowers. The mutant gene *br* (*brevistylis*) produces greatly shortened styles and somewhat misshapen stigmas, with nearly complete ovule sterility. This was one of the first aberrants discovered by de Vries at Hilversum in 1887. The chief interest of these genes is the influence that they had for a time on cytogenetic thinking. *Co* seemed to represent the only exception to the rule that all genes in a circle behave as though they are linked. *Brevistylis*, furthermore, constituted the chief barrier to Shull's acceptance of the chromosome cohesion theory in *Oenothera*. He failed to place this gene in his first linkage group because it segregated independently of other genes in this group. It was placed instead in a second linkage group. The genes that we now know to be situated in the paired chromosome in *lamarckiana* he included in a third linkage group. Because, therefore, there appeared to be three linkage groups but only two chromosome groups, he rejected the chromosome cohesion theory of linkage and placed each of his three linkage groups in a separate but single chromosome. The apparent discrepancy between the number of chromosome groups and the number of linkage groups was explained by Emerson and Sturtevant (1931b) and Cleland and Brittingham (1934) as due to the location of *Co* and *br* so close to the end of the chromosomes in which they reside that crossing over occurs between them and the lethals in essentially 100% of the cells.

The *Co* locus is like most others that lie in the terminal segments in being homozygous in each of the various races. Crossing over shows up only in hybrids in which the associated complexes possess different flower size alleles. The situation in the case of *br*, however, is somewhat different. This mutation causes crippling of the carpels to such an extent that it is difficult to secure progeny by selfing, although the pollen is good. A few progeny have been obtained by selfing, however, all of them *br*. The gene is normally carried along by pollinating *lamarckiana* with *brevistylis*. The hybrid progeny thus produced are normal in appearance but are carriers of *br*. The independent segregation of *br* is shown by the appearance of *br* in the selfed progeny of such a hybrid although it has ⊙12, and by outcrosses to other races producing hybrids with large circles, which, when selfed, throw *br*. It is interesting that laetas sometimes produce *br* plants and at other times velutinas show the character. Since *br* shows 50% crossing over with respect to the lethals, it can lie in either the *gaudens* or the *velans* complex of *lamarckiana* and cannot be considered as belonging to one rather than the other.

Although crossing over, therefore, has been detected in *Oenothera*, the number of cases is remarkably small, particularly in the races themselves, especially when one considers the thousands of cultures that have been grown of these plants.

Both Emerson (1932) and Renner (1942a) have shown that the extent and frequency of crossing over does not differ appreciably between chromosomes that form pairs and those that belong to circles. Apparently *Oenothera* chromosomes fail to spin out in the central region sufficiently for synapsis to occur, irrespective of whether they are members of pairs or circles. Crossing over is therefore confined to the tips, even in paired chromosomes, and it is only the relatively few genes that lie in the pairing segments that, if heterozygous, will show detectable crossing over.

Darlington (1931) has suggested that crossing over may occur exceptionally in the proximal segments. As pointed out earlier (p. 77) he thinks that breaks resulting in segmental interchange may occur at points other than the centromere. As a result, situations could occasionally arise where homologous segments were present in the proximal regions of chromosomes that were otherwise non-homologous. Synapsis between such segments might result in crossing over involving the proximal regions of these chromosomes. Such crossovers would result in exchanges between non-homologous pairing ends, thus simulating reciprocal translocations. Darlington (1931) presents drawings and photographs which purport to show chiasmata in such regions. These figures, however, could also be interpreted as overlaps. Never having observed an interstitial chiasma in *Oenothera*, I am not convinced that

crossing over ever occurs in the proximal regions of the chromosomes in
Oenothera. I believe that changes in segmental arrangement occur only
as a result of reciprocal translocations.

In conclusion, the following points in regard to crossing over in *Oenothera* seem to be clear: (1) crossing over occurs in the pairing ends,
resulting in the chiasmata that hold the chromosomes together at their
ends; (2) most of the genes in the pairing ends have become homozygous
in the races so that crossing over which involves them is not genetically
detectable; (3) crossing over is confined to the pairing ends and does not
take place in the proximal region; (4) a relatively few genes in the pairing
segments may, especially in hybrids, show crossing over, usually with
low frequency, the frequency depending upon whether synapsis does or
does not reach them in a particular complex-combination; (5) crossing
over is most frequently seen in newly made hybrids that are heterozygous
for certain alleles situated in pairing ends. In later generations derived
from such hybrids by selfing, crossing over tends to disappear, owing to
unconscious selection of individuals homozygous for the dominant.

CHAPTER 11

Position Effect

If one were to look for the phenomenon of position effect in higher plants, one would naturally think of *Oenothera*; with such a relatively extensive history of reciprocal translocations, one would think that some evidence of position effect would be found in this genus. However, certain considerations might make one less sanguine about finding evidence for such behavior. In the first place, one should realize that, even in *Oenothera*, translocations are not everyday occurrences. It is only on the evolutionary scale that they have occurred with relatively high frequency. Secondly, if interchanges occur in the middle of heterochromatin, one would probably not expect position effects, and there is good reason to suppose that interchanges in *Oenothera*, at least those that have survived, have occurred most frequently in heterochromatic regions close to the centromeres, or possibly in cases where they are the result of separation of interlocked chromosomes, so close to the ends of chromosomes that no recognized genes lie distal to the point of exchange, and so far removed from heterochromatin that no position effect would therefore be expected.

In over 40 years of growing oenotheras, during which time many thousands of plants have been studied, I have detected a reciprocal translocation genetically only once in my garden, and in this case no position effect was observed (Cleland, 1950b): in 1936, an interchange occurred in a plant of the *grandiflora* of de Vries, as a result of which a plant was produced in 1937 with an altered chromosome configuration. This plant, however, was indistinguishable from normal plants phenotypically and it was not realized that a change had occurred until it was found that hybrids of this 1937 individual did not have the chromosome configurations predicted for them. A check on the chromosome configuration of this parent plant revealed that it had ⊙12 instead of ⊙14. The interchange was later analyzed, and it was found that the *acuens* complex had 1·13 4·14 instead of 1·4 13·14. The altered complex was named *neoacuens*. The *grandiflora* line descended from this plant differed in no respect phenotypically from the original line, and showed, therefore, no evidence of position effect.

It remained for Catcheside (1939, 1947a, b) to discover a clear case of position effect, the only one that has been found in *Oenothera*, or for that matter in higher plants. This occurred in *O. blandina*, a derivative of *lamarckiana. Blandina* is a complex-homozygote and carries in its 3-end the gene P^s (broad red stripes on the flower cone, red papillae on the stem). Catcheside irradiated *blandina* with X-rays and obtained a complex with an altered segmental arrangement, 3·12 4·11 (using the Cleland system; Catcheside designated them 3·11 4·12) instead of 3·4 11·12. This he called *blandina-A*, the normal complex being ^{h}*blandina*. This complex was combined with unmodified ^{h}*blandina*, and the combination, which had ⊙4, 5 pairs, showed certain phenotypic modifications: narrow leaves, easily recognized even in seedling stages, and reduced red pigmentation of the bud cones, the red stripes being narrower, and broken by irregular green stripes running longitudinally through them. Two possible explanations for this mottling effect suggested themselves to Catcheside. It might be the result of a mutation of P^s which occurred simultaneously with the interchange 3·4 11·12 → 3·12 4·11, or it might be a position effect. Catcheside found that the latter was the correct explanation when he pollinated P^s *blandina-A* · P^s ^{h}*blandina* with P^r ^{h}*blandina* · P^r ^{h}*blandina* and backcrossed the resulting P^s *blandina-A* · P^r ^{h}*blandina* to pure P^s ^{h}*blandina*. This yielded, as expected, equal numbers of P^s ^{h}*blandina-A* · P^s ^{h}*blandina* (showing mottling) and P^r ^{h}*blandina* · P^s ^{h}*blandina* (*rubricalyx*), except for one plant which was P^s ^{h}*blandina* · P^s ^{h}*blandina* (normal, unmottled). One could explain the presence of this normal P^s plant only by postulating a mutation of P^r to P^s, which had never before been observed, or by assuming a crossover of the P^s of *blandina-A* and the P^r of ^{h}*blandina*, thus producing a P^s ^{h}*blandina* genome. This would mean that the P^s gene whose effect had been modified when transferred from 3·4 to a 3·12 chromosome, became normal in its effect when returned via crossing over to a 3·4 chromosome. Catcheside adopted the latter hypothesis.

This assumption was based, of course, on only a single case. Consequently, Catcheside performed additional experiments which he published in 1947. He used both P^s and P^r and both gave the mottled effect when they were present in *blandina-A* combined with ^{h}*blandina*. When the backcross mentioned above was carried out (P^s *blandina-A* · P^r ^{h}*blandina* × P^s ^{h}*blandina* · P^s ^{h}*blandina*), seven crossovers were found among the 699 offspring. Both crossover classes were obtained, normal P^s ^{h}*blandina* · P^s ^{h}*blandina* and *rubricalyx* mottled P^r *blandina-A* · P^s ^{h}*blandina*. The crossover percentage, therefore, between the P locus and the interchange point was about 1%.

P residues in the same chromosome end with s (sulfur flowers). In his material, Catcheside had found about 8% crossing over between these two loci, with s distal to P. Various combinations of P alleles with S or s were obtained and tested for position effect. Surprisingly the S locus also showed position effect, although it was as much as eight units of map distance removed from P. The resultant effect was large sulfur patches on yellow petals. For example: P^sS $blandina$-$A \cdot P^rs$ $^hblandina$ (P^sS-$A \cdot P^rs$) $\times$ P^ss $^hblandina \cdot P^ss$ $^hblandina$ ($P^ss \cdot P^ss$) gave the following progeny:

Non-crossovers

    75 $P^rs \cdot P^ss$ (normal rubricalyx sulfur)
112 P^sS-$A \cdot P^ss$ (mottled red and green cones, mottled yellow and sulfur petals)

Crossovers

  4 $P^sS \cdot P^ss$ (normal red, yellow)
19 $P^rS \cdot P^ss$ (normal rubricalyx, yellow)
  6 P^ss-$A \cdot P^ss$ (mottled red and green cones, sulfur)
  1 P^rS-$A \cdot P^ss$ (mottled rubricalyx and green cones, mottled yellow and sulfur)
  1 P^rs-$A \cdot P^ss$ (mottled rubricalyx and green cones, sulfur)

There were 187 non-crossovers in this experiment and 31 single or double crossovers. It will be noted that in all cases where a gene was transferred by crossing over from the 3·12 chromosome to the 3·4 chromosome, its behavior returned to normal. This applies to both P and S.

Catcheside carried out eight such backcrosses and the results were entirely consistent. Out of a total of 1112 progeny, there were 1001 non-crossovers, 109 single and 2 double crossovers, all with the expected phenotypes. The results of these crosses indicate that in this material the P locus is 1·7 units from the translocation break and S is 8·5 units removed. Catcheside's interpretation of the mottling is that the P and S loci have been translocated to the neighborhood of heterochromatin in the 3·12 chromosome.

As mentioned above, Catcheside (1939) found that the translocation-heterozygote, as would be expected, contained a circle or chain of four plus five pairs. He noted that the chromosomes in this chain were not all of the same size (1954). Two were normal in size and morphology, one was shorter than normal with one arm normal and the other reduced in length, and the fourth had one arm that was extra long, with two condensed regions instead of the usual one, separated by a thinner segment. Catcheside, therefore, concluded that the exchange had been unequal.

He identified the enlarged chromosome as the 3·11 (3·12 on our system), the shortened chromosome as the 4·12 (4·11).

Dr. Catcheside has kindly sent us seed of his position effect material and we have carried this on for four generations. We have been able to confirm his finding that all plants heterozygous for the translocation and therefore showing the position effect have a circle or chain of four chromosomes (Fig. 11.1): all plants that are homozygous for 3·4 11·12

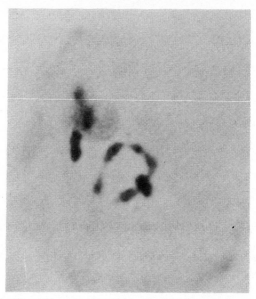

Fig. 11.1. *O. blandina.* Chromosomes of translocation—heterozygote showing chain of four. Ends from left to right are 11, 4, 3, 12, 11. The 3·4 chromosome has a prominent constriction in the 3 arm.

show seven pairs. We have also confirmed the difficulty in obtaining plants homozygous for the interchanged chromosomes. We have attempted to grow plants which are undoubtedly translocation homozygotes but have succeeded in bringing none of them into flower. They have very narrow leaves, grow very slowly and remain indefinitely in the rosette stage. Catcheside did succeed in bringing four plants to the flowering stage, but did not report on the chromosome configuration.

In one respect our cytological results extend as well as support those of Catcheside. As the result of a rather extensive survey of a large number of hybrids involving many races, and with many different chromosome configurations, we have reached the conclusion (not yet published) that the large nucleolus in the subgenus *Oenothera* is attached to arm number

3. On examining the chromosomes of plants showing the position effect it has been found that the nucleolus is apparently attached to the long arm of the enlarged chromosome or to the arm of the adjacent chromosome that is attached to this arm (Fig. 11.1). This would mean that the pairing segment of the long arm of the long chromosome is end number 3, as shown in Fig. 8.7b, and that the lengthened chromosome is 3·12 (3·11 of Catcheside) as stated by Catcheside.

In a small minority of cells a closed circle of four is present in diakinesis; in most cases, however, one finds an open chain. With few exceptions, the chiasma failure that causes the circle to open is at end 11, but I have seen one cell in which the circle was open at the 3 end.

If the mottling effect is due to the fact that P and s have been transferred to the neighborhood of heterochromatin, the break in the 3·4 chromosome must have occurred quite close to these loci. Examination of the translocated chromosomes in diakinesis, however, shows that the 3·12 chromosome has been increased by the addition of a sizeable segment, and the shortened 4·11 has been noticeably reduced. Furthermore, the elongated arm is composed of two clearly distinguishable segments, as Catcheside has shown, a longer proximal segment, and a shorter, but by no means insignificant, distal segment, the two segments separated by a short constriction. This suggests that the position of s and P in the 3·12 chromosome is slightly distal to the constriction. If this is correct, then P and S are farther removed from the end of the 3·4 chromosome than had been previously surmised. Furthermore, a heterochromatic region exists quite far removed from the centromere in 11·12.

The discovery of a case of position effect in *Oenothera* is interesting for several reasons: it is the first clear-cut case of position effect found in higher plants; it involves an extraordinary length of the genetic map (some eight units of map distance); the mottling produces quite large patches, which suggests that when the dominant gene is suppressed in a given cell, it remains suppressed through a series of cell divisions. The patches tend to be larger when the plant is actively growing than later. Finally, this is a case where the translocation that took place was unequal. One chromosome received a longer segment than it donated. This is of interest in view of the fact that circle chromosomes in *Oenothera* are in general alike in size and length of arms. Evidently, in the course of evolution only those individuals whose chromosomes are equal and isobrachial and therefore able to disjoin regularly have survived. Catcheside's material demonstrates, however, that unequal exchanges are possible, although a strain of plants with a circle of four, consisting of chromosomes of different lengths, would probably not survive in nature.

This case is of interest in one other respect. The fact that the interchange complex is unable to produce healthy plants when homozygous suggests that the interchange has either caused the loss of some locus, or an important locus has been so seriously affected by the transfer to a region of heterochromatin that it is unable to function properly. The interchange complex is not fully functional. In homozygous condition it is semi-lethal.

CHAPTER 12

Inconstant Characters

Two characters have been found in *Oenothera* that frequently or regularly show inconstancy of behavior. These are cruciate petals (*cr*) and missing petals (*mp*). The former has been studied for a long time by several investigators, the latter is discussed here in detail for the first time (see Cleland, 1970).

1. CRUCIATE PETALS (*cr*)

Cruciate petals (*cr*) is a condition in which the petals are narrowed to crippled, strap-like structures, usually with irregular edges, and often sepaloid—more greenish than yellow in color (Fig. 12.1).

FIG. 12.1. *O. biennis apetala*, showing cruciate petals.

The character has been known for a long time. It was first noted by Nuttall about 1830 in material from Massachusetts which was described by Don (1832) as *O. cruciata* Nutt. The material which de Vries (1903a) studied originally came from the vicinity of Lake George, New York, and was obtained from D. T. MacDougal. Both of these strains belong to the *parviflora* assemblage* as do several other cruciate strains, including three that we have studied (*cruciata I* and *II* and *Ostreae*) and also the strain known as *atrovirens* Shull and Bartlett (Bartlett, 1914) which furnished many of Renner's data.

Other cruciate strains belong to the *biennis* assemblage including *biennis cruciata apetala* and *biennis cruciata* A_2a, studied by Oehlkers, and a race studied by us from Elgin, Illinois. The *"lamarckiana cruciata"* of Oehlkers was obtained from the Tübingen Botanical Garden and was probably the result of a cross between *lamarckiana* and *biennis cruciata*, having the composition *velans·rubens*.

The character was studied in some detail by de Vries (1901–3, Vol. II, p. 593 *et seq.*). In the material from Lake George, New York, the condition was found to be recessive to normal petals in outcrosses but to breed true in the race itself. In contrast, de Vries studied another strain, of uncertain origin, which he called *cruciata varia*, in which the character was variable in expression. Some plants of this strain showed the trait more strongly than others, and the flowers of a single individual might range from fully cruciate through all degrees of intermediacy, to the completely normal condition. de Vries included races that behaved in this way among his "Zwischenrassen", or "Mittelrassen", which he defined as races that have acquired through mutation a new pangene that competes with the old one, the balance between new and old being rather even, thus resulting in a fluctuation in the degree to which old and new characters are expressed. de Vries' interpretation was supported by Stomps (1913).

Oehlkers was the first to study the genetics of this character intensively (1930a, b; 1935a; 1938). He used a relatively constant race from the Leiden botanical garden (*O. biennis cruciata apetala*) and two inconstant lines, *O. biennis cruciata* A_2a, obtained from Klebahn, and *O. larmarckiana cruciata*, derived from the botanical garden in Tübingen. When the constant race was outcrossed, he found that the character behaved as a simple recessive. In the inconstant lines behavior was highly variable, both within the race and in hybrids. In one of his inconstant races (A_2a), selection yielded positive results (1930a, p. 486): there was one plant in the culture which showed a striking difference between a side branch

* See Chapter 18.

with almost normal petals, and the main shoot which had strongly cruciate flowers. He selfed flowers from both branches, and from the weakly cruciate branch established a weakly cruciate line, from the strongly cruciate one a strongly cruciate strain. The other inconstant line, *lamarckiana cruciata*, at first showed little effect of selection, but in later generations, the effect became greater (Oehlkers, 1935a, p. 167 *et seq.*). As a result, a seemingly recessive trait became in some lines an apparently dominant character as a result of selection.

Oehlkers crossed his cruciate lines to other races and found that the character expressed itself to a different degree in different hybrid combinations. He arranged the various degrees of expression of this character into seven classes. Class I included the normal condition, class VII the completely cruciate situation.* As a result of these studies, he concluded that there exists among the various races a series of alleles, constituting a multiple allelic series. He recognized nine alleles on the basis of their ability to express or suppress the cruciate condition (1930b). These are as follows (Cr_1 is the strongest normal allele, cr_5 the strongest cruciate factor):

Descending degrees of dominance

Cr_1 in ^{h}*hookeri*
Cr_2 in *albicans* (from *suaveolens*)
Cr_2 in *flavens* (from *suaveolens*)
Cr_3 in *albicans* (from *biennis Hannover*)
Cr_4 in *rubens* (from *biennis Hannover*)

Ascending degrees of expression of cruciate condition

cr_1 in *albicans* (from *biennis cruciate A_2a*)
cr_2 in *rubens* (from *biennis cruciata A_2a*)
cr_3 in *gaudens* (from *lamarckiana cruciata*)
cr_4 in *velans* (from *lamarckiana cruciata*)
cr_5 in *albicans* (from *biennis cruciata apetala*)
cr_5 in *rubens* (from *biennis cruciata apetala*)

Oehlkers postulated that this situation had come about because the *cr* locus is highly mutable, thus resulting in multiple alleles. All alleles in the series are mutable, but in stable races such as *hookeri* or *biennis*

---

* It should be noted that Renner also recognized different classes of phenotypic expression, but his series had only five classes and they were in inverse order, class I being purely cruciate and class V wholly normal.

cruciata apetala, mutations do not extend far enough along the series to affect the phenotypic expression, so that these races breed true to the normal and cruciate conditions respectively. In races, however, whose complexes are nearer the center of the series, mutations may alter the phenotype, resulting in the production of more or fewer defective petals.

Oehlkers found a few cases where the results did not seem to conform to expectation on the basis of this pattern of relative strengths. Some of these he attributed to crossing over, others to a change in dominance resulting from mutation.

One case (1930a) which he ascribed to crossing over involved the cross *biennis cruciata apetala* × (*suaveolens sulfurea* × *apetala*) *flavens·rubens* (i.e., *cr-albicans·cr-rubens* × *Cr-flavens·cr-rubens*). He obtained:

> *albicans·flavens* 41 of class I–II, (normal or subnormal)
> 75 of class V–VI, (subcruciate)
> *albicans·rubens*   6 of class I–II
> 22 of class VII, (wholly cruciate)

The *albicans·flavens* obtained in this case should have carried the genes *cr Cr* ($cr_5 Cr_2$ according to his scheme presented in a later paper) and all plants should have been normal or subnormal. Instead, 75 out of 116 plants were sub-cruciate.* The *albicans·rubens* should have been *cr cr* ($cr_5 cr_5$) and therefore strongly cruciate, but six plants were normal or subnormal. Oehlkers ascribed the presence of the unexpected plants to exchange of *cr* from *rubens* to *flavens* in the first case, of *Cr* from *flavens* to *rubens* in the second. However, the configuration of *flavens·rubens* is ⊙12, 1 pair, and the pair is composed of 5·6 chromosomes. According to Renner (1957), *Cr* may be in 5·6. If this is correct, then it is possible that this particular exchange of *Cr* and *cr* can be explained as the result of independent assortment, not of crossing over as postulated by Oehlkers. The location of *Cr*, however, cannot be regarded as determined with certainty.

Other evidence of crossing over involving the *Cr* locus, however, was presented by Oehlkers in 1938. For example, *biennis cruciata* A_2a strongly cruciate × *lamarckiana brevistylis* gave *albicans·velans* plants (⊙14) which had normal flowers. When these were selfed, they produced progeny with the following composition: 37 *Br Cr*, 10 *Br cr*, 14 *br Cr*, 4 *br cr*, a close approach to a 9:3:3:1 ratio. It is known that *Br br* segregate independently of the lethals, so that a 3:1 ratio of *Br* and *br*

---

* A sub-cruciate plant is not one in which all petals are necessarily sub-cruciate. A plant that is intermediate between normal and cruciate will usually show a range of conditions, even within a single flower. In a sub-cruciate plant, the middle of the range will approach but not reach the fully cruciate condition.

was to be expected (Emerson and Sturtevant, 1932). In addition, however, this complex-combination seemed to show a 3:1 segregation of Cr and cr. Since this hybrid had $\odot 14$, independent assortment could be ruled out: the only way for Cr and cr to change complexes was by crossing over. In this material, Cr and cr appeared to show 50% crossing over independently of $Br\ br$ which also showed 50% crossing over.

Not in every case, however, has the Cr locus shown indication of such a high percentage of crossing over. At the other extreme Oehlkers (1938) found that *albicans·gaudens* ($\odot 6$, $\odot 8$) showed no crossovers among 68 plants, and *albicans·^hhookeri* ($\odot 14$) failed to show crossovers among 983 plants. The latter cross was among those which gave results that Oehlkers explained on the basis of change of dominance: among 93 plants of the F_1 of *biennis cruciata apetala* × *hookeri* (*cr-albicans·cr-rubens* × *Cr-^hhookeri*), four *albicans·^hhookeri* plants appeared that were strongly cruciate (Class V) instead of normal. When these were selfed, however, the 983 progeny included 26 *albicans·^hhookeri* which were all cruciate, and 957 *^hhookeri·^hhookeri*, all of which were normal. Oehlkers considered this a case of change in dominance resulting from mutation. The four F_1 plants had the composition *cr-albicans·Cr-^hhookeri*, but *cr* had become dominant. When selfed, they produced *cr-albicans·Cr-^hhookeri* plants which were all *cr*, and *Cr-^hhookeri·Cr-^hhookeri* plants, all of which were homozygous for the normal gene, which had become recessive to *cr* in the F_1.

Another case that was considered to demonstrate a change of dominance involved *lamarckiana cruciata*. At first this strain, in which the character was not expressed in extreme form, did not respond to the process of selection. After a few years, however, it did respond, and Oehlkers developed two lines, one strongly cruciate, the other weakly so. When normal *O. lamarckiana* was pollinated by *lamarckiana cruciata* strong (1938, p. 284), both normal and cruciate plants were found in the F_1 in spite of the fact that all plants should have been $Cr\ cr$. One of the cruciate plants was selfed and yielded an F_2 that again contained normal and cruciate plants—in other words, this F_1 cruciate plant behaved like a heterozygote, with *cr* dominant. One normal and three cruciate F_2 plants were then selfed. The normal plant produced only normals, suggesting its homozygosity for $Cr\ Cr$. One of the cruciate plants also bred true, apparently showing its homozygosity for *cr cr*. The other two cruciate plants produced splitting progenies. Thirty-four of the $44°F_3$ plants were fully cruciate, five plants were fully normal and five plants were intermediate. The cruciate gene, therefore, behaved as though it was a dominant or partial dominant, although in most outcrosses involving *lamarckiana cruciata*, it behaves as a recessive.

Oehlkers' general conclusion from his work was that the *cr* locus is highly mutable. As a result, multiple alleles have developed among the oenotheras which account for the varying degrees of expression or repression of the cruciate character. In strains which show inconstancy or fluctuation in the degree of expression of the character, the complexes show an intermediate balance. Oehlkers also considered it likely that somatic mutations occur with high frequency in such forms, thus bringing about changes in dominance. Such mutations can occur, according to Oehlkers, in homozygotes, as well as heterozygotes.

Renner took up the study of *cr* in the 1930's and carried on this work for a long period, his first paper appearing in 1937, the last in 1959. So far as the facts are concerned, Renner confirmed Oehlkers' work quite fully. He differed somewhat, however, in his interpretation of the physical basis for the phenomena observed. He concluded that mutations affecting the *cr* locus occur only in heterozygotes and he adopted a modification of Winkler's (1930, 1932) gene conversion hypothesis to account for the behavior of *cr*. Winkler developed his hypothesis as an alternative to the concept of crossing over, and related it only to meiosis. His suggestion was that a gene in a heterozygote can cause its allele to mutate to its own condition. Renner transferred this concept to the soma and adopted as a working hypothesis the idea that a gene in a heterozygote can "convert" its allele into its own condition. He did not commit himself as to whether this can also occur in meiosis. Conversion can occur in either direction—*Cr cr* can become *cr cr* or *Cr Cr*. In either case, a heterozygotic cell becomes homozygotic. Since conversion may occur here and there in a plant, the result may be the production of a mosaic. (Mosaics can also result from partial conversion producing an intermediate genic balance.) Some regions may become *cr cr*, and the flowers produced in such regions will be cruciate. Others may become *Cr Cr* and normal flowered.

Examples that Renner gave to illustrate gene conversion include the following (Renner and Sensenhauer, 1942, p. 584, no. 94–100):

(a) *cr-pingens·Cr-rubens* × *cr-velans·cr-gaudens* gave:
   pingens·gaudens   (378 normal, 13 cruciate or intermediate)
   pingens·velans   (30 normal, 3 cruciate or intermediate)

These should all have been cruciate. Renner suggested that the *Cr* of *rubens* had converted *cr* of *pingens* into *Cr* in most of the cells of the female parent.

That the *Cr* gene had not been transferred from *rubens* to *pingens* in *pingens·rubens*, but rather the *cr* in *pingens* had been transformed into *Cr* by the *Cr* of *rubens* in most of the cells of *pingens·rubens* was indicated by the reverse cross, which showed that *rubens* still retained its *Cr* (*loc. cit.* no. 107):

    (a) *cr-velans·cr-gaudens* × *cr-pingens·Cr-rubens* gave:
        velans·rubens (157 normal, 7 cruciate, 8 intermediate)
    (b) *atrovirens* (*cr-pingens·cr-flectens*) × *suaveolens* (*Cr-albicans· Cr-flavens*) produced *cr-pingens·Cr-flavens* with normal petals. This was backcrossed to *atrovirens* (Renner, 1937b, p. 116):

    cr-pingens·Cr-flavens × *atrovirens* (*cr cr*) gave:
      pingens·flectens (29 normal, 3 cruciate, 23 intermediate)
      flavens·flectens (mostly normal)

The *pingens·flectens* resulting from this backcross should have been *cr cr* and should have had cruciate petals. Renner ascribed the presence of *Cr* in some of these plants to partial or complete conversion of the *cr* of *pingens* in *pingens·flavens* to *Cr* in the presence of the *Cr* of *flavens*. If this was the case, then *pingens·flectens* plants with normal petals should have had the formula *Cr-pingens·cr-flectens*. When, however, such a plant was again backcrossed as female to *atrovirens*, the resultant *pingens·flectens* were not all *Cr*, as would be expected, but about a quarter of the plants were cruciate (Renner, 1937b, p. 112, no. 896). Renner suggested that the *Cr* in *pingens* was in some cases converted back to *cr* under the influence of the *cr* of *flectens*. The segregations and unexpected appearances of *Cr* or *cr* cannot be ascribed in this case to independent assortment. *Pingens·flavens* has $\odot$12, 1 pair, the pair being 11·12. *Cr*, however, is probably not in this chromosome, but possibly in 5·6, so it would not be able to segregate independently of the lethals; nor would one expect its transfer from one complex to the other in such wholesale fashion on the basis of crossing over. *Pingens·flectens* has $\odot$14, giving no opportunity for independent assortment.

Although Renner claimed that the initial conversion can occur only in heterozygotes, certain cases suggested to him that a converted gene in a homozygote is often in a labile condition and can revert to its original status, so that what appear to be reverse mutations can occur in homozygotes. For example, (*atrovirens* × *biennis* München) *cr-pingens·cr-rubens* (cruciate), in which the *Cr* of *rubens* has been converted to *cr*, gave when selfed occasional plants with normal flowers, such individuals appearing also in subsequent generations (Renner and Sensenhauer, 1942, p. 575, nos. 76, 84–86). On the other hand, a gene that has converted another can also become labile as a result. For example, if the *cr-pingens·cr-rubens* just mentioned was pollinated by *biennis* (containing *Cr-rubens*), the resulting plants were wholly *Cr*; *cr* had been so "shocked" as a result of converting Cr to cr that it was now suppressed by a second exposure to Cr, i.e., unable any longer to convert Cr (*loc. cit.*, nos. 89, 90). On the other hand, if the normal-flowered *cr-pingens·Cr-rubens* thus obtained was used to pollinate *cr-pingens· cr-flectens* (= *atrovirens*) the progeny was made up mostly of cr plants (no. 104): in this case, it was the Cr that was "shocked" by its contact with the cr of *pingens*, so that it was no longer able to resist conversion by cr (Renner, 1959a, p. 284).

Renner agreed with Oehlkers that differences in strength or "potency" exist among the genes of different races. Some are more able to bring about conversion than others, or to resist conversion. Among the races that he analyzed in detail, the cr of *biennis cruciata* and of *atrovirens* appeared to be equal in strength; the cr of *lamarckiana cruciata*, however, was definitely weaker. The existence of different levels of potency he ascribed to stepwise conversion. The weakness of one allele or the resistance of another to change may result in only partial conversion.

He also found that shifts in potency may occur in the course of time. In 1926 *cr-pingens·Cr-rubens* × *cr-pingens·cr-flectens* gave almost wholly normal flowered *pingens·flectens* (Renner, 1937b, p. 117). In 1953, however, the same cross yielded the following *pingens·flectens* offspring: 173 cr, 17 intermediate, and 27 Cr; i.e., cr had become relatively more potent, Cr less so. (Renner, 1958c, p. 381.)

Renner's and Oehlkers' ideas differed, therefore, primarily in one respect: according to Oehlkers, cr and Cr are highly mutable genes that can change in either direction and in either a homozygote or a hetero-zygote; according to Renner, "conversion" can occur initially only in heterozygotes. A heterozygous cell may become homozygous, not the reverse, although a converted gene, because it has become labile, may revert to its original condition. Oehlkers claimed that homozygotes could also become heterozygous, but if their Cr or cr was of maximum strength such mutation might not produce a phenotypic effect.

Because of the presence of circles and balanced lethals, it is not easy to determine in every case whether a plant is homozygous or heterozygous for a given trait. This system prevents recessives from showing up, does not permit Mendelian segregation after selfing. One must resort to out-crossing by which both complexes of a race are combined with a tester complex which does not carry the trait. But tester complexes must belong to isogamous races if they are to be combined with both complexes of a heterogamous race. The possibility of the presence of different modifying factors in associated complexes introduces a further cause of uncertainty. It is not always easy, therefore, to discover whether mutations or con-versions are confined to heterozygotes for the cr locus, or whether they may also occur in homozygotes.

Apropos of the possibility of the existence of modifying factors, Renner found that certain other genes do indeed exercise an influence on the stability of *cr*. For example (1957, p. 387), if a *lamarckiana* (*gaudens · velans*) derived from a cruciate *albicans · gaudens* × a cruciate *albicans · velans* has small flowers (*Co Co*), it is wholly cruciate and stable. If, however, it has intermediate-sized flowers (*Co co*), it is somewhat less decidedly cruciate, and if it has large flowers (*co co*), its flowers are mostly normal, though with occasional slight defects in the petals. Even the rare large-flowered cruciate individuals produce progeny that are more often normal than cruciate (in the case cited, 127 normals and 29 cruciate). According to Renner, a *cr* that has been derived from *Cr* through conversion is capable of reverting to *Cr* in the presence of the large-flowered gene (*co co*) but remains stable in the presence of *Co Co*: this in spite of the fact that *cr* and *co* are probably situated in different chromosomes. Renner placed *cr* tentatively in chromosome 5·6 while *co* is either in 13·14 (Harte) or 11·12 (Catcheside). This apparent case of mutation in a homozygote for *cr* was another example of reconversion, in the presence of another gene, of a previously converted, and therefore labile gene, thus changing a homozygote back into a heterozygote.

The gene *R* (red midribs) also affects the behavior of the *Cr* locus. For instance (Renner 1952, p. 369), a cruciate *hookeri*, derived by segregation from the hybrid *hookeri* × *biennis cruciata*, is stable when small flowered. If it is large flowered, it becomes unstable, with petals intermediate between cruciate and normal, unless *R* (in chromosome 1·2) is present. *R* stabilizes the effect of the cruciate gene just as does *Co*. It prevents the reconversion of labile *cr* in the presence of *co*. Again, *R* is in a different chromosome from either *cr* or *co*.

As to the mechanism by which conversion takes place, Renner offered no suggestions. Goldschmidt (1958), in reviewing this situation in *Oenothera*, suggested the possibility of position effect resulting from unequal crossing over. This, however, would not account for the variegation within an individual. Unequal crossing over, as in the case of *Bar-eye*, in *Drosophila*, produces an S-type (stable) position effect, as opposed to the V-type (variegated) position effect found when a gene is translocated to a region of heterochromatin. Presumably both types of alteration take place at meiosis, although the pairing of homologous chromosomes in the dipteran *Drosophila* has been shown to permit occasional somatic crossing over (Stern 1936). In *Oenothera*, the mosaicism of *cr* is a somatic phenomenon, unrelated to crossing over which does not occur in the soma in *Oenothera*, where homologues are unpaired. It is also unrelated to the V-type position effect resulting from translocation, as described by Catcheside in *O. blandina* (see Chapter 11).

Summing up the work of Oehlkers and Renner, they found the cruciate character to be a highly variable one. In some races all flowers are fully cruciate, in other races they are wholly normal, in still others a mosaicism is found in which normal and partially to completely cruciate flowers may be present. The cruciate gene *cr* behaves as a recessive in many crosses but appears to behave as a dominant in others. Still other hybrids show a mosaicism similar to that seen in inconstant races: it is not unusual to find the entire range from normal to fully cruciate flowers on the same hybrid plant or even on the same branch, on the same day. Two hypotheses have been put forward to explain this behavior. Oehlkers claimed that the locus is highly mutable. Either allele can mutate at any time, either in a heterozygote or a homozygote. Renner considered that mutations occur as a result of gene conversion, where one allele of a heterozygous pair converts the other to its own state, thus producing a homozygous condition which may be either stable or labile; or in other instances one allele merely reduces the potency of the other, in which case heterozygosity is retained but dominance has been altered. The two hypotheses differ in respect to the question as to whether the initial mutation can occur in the homozygous condition, or whether it is confined only to cells that are heterozygous. This is not easy to determine in *Oenothera* because of the presence of chromosome circles and lethals.

Before attempting to evaluate the hypotheses of Oehlkers and Renner we should call attention to the fact that the cruciate character has been found in a number of races of *Oenothera* in addition to those studied by Oehlkers and Renner. At least four such races from the United States have been grown in our garden.

Ostreae was collected in 1927 by A. H. Sturtevant near the Oyster Pond at Falmouth, Massachusetts. We grew it from 1931 until 1955, but it was later lost. It was a member of the *parviflora II* assemblage (see Chapter 18), rather close in phenotype to *muricata* (*syrticola*), but with more anthocyanin in midribs, stems and buds. It was a complex-heterozygote with ⊙14. Its petals were often missing, but when present were strongly cruciate. We crossed this form with a number of other races and found that the beta *Ostreae* complex carried a strong *cr* allele, the alpha complex presumably carried a weak *cr* which was unable to function in most outcrosses. The missing petal character (*mp*) was carried by the alpha complex (see Table 12.1).

Cruciata I was a strain grown in our garden for a brief period (1934–1938). It was collected by Dr. R. H. True near Waterbury, Vermont, and was also a member of the *parviflora II* group. Its flowers were 100% cruciate, but did not show the missing petal character. When used as female parent, its alpha complex produced normal-petalled progeny in combination with [h]*hookeri*, [h]*Devil's Gate*, *punctulans*, β *Omaha*, *maculans*

TABLE 12.I

Status of cruciate and missing petal characters in hybrids of Ostreae

Hybrid	Complex-combination	Chromosome configuration	cr	mp
Ostreae × (blandina × suaveolens)	fascians·ʰblandina	⊙8	Petals normal	Not observed
	fascians·flavens	⊙4, ⊙6	Petals normal	Not observed
Ostreae × franciscana Sh.	fascians·ʰfranciscana	⊙6	Petals normal	Not observed
Ostreae × grandiflora of de V.	fascians·acuens	⊙4, ⊙8	Petals normal	Not observed
	fascians·truncans	⊙14	Petals normal	Not observed
Ostreae × Haskett	fascians·β Haskett	⊙12	Petals normal	Not observed
Ostreae × Johansen	fascians·ʰJohansen	⊙8	Petals normal	Not observed
Ostreae × (lamarckiana × blandina)	fascians·ʰblandina	⊙8	Petals normal	Not observed
	fascians·gaudens	⊙6	Petals normal	Not observed
Ostreae × Leonard	fascians·β Leonard	⊙4, ⊙8	Petals normal	Not observed
Ostreae × Nobska	fascians·aenescens	⊙14	Rarely defective	Strong expression
Ostreae × Oakesiana	fascians·denudans	⊙14	Petals normal	Weak expression
Ostreae × Palmer Lake	fascians·β Palmer Lake	⊙4, ⊙8	Petals normal	Not observed
argillicola × Ostreae	ʰargillicola·β Ostreae*	⊙4, ⊙6	Often cruciate or intermediate	Not observed
cruciate I × Ostreae	α cruciate I·β Ostreae	⊙6, ⊙6	Cruciate	Not observed
grandiflora × Ostreae	truncans·β Ostreae	⊙12	Cruciate	Not observed
muricata × Ostreae	rigens·β Ostreae	⊙14	Subcruciate	Not observed
Nobska × Ostreae	pubens(?)·β Ostreae	⊙14	Subnormal	Not observed
Omaha × Ostreae	α Omaha·β Ostreae	⊙14	Some plants cruciate Some plants intermediate	Not observed
shulliana × Ostreae	jugens·β Ostreae	?	Some plants cruciate Some plants intermediate	Not observed

* Sturtevant, who named the alpha complex of Ostreae "*fascians*," did not name the beta complex.

and β *Iowa I*. In combination with *truncans*, five plants had normal petals, three plants cruciate flowers and six plants were mosaics, showing various levels of intermediacy. With *jugens* all plants were intermediate, some approaching normality, others approaching the cruciate condition. With β *Ostreae*, all plants were cruciate. It is clear that α *cruciata I*, unlike α *Ostreae*, had a *cr* allele, which in some hybrids was unable to express itself but in other cases produced a threshold situation, or assumed a dominant role. The segregation observed among the α *cruciata·truncans* plants might have been interpreted by Renner and Oehlkers as the result of shifts in prepotency or dominance. The α *cruciata I·β Ostreae* hybrid was, of course, homozygous for *cr*.

Only one successful cross was made using *cruciata I* as male. The resultant α *Waterbury·β cruciata* was cruciate to sub-cruciate, thus suggesting, since *Waterbury* has normal petals, that β *cruciata I* also carried a *cr* allele. The evidence is insufficient to compare the strengths of the *cr* alleles in alpha and beta *cruciata I*. They may have been identical.

Cruciata II is a strain of unknown origin. It appeared on a compost pile in our experimental garden in 1952 and is still being grown (nine generations). It is definitely a *parviflora II* type and since this assemblage is not found in the midwest, it apparently came in on material derived from the *parviflora* range. There is reason to believe, on other grounds, that some of the other material that found its way on to this particular compost pile came from Vermont, so there is little doubt that this strain, like *cruciata I* and de Vries' *cruciata*, came from that region. *Cruciata II* is like *Ostreae*, but unlike *cruciata I*, in possessing the missing petal character. When *cruciata II* is used as female, the α *cruciata* complex fails in all cases so far tested to induce the cruciate character (see Table 12.IV). The beta complex comes through the egg in some crosses and its influence differs in different combinations. In β *cruciata·punctulans* the petals are normal, but in β *cruciata·velans* they are cruciate. In β *cruciata· truncans*, petals are cruciate or semi-cruciate. In β *cruciata·*[h]*franciscana*, some plants have normal or slightly abnormal petals, others are cruciate or sub-cruciate. The latter, when selfed, yield [h]*franciscana·*[h]*franciscana* plants which are normal flowered, and β *cruciata·*[h]*franciscana* plants which are like the F_1 in being cruciate or sub-cruciate.

The alpha complex, therefore, appears not to carry *cr*, (it does carry the *mp* gene). The beta complex carries a cruciate allele that has been able to suppress or weaken the effect of associated *Cr* genes in three of the four combinations so far studied as well as in the race itself. Its *cr* is probably comparable in strength with that in α *cruciata I*.

The fourth cruciate strain that we have studied differs from the others in not belonging to the *parviflora* assemblage. It is a *biennis* with somewhat uncertain affinities, in that its *strigosa* complex seems to be related more closely to the alpha than to the beta strigosas (see Chapter 18). This material was collected in 1952 near Elgin, Illinois, by Robert Long. Its petals are strongly cruciate or sub-cruciate but it does not show the missing petal character (although petals are occasionally missing in some of its hybrids).

When used as female parent, *Elgin* transmits only the alpha (*strigosa*-like) complex. This carries a *cr* allele which is as strong as any we have studied: when combined with β *Bestwater I, punctulans, velans,* [h]*franciscana,* β *Iowa II,* β *Iowa VI* and *maculans*, the hybrids are cruciate to sub- or semicruciate; with *elongans* of *cockerelli* and with α *Iowa VI*, flowers are normal, or slightly nicked along the edge.

When used as male parent, the alpha complex is transmitted much more frequently than the beta. In four cases, identical complex-combinations were obtained when α *Elgin* came through both sperm and egg (with β *Bestwater I, velans,* [h]*franciscana* and α *Iowa VI*). In three of the four the results were identical in the reciprocal hybrids, in the fourth there was somewhat more tendency toward cruciateness in *Elgin* cytoplasm. On the whole, cytoplasm seems to have a minor effect. α *Elgin* derived through the pollen was combined with additional complexes and gave the following results: cruciate or sub-cruciate flowers with α *Bestwater I*, *gaudens*, *curvans*, *rigens*, α *Paducah* and *jugens*; normal with α *Iowa XII*, *truncans* and *curtans*; mostly normal, but with cruciate or intermediate sectors, with *excellens*, α *Iowa II*; a splitting progeny including normal, cruciate and intermediate plants with α *Camas*.

The beta (*biennis*-like) complex produced wholly normal flowers in almost all combinations (with *excellens*, *punctulans*, *velans*, [h]*franciscana*, *acuens*, *truncans*, α *Iowa I*, *rigens*, and *jugens*). In one case, however, there were indications of the presence of *cr* in that β *Bestwater I* · β *Elgin* showed slight nicking of the margins. In general we may conclude that α *Elgin* carries a *cr* gene, and β *Elgin* may carry a very weak *cr* which is recessive to essentially all of the *Cr* genes with which it has become associated.

We should now mention the one instance in which the cruciate condition is known to have arisen spontaneously, probably as the result of mutation. Interestingly enough, however, this new character (designated *sep*) is considered not to be allelic with *cr* (Rossmann, 1963a, b).

This character appeared first in the F_2 of the cross *O. coronifera* × *purpurata*, the hybrid having the composition *quaerens* · [h]*purpurata* ($\odot$8) (Rossmann, 1963b). Neither of the parental races has ever shown the cruciate character nor has it appeared elsewhere in any of the numerous outcrosses that have been made using these races. Furthermore, although the above cross was made several times, using different individuals of *coronifera*, the cruciate character appeared in the progeny of only one. For these reasons Renner, in whose garden the character appeared in 1957, and his student Rossmann, who undertook the study of the trait, considered it to have arisen through mutation rather than as a result of crossing over or independent assortment.

Coronifera is a rather large-flowered form closely related to *lamarckiana*. Its complexes are *quaerens* and *paravelans*, its configuration $\odot$12, 1 pair. *Paravelans* is nearly identical genically with *velans*, and has the same segmental arrangement, but *quaerens* gives $\odot$8, 3 pairs, with *gaudens*. *Purpurata* is a homozygous race, possessing in double dose the presumably ancestral arrangement of chromosome segments, the commonest found among the hookeris. Neither race, therefore, is closely related to the parvifloras or to the *biennis* groups in which cruciate flowers had previously been found.

In the 1957 culture in which the character first appeared, almost one third of the plants that flowered were cruciate. A selfed line was established which has bred entirely true to an extreme cruciate condition, no trace of normal petals having been observed. In outcrosses, however, the trait behaves as a strict recessive. In a total of 51 different crosses, the hybrids have shown no indication of cruciate or crippled petals (Rossmann, 1963b).

The chromosome configuration of the F_1 in whose progeny the character first appeared was $\odot 8$, 3 pairs, and this same configuration was found in both the cruciate and non-cruciate members of the F_2, as well as in all subsequent generations of cruciate plants. This indicates that the appearance of the trait was not associated with a change in chromosome configuration. The locus of the trait, therefore, was not in the circle but in one of the pairs, namely, in $1\cdot2$, $5\cdot6$ or $7\cdot10$. Since almost a third of the F_2 population showed the character when it first appeared, the F_1 individual from which these were obtained by selfing must have been heterozygous (*Sep sep*), and the mutation no doubt occurred either in the ovule of the *coronifera* parent from which this particular F_1 plant came, or else it occurred in the pollen mother cell of *purpurata* which produced the pollen grain involved in this particular fertilization. Assuming that *Sep* and *sep* were in a free pair in the F_1, free assortment would account for the large number of *sep sep* plants that appeared in the F_2.

That $5\cdot6$ is one of the possible sites of this locus is possibly significant in view of the fact that it is this chromosome that was considered by Renner to carry the locus of *cr*. However, Rossmann states without giving details (1963b, p. 22) that "Kreuzungen mit cruciaten Arten zeigen, dass sepaloid und cruciat nicht allel sind", which presumably means that such crosses give normal petals. We may be dealing, therefore, with a case of pseudo-allelism.

In order to explain and evaluate the facts and hypotheses that have been put forward by Oehlkers, Renner and their students, it is necessary to mention work on petal embryology and anatomy done by Renner's student Kowalewicz (1956) and confirmed by Rossmann (1963b). This work helps us understand the effect that the genes at the *cr* and *sep* loci have on petal structure. These authors find that *cr* and *sep* are not merely inhibitors of petal development—their effect is not the result of failure to mediate a step in the biochemical sequence leading to the growth of the petal. Instead, they play a positive and competing role in that they further the development of a type of structure quite different from, and in some ways more complex than, that of the normal petal, one that resembles in certain respects the sepal rather than the petal. A petal in *Oenothera* begins as a structure several cells thick in which normal chloroplasts, with a normal complement of chlorophylls and carotinoids, develop. This structure expands into a thin sheet, three (and toward the edges two) cells thick, in which the chlorophyll has almost completely disappeared, the carotenes have been oxidized into xanthophylls, and the plastids have swelled and often burst. Stomata and typical mesophyll structures are absent.

The structure produced under the influence of *cr* and *sep*, however, is fundamentally different. It remains several cells thick throughout, its chloroplasts and their pigments tend to remain intact, and it develops typical stomata and hairs on its outer, or morphologically lower, surface. In races in which the sepals are red pigmented, cruciate petals may also develop anthocyanin pigmentation on the outer surface. We are dealing therefore, with genes which have positive but conflicting roles.

Kowalewicz also made the important finding that a petal that is intermediate in its expression of the cruciate factor is not, in many cases, composed of cells whose structure is intermediate between those of typical petals and sepals. Rather, it is a mosaic, largely made up of islands of typical petaloid cells interspersed with islands of typical sepaloid cells. In certain islands of cells one allele or locus is apparently in full command, in others the other allele or a different locus is dominant. The extent to which petaloid, sepaloid and intermediate areas are present depends upon the material, and within an individual plant, upon its total genic composition. At one extreme is Rossmann's material in which no trace of petaloid tissue is present, and the petals are almost indistinguishable, microscopically and macroscopically, from the sepals. At the other extreme, petals in some hybrids or races may be almost entirely petaloid, showing only isolated areas of sepaloid tissues on the margins, resulting in nicks or indentations. The amount of intermediate tissue also varies in different materials.

As we have seen, both Oehlkers and Renner, in attempting to interpret the variable and frequently anomalous behavior of the alleles at the *cruciata* locus, postulated an actual change in genic structure, a process of gene mutation, either spontaneous, or induced by the presence of the associated allele. In support of an hypothesis of genetic mutation was cited the tendency for apparent alterations to be transmitted to succeeding generations—for strongly cruciate sectors of a plant to produce cruciate offspring, for normal-flowered sectors to produce normal progeny, and for flowers that are intermediate in structure to produce intermediate or splitting descendants. Many of Renner's data, however, show that these tendencies are not absolute—e.g., while the progeny from cruciate sectors are predominantly cruciate, there may be a minority of plants with intermediate or normal flowers—and vice versa. Renner was forced to assume, as explanation for these exceptions, that converted genes often become weakened or labile and can revert to their original condition.

There is an alternative, however, to concepts involving a high level of mutability at the *Cr* locus which seems to be worth consideration.

It seems possible that appearance of the cruciate condition is not the result of frequent and recurrent mutations at a given locus, but is rather a case where a gene that plays a key role in sepal formation has mutated in such a way that it is able to function in cells of petal primordia as well as in those of sepal primordia, under a locally determined set of conditions. When this gene becomes active in cells of the petal primordium, it tends to suppress genes that are normally responsible for development of petal cells at this location. The result is that islands of sepaloid tissue develop within the petal, the relative amount of sepaloid and petaloid tissue in a given petal depending upon the number of cells in which conditions permit or stimulate the sepaloid gene in question to become active.

We should emphasize the fact that the islands of cells in semicruciate petals are frequently purely petaloid or purely sepaloid in structure. In general, an all-or-none reaction is found in such cells. This calls to mind the work of Sonneborn (1957) on *Paramecium*, and Nanney (1963) on *Tetrahymena*, who describe, in the case of mating type and certain serotypes an all-or-none condition to which Nanney has applied the term "mutual exclusion". Turning on a particular biochemical sequence in a cell completely turns off alternative sequences. In some cases, Nanney found the competition to be between non-allelic loci, in other cases alleles compete in this manner, giving rise in the former case to "non-allelic repression", in the latter case to "allelic repression". Since it is unlikely that a single locus is responsible for the initiation of both sepals and petals, the situation in *Oenothera* is no doubt one where one locus suppresses a different locus.

We suggest, therefore, that the *Cr* locus is one whose activation is basic to the sequence (no doubt involving a succession of genes) that initiates and develops a sepal. In the case of *cr* we are dealing with a gene that has mutated in such a way that its potency is enhanced to the point where it can become active, not only in sepal primordia but also under certain conditions in petal primordia. When it becomes active in petal primordia it may suppress or turn off the genes that normally bring about differentiation of petal structure. It does not turn off the regulator that initiates petal formation—petals are still formed—but the way in which petal cells develop can be altered when *cr* becomes active.

It may seem anomalous at first sight that a gene with increased potency appears often to be recessive to its normal allele *Cr*: *cr* however, is not in competition with *Cr*; on the contrary, from the standpoint of sepal development it is quite normal. Both *Cr* and *cr* presumably initiate normal sepal development; *cr*, however, because of its greater potency, is brought into competition with an entirely different locus, one that is involved in the differentiation of normal petal structure. This is a case, therefore, of "non-allelic repression", but it presumably takes a double dose of *cr* to bring about substitution of sepaloid for petaloid tissue in cases where *cr* alleles seem to be recessive to *Cr*. There are *cr* alleles, however, that appear to dominate over *Cr*, in which case they are apparently able to produce their effect in single dose. Cruciate genes in different strains, therefore, differ in strength—in their ability to suppress genes responsible for the development of normal petaloid structure. That different mutations have occurred in different materials is not surprising. It is not likely that alleles in groups as unrelated as *biennis* and *parviflora* trace back to a single origin—and it is certain that Rossmann's *sep* has arisen as an independent mutant.

The ability of *cr* to express itself in the petals is also influenced by the presence of modifying factors. Thus, *co* (large flowers) tends to restrict the activity of *cr* in the petals, whereas *R* tends to enhance it. The degree to which *cr* produces its effect, therefore, depends upon two things—the level of potency of the particular mutant allele of *cr*, and the special mix of modifying factors that together make-up the milieu in which *cr* functions in a plant.

It is possible, of course, that somatic mutations may occur from time to time, producing plants that are genetic mosaics for *cr* alleles. Both Oehlkers and Renner found a few cases where selfing of a normal flower on a branch on which most of the flowers were normal led to the establishment of a weak cruciate line, whereas a cruciate flower from a largely cruciate branch on the same plant produced a strong cruciate line. Such cases are apparently rare, however, and do not account for most of the mosaicism based on the *cr* locus.

The fact that zones or layers of cells intermediate in structure are sometimes found especially where petaloid and sepaloid tissues impinge, might seem to argue against the hypothesis of competition and mutual exclusion. We may be confronted, however, with a situation somewhat analogous with that found by Nanney in *Tetrahymena* (1963). Macronuclei are clusters of small but genically complete nuclei. Nanney found that if one of these constituent nuclei in a new and developing macronucleus becomes "differentiated", i.e., one mating type allele is turned on to the exclusion of others, this influences the surrounding nuclei to turn on the same allele, presumably as the result of diffusion of an activating substance. In like manner, in *Oenothera* it is conceivable that not all of the cells of an island of sepaloid or petaloid cells have been derived by division from a single cell in which one of the competing loci has been activated, but that as yet undifferentiated cells are influenced to develop in the petaloid or sepaloid direction as the result of substances that have diffused out from the initially activated cells. If such substances are formed in sepaloid and petaloid centers of development, it is possible that some cells will be so placed that they will receive both substances in sufficient quantity so that they will develop in a manner intermediate between those of typically petaloid and sepaloid cells. In such cells, the principle of mutual exclusion will not hold, but we are still dealing with the regulation of gene action rather than with frequently recurring mutations.

Kowalewicz interpreted the presence of such intermediate zones as evidence in favor of Renner's conclusion that gene conversion may occur in step-wise fashion—e.g., *Cr* is converted only partially toward the *cr* condition in cells which are intermediate in structure, and completely to *cr* in a sepaloid area. An interpretation based on the presence of diffusible substances that cause particular genes to be turned on seems more plausible. It is true that we are suggesting a process at the cellular level which Nanney found at the nuclear level but this does not reduce the plausibility of this hypothesis.

It seems reasonable, therefore, to postulate that the behavior of the *cr* locus is the result, not of constantly recurring alterations of genic structure, but of an original mutation that has extended the range of activity of a gene basic to sepal development and brought it into competition with genes responsible for petal formation.

2. MISSING PETALS (*mp*)

The other character in *Oenothera* that follows a mosaic pattern is the missing petal character. This is a condition where one or more petals in a flower are completely lacking. A given flower may have four, three, two, one or no petals (Fig. 12.2). When a petal is missing, those that are present do not as a rule adjust their positions to achieve a uniform spacing. Thus, where three petals are present, they are spaced 90° apart, not 120°, so that one position remains vacant. When two petals are absent, the two that remain may occupy adjacent positions, or opposite positions, the former being somewhat the commoner. Petals that are present are normally perfect, although in two races that we have studied this character is associated with the cruciate character (in *cruciata II* and *Ostreae*), in which case petals are defective. It is rare that all flowers lack all of their petals. Usually one observes on a plant a mosaic of flowers with differing numbers of petals. All conditions may be found on the same plant and even on the same branch.

As we shall see, rare exceptions have been found. A very few flowers have been observed where one or more petals, instead of being completely absent, are present as diminutive structures of more or less normal width but only 1–2 mm in length. In a very few other flowers, a single petal has appeared as a long slender filament slightly expanded at

FIG. 12.2. The missing petal character. Flowers with 4, 3, 2, 1 and 0 petals. *O. (muricata × Magnolia) rigens · β* Magnolia.

the tip, but less than 1 mm broad at its broadest. The length in this case is usually that of the normal petal. Another exception to the rule was seen in a few cases where some flowers on a plant were truly trimerous (see below, p. 151).

Table 12.II lists the materials in which *mp* was initially found.

Our first encounter with this character involved the race known as *Ostreae* (see p. 138), in which the character showed very strongly. As seen in Table 12.I, however, only two hybrids derived from *Ostreae* as female have shown any evidence of *mp*. One hybrid showed the character quite strongly, and one very infrequently. It should be pointed out,

TABLE 12. II
Initial appearance of missing petal character (mp)

Name	Species	Level of expression	Chromosome configuration
Races			
Ostreae	parviflora II [a]	Strong	⊙14
cruciata II	parviflora II [a]	Strong	⊙14
muricata	parviflora II [a]	Weak	⊙14
Galeton	parviflora II [a]	Weak	⊙14
Hybrids			
muricata × Warrenton	parviflora II × biennis I [b] (rigens·β Warrenton)	Strong	⊙14
muricata × Magnolia	parviflora II × biennis I [b] (rigens·β Magnolia)	Strong	⊙14
muricata × Petersburg	parviflora II × biennis I [b] (rigens·β Petersburg)	Strong	⊙12, 1 pair
Elma V × Iowa II	biennis II [c] × strigosa [d] (α Elma V·β Iowa II)	Strong	⊙4, ⊙6, 2 pairs
Elgin × Bestwater I	(strigosa × biennis I) × biennis I (α Elgin· β Bestwater I)	Weak	⊙12, 1 pair
Elgin × cockerelli	(strigosa × biennis I) × strigosa [d] (α Elgin· elongans)	Weak	⊙14
Elgin × Iowa II	(strigosa × biennis I) × strigosa [d] (α Elgin· β Iowa II)	Strong	⊙14
Elgin × shulliana	(strigosa × biennis I) × biennis II (α Elgin· maculans)	Intermediate	⊙14
Iowa II × Elgin	strigosa × (strigosa × biennis I)(α Iowa II· α Elgin)	Weak	⊙6, 4 pairs
muricata × Elgin	parviflora II × (strigosa × biennis I) (rigens· α Elgin)	Strong	⊙10, 2 pairs
shulliana × Elgin	biennis II × (strigosa × biennis I) (jugens· α Elgin)	Strong	⊙6, ⊙8

[a] parviflora II = parviflora L. subsp. parviflora Munz
[b] biennis I = biennis L. subsp. centralis Munz
[c] biennis II = biennis L. subsp. caeciarum Munz
[d] strigosa (Rydb.) Mack and Bush subsp. canovirens (Steele) Munz

however, that in most cases these crosses were made to forms with large flowers, which factor in general prevents *mp* from being expressed. When *Ostreae* was used as the male parent, no hybrids showed the missing petal trait, although all were to a greater or lesser extent cruciate. From these results, it appears that α *Ostreae* (= *fascians*) carried a weakly mutated *mp* allele, and probably a weak allele for *cr*; β *Ostreae* carried a strong cruciate factor, and a weakly mutated *mp* gene.

The second case of missing petals appeared in 1938, in the hybrid (*muricata* × *Warrenton*) *rigens·β Warrenton* (⊙14). This F_1 hybrid between a *parviflora II* and a *biennis I* race (see Chapter 18) showed the frequent occurrence of flowers in which one or more petals were missing. A plant of this culture was selfed and gave a small F_2 family (five plants), four of which had only normal flowers, the fifth again displayed the missing petal character. This case was not studied further, but it is interesting that a plant with ⊙14 produced a splitting progeny.

In 1939 another hybrid combination revealed a similar situation. This was (*muricata* × *Magnolia*) *rigens·β Magnolia* (*m* × *M*). The maternal parents of this and the previous cross belonged to the same race, and the two races used as male parents were closely related *biennis I* races, their beta complexes differing in segmental arrangement by only a single interchange. The latter hybrid (*m* × *M*), which had ⊙14, was selfed, and the F_2, in contrast with the first cross, bred true to the missing petal character, i.e., all plants were mosaics, whose flowers ranged from the four-petalled to the zero-petalled condition.

This hybrid, with its ⊙14, has been carried on in selfed line for 16 generations and has bred true in every respect. The missing petal situation remains as it did in the F_1, although attempts have been made to select for petal-less and for four-petalled flowers by consistently selfing flowers with these conditions.

A typical analysis of a culture is one made in 1952. On a given day, chosen on the basis of convenience, the 20 plants showed the following:

Number of petals per flower	4	3	2	1	0
Number of flowers	14	23	58	69	67

The number of petals on these 231 flowers represent 33·5% of the full number of petals that would have been present had all possessed their full complement of four petals each. Of the total number of flowers, 29% had no petals, 31% had a single petal, and only 6% had all four petals. There was some tendency for flowers on the same spike to be similar in structure. Thus, on another day 66 spikes which had more than one flower were examined; on 22 of these, the flowers were of the same class (i.e., all lacked petals completely, or had a single petal, etc.); on 29, flowers differed by not more than one class (e.g., they had zero or one petal, one or two petals, etc.); on 15, the differences were greater than this. On only one spike was the maximum difference found, i.e., flowers with zero and four petals.

This hybrid complex-combination, which has been carried on as a true-breeding race for many generations, has been crossed, both as male and female parent, with other races with the results shown in Table 12.III. In 17 hybrid complex-combinations, no trace of missing petals has been seen; in 16 combinations, the character has appeared in some degree, for the most part only rarely but in one case (*cruciata II* × (*m* × *M*) α *cruciata·β Magnolia*) in essentially 100% of the flowers throughout most of the growing season.

A fourth case of missing petals was found in another hybrid which once more involved *muricata* as the female parent. In 1951, the F_1 hybrid (*muricata* × *Petersburg*) *rigens·β Petersburg* was grown. *Petersburg* is a *biennis I* race from Virginia, but its beta complex is not closely related segmentally to the other beta *biennis I* complexes so far analyzed, or for that matter, to any complex hitherto examined. This F_1 (*m* × *P*) showed the missing petal character to a marked degree. The hybrid has bred true in seven succeeding generations, the missing petal character being retained at full strength. The hybrid has ⊙12, 1 pair, the paired

TABLE 12.III

Status of the missing petal character in derivatives of the hybrid
muricata × *Magnolia* (*m* × *M*)

Hybrids in which petals may be missing[a]	Hybrids with no missing petals
biennis apetala × (*m* × *M*) albicans·β Magnolia (rare)	*blandina* × (*m* × *M*) ʰblandina·β Magnolia
muricata × (*m* × *M*) rigens·β Magnolia	*biennis* Hannover × (*m* × *M*.) albicans·β Magnolia
cruciata II × (*m* × *M*) α cruciata·β Magnolia (almost 100%)	*franciscana* de V. × (*m* × *M*) ʰfranciscana·β Magnolia
Galeton × (*m* × *M*) α Galeton·β Magnolia (*ca.* 10% of flowers defective)	*Waterbury* × (*m* × *M*) α Waterbury·β Magnolia
(*m* × *M*) × *muricata* rigens·curvans (rare) β Magnolia·curvans (rare)	(*m* × *M*) × *blandina* rigens·ʰblandina β Magnolia·ʰblandina
(*m* × *M*) × *chicaginensis* rigens·punctulans (rare) β Magnolia·punctulans (rare)	(*m* × *M*) × *Ft. Lewis* rigens·velans β Magnolia-velans
(*m* × *M*) × *Galeton* rigens·β Galeton (rare)	(*m* × *M*) × *shulliana* β Magnolia·maculans rigens·maculans
(*m* × *M*) × *Iowa II* rigens·β Iowa II (*ca.* 4% of flowers defective)	(*m* × *M*) × *suaveolens* β Magnolia·flavens rigens·flavens
(*m* × *M*) × *Iowa VI* rigens·β Iowa VI (rare) β Magnolia·β Iowa VI (rare)	(*m* × *M*) × (*cruciata II* × *franciscana* de V.) rigens·ʰfranciscana β Magnolia·franciscana
(*m* × *M*) × *shulliana* rigens·jugens (rare)	(*m* × *M*) × (*franciscana* de V. × Elgin) rigens·ʰfranciscana β Magnolia·ʰfranciscana
(*muricata* × (*m* × *M*)) × *biennis apetala* β Magnolia·rubens (rare) rigens·rubens (rare)	(Ft. Lewis × Elgin) *gaudens·*α × (*cruciata II* × (*m* × *M*)) α Elgin·β Magnolia
(*cruciata II* × (*m* × *M*)) × (*Ft. Lewis* × *Elgin*) α cruciata·α Elgin α cruciata·gaudens	

a Unless otherwise indicated, percentage of missing petals approaches that observed in (m × M).

chromosome being 8·14. An analysis made in 1952 is typical of all plants in all generations. On August 21 of that year, 108 spikes had 199 flowers in bloom. The number of petals per flower was as follows:

Number of petals per flower	4	3	2	1	0
Number of flowers	34	43	47	42	33

Thus, only 17·1% of the flowers had a full complement of petals and only 50·8% of the number of petals to be expected in 199 flowers were present (75% of the petals on the defective flowers were missing).

The same tendency found in $(m \times M)$ was observed in $(m \times P)$, namely, a tendency for flowers on the same spike to resemble one another in respect to petal number. There were 56 spikes with two or more flowers on this particular day. On 22 of these, the flowers were of the same class (i.e., had the same number of petals), on 20 spikes, the flowers belonged to adjacent classes. In only 14 of the 56 tips was the spread wider.

Except for reciprocal crosses with *blandina*, in which missing petals proved to be recessive, no outcrosses to other races were made with this hybrid. It was, however, backcrossed to *muricata* in both directions. *Muricata* $\times (m \times P)$ F_2 gave a repetition of the F_1 hybrid (*rigens · β Petersburg*) in 77 of 78 plants (three pollinations). These were indistinguishable from $(m \times P) \times$ s. The single exception was a metacline (*curvans·β Petersburg*, with the expected $\odot 4$, $\odot 6$, 1 pair). Both classes showed missing petals. The reciprocal cross $(m \times P)$ $F_2 \times muricata$ (made twice) yielded twin hybrids. One class was supposedly *rigens · curvans* (= *muricata*) reconstituted, but the progeny from one of the two pollinations was not exactly typical *muricata*, the leaves being narrower and less hairy, buds more slender, sepal tips more strongly subterminal, the internodes longer. The F_2 $(m \times P)$ plant used in this cross probably had the *β Petersburg* or the *rigens* 8·14 chromosome in double dose rather than one from each parent. No petals were missing in the *rigens·curvans* plants. The other backcross class was *β Petersburg · curvans* ($\odot 4$, $\odot 6$). The missing petal character was evident in this combination. For instance, on one day, 1157 flowers were scored; of these, 418 showed the absence of one or more petals (36%). Beta *Petersburg*, therefore, is able to express the missing petal character in cooperation with either *rigens* or *curvans* of *muricata*.

A most interesting feature associated with the hybrid *β Petersburg · curvans* was the appearance in 1956 of trimerous flowers. On two plants out of 26 all flowers on the central shoot for a considerable distance were completely trimerous, with three sepals, three petals, six stamens and three carpels. Sepals and petals were spaced 120° apart. There was one flower with an intermediate condition—three well developed and one reduced sepal tip, two good and one split petal, seven stamens and three stigma lobes.

Four attempts were made to establish trimerous lines by selfing trimerous flowers. In no case was the progeny wholly trimerous, although a small proportion (5–10%) of the flowers were trimerous in all cases. In every case, and in both generations, the plants involved had the $\odot 4$, $\odot 6$ expected of β *Petersburg·curvans*, and in all individuals there were flowers with missing petals. The evidence is against the supposition that trimery was the result of a new gene mutation.

The three cases of missing petals that have just been described were all hybrids that had *muricata* as their maternal parent. Missing petals had not been observed in *muricata* itself prior to the discovery of these hybrids. A subsequent careful daily check of the *muricata* plants in our field, however, revealed an occasional flower with a missing petal; in fact in some years, the character has expressed itself fairly strongly. This was especially true in 1959, when on a certain day all ten plants in the culture showed the character. On this day (July 29), 170 (65·4%) of the 260 flowers examined lacked at least one petal, 15 flowers had no petals, and the total number of petals (693) was 66·6% of the 1040 expected on 260 flowers. This was almost as strong an expression of the character as is normally exhibited by *muricata* × *Petersburg*. Significantly, the effect became less striking later in the summer.

A fifth form to show *mp* is *cruciata II*, which has been mentioned earlier as showing the cruciate character (p. 139). The race shows the *mp* character quite strongly. For example, on one day in 1955, only 14% of the 99 flowers scored had four petals, 17% had no petals, and the total number of petals on these flowers was exactly 50% of what should have been present (40·5% in the case of the flowers which showed the trait). A number of crosses were made involving this race, resulting in 15 hybrid combinations. Twelve of these received the alpha complex from *cruciata II*, three the beta complex. It will be seen from Table 12.IV that eight of the twelve hybrids of α *cruciata II* showed no trace of the *mp* character. Included in this group were all of the combinations involving complexes from large-flowered races (*lamarckiana, franciscana, grandiflora* and *suaveolens*). Four α *cruciata II* combinations, however, showed *mp* quite strikingly, namely, *cruciata II* × *Iowa II* (α *cruciata*, β *Iowa II*), *cruciata II* × *Iowa VI* (α *cruciata·* α *Iowa VI*), *cruciata II* × (m × M) (α *cruciata·* β *Magnolia*), and *cruciata II* × *chicaginensis* (α *cruciata· punctulans*).

Only three hybrids have been obtained that involved beta *cruciata II*. None of these showed the *mp* trait.

The sixth case of missing petals was found in 1956 in a hybrid, but one

TABLE 12.IV

The cruciate and missing petal characters in hybrids of *cruciata II*

Hybrid	Year grown	Complex-combination	Chromosome configuration	Status cruciate character	Status of missing petal character
cruciata II × *chicaginensis*	1956	α cruciata II · punctulans	⊙14	Petals normal	Often missing
cruciata II × *Ft. Lewis*	1956	α cruciata II · gaudens	⊙6, 4 prs.	Petals normal	No missing petals
		α cruciata II · velans	⊙8, 3 prs.	Petals normal	No missing petals
		β cruciata II · velans	⊙14	Petals cruciate	No missing petals
cruciata II × *franciscana* de V.	1956	α cruciata II · ʰfranciscana	⊙6, 4 prs.	Petals normal	No missing petals
		β cruciata II · ʰfranciscana	⊙14	Petals normal to cruciate	No missing petals
cruciata II × *grandiflora*	1956	α cruciata II · acuens	?	Petals normal	No missing petals
		α cruciata II · truncans	?	Petals normal	No missing petals
		β cruciata · truncans	⊙12, 1 pr.	Petals cruciate or semi-cruciate	No missing petals
cruciata II × *Iowa I*	1956	α cruciata · β Iowa I	⊙14	Petals normal	No missing petals
cruciata II × *Iowa II*	1956	α cruciata · β Iowa II	⊙4, ⊙10	Petals normal	Almost no petals present
cruciata II × *Iowa VI*	1956	α cruciata · α Iowa VI	⊙12, 1 pr.	Petals normal	Very few petals present
cruciata II × *shulliana*	1956	α cruciata · maculans	⊙4, ⊙6, 2 prs.	Petals normal	No missing petals
cruciata II · × *suaveolens*	1956	α cruciata · flavens	⊙4, ⊙6, 2 prs.	Petals normal	No missing petals
cruciata II × (*muricata* × *Magnolia*)	1958	α cruciata · β Magnolia	⊙14	Petals normal	Almost no petals present

which did not involve *muricata*. The cross *Elma V* (a *biennis II* race) × *Iowa II* (a *strigosa*)* was made in 1955. Neither race had shown missing petals, nor have they shown this character since. The F_1, however, showed the character quite strongly.[†] The hybrid was selfed and has been carried through to the F_4. Despite its configuration of ⊙4, ⊙6, 2 pairs, all plants examined in all generations have shown this configuration, and they have also continued to show a constant lack of petals. Table 12.V summarizes the results of counts made in the generations

TABLE 12.V

Missing petals in *Elma V* × *Iowa II*

Number of flowers examined	Number of flowers lacking one or more petals	Number of petals expected (4 × number of flowers)	Number of petals found	Percentage of petals missing	Percentage of petals missing in defective flowers
1st cross					
F_3 1638	487 (29·7%)	6552	5638	14	46·9
F_4 1371	296 (21·5%)	5484	4908	10	51·5
2nd cross					
F_1 1595	442 (24·1%)	6380	5474	14·1	51·3

indicated. From the table it is seen that on the average about a quarter of the flowers were defective, and in the defective flowers about half of the petals failed to appear.

This hybrid strain was of interest because of the presence of a very few trimerous flowers and a number of petals that were much reduced in length, some of them mere stumps, others narrow and small but yellow and petal-like in structure. In a few flowers, petals were cruciate and sepaloid.

A seventh case of missing petals has been found in a wild strain— *Galeton*, a *biennis I* from Potter County, in northern Pennsylvania. This was first grown in 1953 and the cultures of the first and second generations, grown the following years, had only normal flowers. However, the second generation was repeated in 1955 and in that year many flowers early in the season showed the character. As the season wore on, the proportion of defective flowers diminished until toward the end of the season all flowers were normal.

* See chapter 18.

† To make sure that no mistake had been made the cross was repeated in 1958, with the same result (results incorporated in Table 12.V).

In subsequent generations the character has continued to appear in this race, but at a low level. Most plants have shown a few defective flowers in the earlier part of the season, but few or none later in the summer. The race has ⊙14 and is a true-breeding complex-heterozygote.

Galeton has been crossed with eleven other races, but as might be expected from the low level of expression of the character in the race itself, almost none of its hybrids has shown the missing petal character. The only exceptions are reciprocal hybrids involving (*muricata* × *Magnolia*), as would have been expected.

Finally, mention should be made of the fact that *mp* has been found in several hybrids of *Elgin*, a cruciate race referred to above (p. 140). This race has been grown for six generations and has itself shown no trace of the character; the trait has, however, appeared in six of the 21 hybrid combinations obtained from this race. In five of these cases, neither parent has itself shown the trait, but in the sixth case, one of the parents was *muricata* in which flowers occasionally lack petals. The hybrids of *Elgin* which show the *mp* trait are included in Table 12.II.

Reviewing the *mp* situation as a whole, an interesting correlation becomes apparent with respect to the distribution of the character. In every case where it has appeared, at least one complex is of the *strigosa* type (see Chapter 18). The alpha complexes of *parviflora II* and *biennis II* are *strigosa*-like, as are the beta complexes of *biennis I*. Consequently, all four of the races, and one of the hybrids (*Elgin* × *shulliana*) listed in Table 12.II have one *strigosa* complex; and the other ten hybrids have both complexes of this type. Furthermore, eleven of the 17 hybrids listed in Table 12.III which have shown the *mp* character have two complexes of the *strigosa* type, while the other six each have one. In addition, all three hybrid combinations mentioned above which involve *cruciata II* (α *cruciata II*·β *Iowa II*, α *cruciate II*·α *Iowa VI* and α *cruciata II*·*punctulans*) include two *strigosa*-like complexes.

Complexes of the *strigosa* type are descended from two original sources (see Chapter 18) which we have called Populations 3 and 4. Those derived from Population 3 include the beta complexes of the strigosas and the *biennis I* races, and the alpha complexes of the *biennis II* and *parviflora II* races. Population 4 has produced the alpha complexes of the strigosas, and is also present in α *Elgin*, probably as the result of a cross between a *strigosa* race as female with a *biennis II* race as male. With the exception of α *Elgin* and α *Iowa VI* which have come from Population 4, all of the *strigosa* complexes which have been found to carry *mp* are descended from Population 3.

This fact suggests that the *mp* trait has originated in Population 3, and that *mp* alleles may now be quite widely distributed among complexes derived from this source. α *Elgin* and α *Iowa VI* have probably received *mp* from associated beta *strigosa* complexes by crossing over.

In view of the fact that flower size affects the frequency of appearance of the cruciate character, as mentioned above, it is interesting to note that we have found in our work no races or hybrids containing large-flowered complexes that have shown any trace of the missing petal character. Large flower size tends to prevent the expression of both cruciate petals and the missing petals character.

External conditions in some cases may also affect the degree of expression of the character. Most striking is the effect of time of flowering. The trait usually shows more strongly early in the summer, at the beginning of the flowering season. Toward the end of the season, petals become more numerous. This was most strikingly seen in the hybrid combination in which the absence of petals was most strongly expressed, i.e., in *cruciata II* × (*muricata* × *Magnolia*) α *cruciata*·β *Magnolia*. In this hybrid, both of whose parents showed the missing petal trait, one could scarcely find a petal during the earlier half of the season. Toward the end of the season, however, they became rapidly more numerous. On September 6, 10% of the petals were present that might have been expected on 431 flowers; on September 20, 30·1% of those expected on 750 flowers were present, and on September 24, 82% were found on 340 flowers. Whether this rapid loss of the *mp* expression is the result of an altered biochemical condition due to aging, or of reduced day length is not known, but obviously the degree of expression is influenced by factors other than the *mp* locus.

Certain other facts of importance in respect to this character should be mentioned. The effect of *mp* is purely local—it affects single petal sites only, no other structure is involved and in most flowers not all the sites are affected. Furthermore, with very few exceptions, the effect is an all-or-none one. A petal is totally missing, or if present is fully normal, except where the *cr* factor is also present and able to express itself. (In the example just cited, *cr* was brought into the α *cruciata II*·β *Magnolia* hybrid from the mother, but was suppressed by the paternal complex so that petals, when they appeared, were fully normal.) Furthermore, unlike the *cr* factor, *mp* gets in its work at the very beginning of the development of the petal. Petal development is completely suppressed. Unlike *cr* also, the effect of *mp* is a purely negative one—it blocks development completely rather than bringing about the substitution of a different type of development for the normal process.

We have not made a histological study of the beginning stages of petal formation in *mp* forms, but we have made careful dissections under the microscope, and find no trace of primordia at any stage in those locations where petals fail to develop. Primordia do not develop and then abort. They are never produced.

When the *mp* character was first discovered, we thought that this might be an extreme form of the cruciate condition, reduction of the petal being carried to the point of complete elimination. It was thought therefore, that crosses of plants showing missing petals with those showing degrees of cruciateness would produce progeny with either an extreme form of the cruciate condition or a strong tendency toward the elimination of petals. This proved, however, not to be true in most cases. For example, five crosses were made using *muricata* × (*muricata* × *Magnolia*), which shows the *mp* character strongly, and *biennis apetala*, which is strongly cruciate, the latter being the male parent. In three of these, the progenies were almost entirely normal-flowered, both the *rigens·rubens* and *β Magnolia·rubens* combinations. In the other two crosses, four plants out of a total of 28 showed a strong tendency toward the cruciate condition, the other plants had entirely normal flowers, or subnormal to semicruciate flowers. Almost no flowers lacked petals. An analysis of the flowers scored is given in Table 12.VI. The preponderance in this case of flowers with the full number of normal petals indicates that *cr* and *mp* are not alleles unless we are dealing with a case of pseudo-allelism (it is not known to which chromosome *mp* belongs).

The reciprocal of this cross, *biennis apetala* × (*m* × *M*), yielded progeny all of which were *albicans·β Magnolia*. With respect to the cruciate character, nine plants had wholly normal flowers, three had various degrees of intermediacy, and 16 had cruciate or sub-cruciate flowers.

TABLE 12.VI

Analysis of flowers scored in five crosses of (*muricata* × (*muricata* × *Magnolia*)) × *biennis apetala* yielding rigens·rubens and β Magnolia·rubens

Character shown	1 (15 plants)		2 (15 plants)		3 (12 plants)		4 (14 plants)		5 (13 plants)		Total	
	rigens	β	rigens	β	rigens	β	rigens	β	rigens	β	rigens	β
Normal	100	532	320	76	135	472	228	287	348	329	1131	1606
Subnormal to intermediate	1	8	5	0	3	5	118	188	66	106	193	307
Subcruciate to cruciate	0	1	0	0	0	0	19	162	0	8	19	171
mp	0	1	0	0	1	3	5	4	2	2	8	10

There was thus a striking segregation among the F_1 offspring. This, however, was unrelated to chromosome configuration: two plants with wholly normal flowers and two with wholly cruciate flowers were examined cytologically. All four plants had the ⊙4, ⊙8, 1 pair, expected in *albicans·β Magnolia*. All F_1 plants were alike, therefore, in having the full chromosome complements of *albicans* and *β Magnolia*. Unless *cr* is in 11·10, the segregation cannot be explained in terms of independent assortment. Since the evidence points to *cr* being in ends 5 or 6 this is probably not the correct explanation. The segregation can be explained in terms of gene conversion but can as readily be ascribed to differences in the control of the *cr* locus—in some plants, *Cr* was turned on, in others *cr*, and in still others, regulation varied in different parts of a plant or even in different petals of a flower.

The *mp* character was almost completely suppressed in this case, only ten flowers in 3207 showing the lack of a petal or its reduction to minute size. The results of this cross, therefore, do not support the suggestion that *mp* and *cr* are alleles unless they are pseudo-alleles.

In only one cross between a cruciate plant and one showing the missing petal character have the flowers shown one of these traits predominantly, namely in the *cruciata II* × (*m* × *M*) *α cruciata·β Magnolia* mentioned above, where petals are usually missing. In this case, however, the missing petal trait was present in both parents, the cruciate character in only one. When petals appeared in the hybrids (toward the end of the season), they were entirely normal, suggesting that *cr* and *mp* are independent traits and not different degrees of expression of a single character.

The behavior of *mp* is similar to that of *cr* in that both exhibit a high frequency of mosaicism. The missing petal character, however, has not been found in as extreme a form as the cruciate character. Several races are known in which all flowers are cruciate, but the only cases of missing petals that approach this level are *cruciata II* × (*m* × *M*) and *cruciata II*× *Iowa II*. Except for the fact that *mp* rarely reaches the high level of expression often reached by *cr*, however, it behaves in much the same way as this locus. The question arises, therefore, as to whether it can be explained in the same way, on the basis of control of gene activity rather than in terms of constant mutability or conversion of genes at this locus.

We have suggested that the cruciate condition is the result of a mutation in a gene basic to sepal differentiation, which permits the development of sepaloid tissue in larger or smaller regions of the emerging petal, often to the exclusion of the sequence which leads to the formation of petaloid tissue. In the case of *mp*, however, petaloid tissue is not replaced with something else—it is simply eliminated altogether. It seems that we may be dealing here with a gene basic to petal initiation that has mutated in such a way that its potency is cut to a low level. As a result, its antagonist is in some cases unable to carry alone the entire burden of synthesis of some essential ingredient, although in other combinations a single gene is able to overcome the handicap of being associated with a defective allele. In many combinations essentially a threshold situation is set up so that minor fluctuations in the milieu are able to determine whether the petal will be initiated or not. At one petal site sufficient substrate is present to initiate or carry forward the necessary biochemical sequence; at another site, an essential ingredient falls below the critical level and petal development is inhibited. The *mp* character, which involves the total elimination of a structure, is best explained, therefore, on the basis of a gene with reduced potency. The cruciate gene, on the other hand, seems to have taken on increased potency, since it is able, in petal primordia, to suppress the genes that normally produce petaloid tissue, and to substitute for them the sequence, actually more complex, that leads to the development of sepaloid tissue.

The location of the *mp* factor is unknown; it is not even known whether all cases of missing petals are owing to mutation at the same locus. The fact, however, that *mp* is always associated with the presence of a complex derived from Population 3 suggests a single origin.

CHAPTER 13

Self-Incompatibility

The first case of self-incompatibility of the *Nicotiana* type in the genus *Oenothera* was found in *O. organensis* by Sterling Emerson (1938). According to East (1940), Emerson had some evidence of incompatibility also in *O. (Anogra) californica*, *O. (Lavauxia) acaulis* and *O. (Pachylophus) caespitosa* var. *marginata* and var. *montana*. This condition was also reported by Gates (1939) in *O. (Megapterium) missouriensis*; and Hecht (1944) found it in *O. (Anogra) latifolia, pallida, runcinata*, and *trichocalyx*, as well as in *O. (Raimannia) heterophylla* and *rhombipetala*. I have found it in another *Raimannia (O. laciniata grandiflora)* (=*O. grandis*). Hagen (1950) confirmed its presence in *O. trichocalyx, missouriensis*, and *caespitosa* var. *marginata* (also var. *montana*). He did not find it in the strain of *O. acaulis* which he studied, but did find it in *O. (Hartmannia) speciosa, O. (Salpingia) greggii* and *hartwegii* and in *O. (Sphaerostigma) bistorta* var. *veitchiana*. Crowe (1955) failed to find self-incompatibility in *O. (Lavauxia) acaulis* and *O. (Anogra) trichocalyx*. All told, this character has been found in nine of the subgenera of *Oenothera*.

The fact that different authors have differed regarding its presence in *O. acaulis* and *O. trichocalyx* may not be due to inaccuracy of observation but to differences existing within these species. Another example of this has been found in our garden. *O. heterophylla* is, as a rule, self-incompatible, as shown by Hecht and exemplified by our strain from Athens, Texas. I have another strain, however, very similar in appearance, from Gainesville, Georgia, which is self-compatible. Both strains have small circles or wholly paired chromosomes.

As a matter of fact, self-incompatibility in *Oenothera* is always associated with small circles or none. It is also a striking fact that, wherever the situation has been investigated, the number of S-factors in a population is very large (Emerson 1940; Cleland 1960b). Emerson, for instance, found 45 alleles of the S-gene in *O. organensis*. The presence of many alleles in a population creates a very efficient system for maintaining hybrid vigor through the prevention of inbreeding. If only a few S-factors were present in a population, the chances of pollen falling on incompatible stigmas would be relatively high, and the necessity of having to depend upon pollen from plants with different S-factors would constitute a major hazard to successful reproduction. It is interesting that the genus *Oenothera* has two drastically different systems for ensuring heterozygosity, the S-factor system, and the system of balanced lethals in large circles coupled with self-pollination. Both systems maintain a high level of hybrid vigor, both have advantages and disadvantages. The S-factor system involves a degree of uncertainty since it depends on the chance visit of insects or hummingbirds. The balanced lethal system avoids this uncertainty but at the cost of partial sterility resulting from the lethals— a disadvantage, however, that is compensated for in most complex-heterozygotes by the increased certainty of pollination due to the self-pollinating habit.

In the subgenus *Oenothera*, *O. organensis* is the only species in which balanced S-factors have been discovered. *O. organensis*, however, is not a typical member of this subgenus. In phenotype it resembles *Raimannia* more closely than the other oenotheras, and in Munz's monograph (1965) has been placed in this subgenus. It does not cross readily with other members of the subgenus *Oenothera*, and with only slightly less difficulty does it cross with *Raimannia*. Emerson (1938) failed to obtain offspring following crosses with five species of *Raimannia*. Hecht was unable to obtain hybrids with *O. (Raimannia) rhombipetala*, but did obtain viable seeds with another strain of *Raimannia* (*longiflora*), which produced yellow seedlings (Hecht, 1950). I have succeeded in getting viable hybrids using *organensis* as female crossed with *affinis* La Plata, *affinis* Villa Ortuzor and *mollissima* Toledo II, all raimannias. All attempts to cross *Raimannia* with *organensis* as male, however, have failed, hybrid seedlings being colorless in cases where seed was set. In crosses with the subgenus *Oenothera* the results have been comparable. A number of attempts have been made in our garden to cross *organensis* with North American members of this subgenus. In all but two cases, seeds have produced white or yellow seedlings which succumbed early. One success-ful cross was with *grandiflora* of de Vries as female. The complex *truncans* of this race gave a successful hybrid with *organensis*, with a chromosome configuration of ⊙14. The hybrids showed many features of *organensis* including the very long hypanthium, red sutures on the bud cones, and foliage characters. The *acuens* complex produced yellow seedlings that soon died. The other cross was with *hookeri* of de Vries, used as female. The plants were somewhat chlorotic as seedlings but recovered fully and became robust plants with unmistakable features of *organensis* combined with *hookeri* characters. The plants were almost completely pollen sterile, but chromosome configurations of ⊙12, 1 pair, were obtained.

O. organensis, therefore, is at the most to be considered an outlying member of the subgenus *Oenothera*. It has been included in this subgenus principally because of its angular seeds. It is included in the present discussion because of its uncertain taxonomic position and also because of the bearing of its behavior upon problems relating to incompatibility phenomena as exhibited by North American members of the subgenus *Oenothera*.

The *organensis* population is confined to a single valley in the Organ Mountains in southern New Mexico. When first studied it consisted of fewer than 1000 plants. Since then the number has dwindled to a very few individuals, which are perennial in habit. Of the 45 alleles found by Emerson all but a few have unfortunately been lost.

The action of the S-factor is gametophytically controlled; i.e., an S-factor will function only in those pollen grains that carry it. In a plant with S_1 and S_2, half the pollen will carry S_1, half S_2, and the reaction of a given pollen grain depends upon the S-factor that it carries.

Incompatibility is brought about by a reaction between the pollen tube and the stigmatic or stylar tissue into which it attempts to grow. A pollen grain with a given S-factor is unable to produce a functional tube in a tissue whose cells contain the same S-factor (Fig. 13.1). In *organensis*, Lewis has found that tubes with different S-factors will grow different distances in incompatible styles. He also found (1952) that incompatible tubes will grow farther at 15°C than at 31°C (the reverse is true of tubes in compatible styles).

In view of the large number of S-alleles in the natural population of *organensis*, it is surprising that their mutation rate is very low. Lewis (1948, 1949, 1951) found the spontaneous rate to be in the neighborhood of one per million pollen grains. Induced mutations following 500–700 r of X-rays varied in frequency depending upon the stage at which treatment was applied, but in all cases they were very rare. If irradiation took place during metaphase I or at the time of gene replication, the rate in some cases reached a level of 25 per million grains, but at other stages it only slightly exceeded the spontaneous rate. Different alleles differed in their rate of mutation. Some have never shown a mutation, others have attained an overall average frequency of five per million grains (Table 13.I). Even at its highest level, this is a very low incidence of mutation for a locus that would have to be highly mutable, it would seem, to have developed as many alleles as are found in the natural population.

Lewis and his co-workers found several additional facts of interest related to mutation of the S-locus in *O. organensis*. In the first place, all mutations have affected the pollen only, not the style (Lewis, 1951). For example, a pollen grain containing a mutated $S_6(S_6')$ can develop a pollen tube and bring about fertilization on a style containing unmutated S_6. On the other hand, stylar cells in which the mutated S_6' is present are able to react with unmodified S_6 pollen and inhibit pollen tube growth.

TABLE 13.I

Mutation of S-alleles in *Oenothera organensis* (Lewis 1951)

	Revertible mutations			Permanent mutations		
		Rate per million pollen grains			Rate per million pollen grains	
Allele	Number of mutations	Spontaneous	Induced	Number of mutations	Spontaneous	Induced
S_4	2	0	5·2	2	0·9	2·1
S_6	6	0·01	4·1	9	0·19	5·4

In other words, S_6' pollen on an S_6 style gives a compatible result, but S_6 on an S_6' style gives an incompatible reaction. All mutations so far found have affected pollen activity only, none has affected stylar activity. This fact caused Lewis initially to postulate a dual structure for the S-gene, one cistron concerned with pollen activity, the other with stylar activity. When the former mutates, the latter remains unchanged.

A second fact of interest found by Lewis was that not all mutations were permanent (1951). Some soon reverted to their original condition. As a result, it was possible to obtain homozygotes for S-factors, which ordinarily is impossible. For example, an S_6' pollen grain was able to grow in a style having S_6 and to affect fertilization, producing a zygote with the genes S_6S_6'. Early in embryonic development, however, the S_6' gene reverted to S_6 in this plant. As a result, most of the plant became S_6S_6 in composition and all of its pollen and stylar cells were typical S_6, resulting in self-incompatibility. Had the mutated gene remained in its mutated condition, the plant would have been S_6S_6' and would have been self-compatible.

A third fact of great interest is that no mutations have been obtained in which an S-factor has been converted into another functional S-factor (Lewis, 1951). Such mutations as have occurred have eliminated inhibitory activity in the pollen, but none has resulted in the production of a new active S-factor. The one type of mutation so far obtained, therefore, is not the type that has produced the multiplicity of S-alleles present in the natural population. The cause of such multiplicity is still a mystery.

A fourth fact of importance should be mentioned. When an S-factor mutates, the pollen grain in which it comes to reside does not lose its viability or its ability to function on a stigma. It is still a vigorous and functional pollen grain, capable of germination on any *organensis* stigma including those on the same plant.

The behavior of S-factors in preventing self-pollination naturally brings to mind the antigen-antibody phenomena so frequently found in animals. Lewis reasoned that the S-factor behavior might be of this kind. His first hypothesis was that one cistron produces an antigen in the pollen grain, another cistron produces in the style an homologous substance capable of complexing with the antigen and thus inhibiting its action. Perhaps the antigen is an enzyme necessary for pollen tube growth. The stylar substance is then an anti-enzyme which inactivates the enzyme and thus inhibits growth of the pollen tube.

Lewis therefore attempted to determine whether different S-factors in the pollen are in reality antigenic in nature, i.e., capable of inducing antibodies when injected into rabbits. His earlier attempts (1960) in this direction proved promising, and in 1962 Mäkinen and he succeeded in obtaining strong precipitin reactions, using the Ouchterlony technique, suggesting that there is a different and serologically highly specific substance associated with each different S-factor.

Lewis was able to obtain these results with extracts both from macerated pollen grains and from whole grains, showing that the antigen is capable of diffusing out of the pollen grain or else is at the surface of the grain and presumably of the pollen tube.

Comparable experiments were not performed with stylar extracts in the case of *Oenothera*, but Linskens (1958, 1959, 1960) has carried out such experiments in *Petunia*. He prepared antisera, using rabbits, for different S-factors in the pollen, then tested extracts from styles against these antisera. He got very strong reactions where the stylar extract was from the same plant that furnished the pollen from which the antiserum was made. The reaction was less strong where pollen and stylar parent shared only one S-factor, and there was no reaction at all where pollen and styles contained no S-factors in common. Thus it was clear that different S-alleles produce substances that are antigenically active and very specific. Self-pollination brings homologous substances together which complex and thus result in the failure of pollen tubes to grow. Linskens identified these substances in *Petunia* as glycoproteins, using radioactive tracers in his determinations. He found that these do not appear in the style until after pollination. When pollination is by incompatible pollen, there appear two kinds of glycoproteins, which he termed the X and Y fractions, but when pollination is by compatible pollen, a single and different glycoprotein is formed, the Z fraction, which contrary to the other fractions moves toward the cathode during electrophoresis. The protein component of the Z fraction is formed by the stylar tissue, the carbohydrate fraction is formed by both pollen and style, mostly by the latter. The X and Y fractions are formed primarily by the style.

Because pollen and style produce different substances which, in the case of incompatibility, interact; and because mutations that occur in the pollen do not alter stylar behavior; furthermore, because mutation which eliminates the serological activity of the pollen does not affect its ability to develop tubes and bring about normal fertilization, Lewis visualized the S-locus as being compound in structure, containing two parts or cistrons. At first he thought, as mentioned above, that one cistron is responsible for the activity of the pollen, the other for the stylar activity. Later he preferred the concept that one cistron controls antigenic activity, the other mediates the growth and functioning of the pollen grain and pollen tube.

The results of Lewis' and Linskens' work seem to make it clear that S-factors are associated with the production of serologically active substances and that the incompatibility reaction is comparable to the antigen-antibody phenomenon found in animals. Antigenic compounds are generally conjugated proteins, the serological specificity being related to the prosthetic group. The protein may be enzymatic in nature, in which case a conjugated protein can be both enzymatic and antigenic. It would seem that an S-factor has the role of adding to an enzyme which is essential for the germination of the pollen grain or the growth of the pollen tube a serologically active element. In the pollen grain this takes one form, in the cells of the style an homologous substance is produced, and the two substances interact when brought together, thus binding the associated enzyme in such a way that it is no longer active. Whether it is necessary to postulate that the S-locus is responsible for both the specificity factor and the enzyme may be questioned. It is perhaps more likely that the enzyme is under the control of some other locus and that all the S-locus does is to mediate the formation or the attachment of a serologically active prosthetic group. In this case, it is not surprising that all S-factor mutations so far found have affected only the serological activity of the pollen grain and have left intact its normal metabolic machinery.

In the subgenus *Oenothera*, balanced S-factors are found only in *O. organensis*. The work of Steiner (1956, 1957, 1961, 1964) and Steiner and Schultz (1958), however, suggests that S-factors may still be widespread in the group, but in an unbalanced condition, in which they serve as one form of pollen lethal. Most members of the subgenus *Oenothera* are heterogamous complex-heterozygotes, in which one complex (the alpha) is transmitted largely or wholly through the egg, the other (beta) through the sperm. In some forms, however, the alpha complex is transmitted occasionally through the sperm (and/or the beta complex through the eggs). As a result, it is possible to make crosses that will yield alpha·alpha combinations. Steiner made a number of such crosses within the midwestern assemblage known as the *biennis I* group (see Chapter 18). Every alpha·alpha that he obtained, however, was self-sterile, the pollen tubes either failing to develop, or growing only a few millimeters following self-pollination. When pollen containing one of these alpha complexes was placed on the stigma of a plant which did not have this complex, it produced fertile plants. Steiner suggested that present-day *biennis I* races have been derived from crosses that took place between individuals of originally isolated populations, one of which had a balanced S-factor situation, the other had no incompatibility factor. Because the two populations had experienced different histories of interchange, occasional hybrids between them had ⊙14. Such a hybrid would have an unbalanced S-factor situation. Half its pollen would carry the S-factor, the other half would have no such factor. Its stigmas, however, would contain the S-factor, as a result of which, pollen with the S-factor would be unable to produce pollen tubes following self-pollination; consequently, the S-factor would become a male gametophytic lethal (Fig. 13.1). If the complex possessing the S-factor was strong enough to overcome its competitor in the competition for the production of embryo sacs, a balanced lethal situation such as now characterizes complex-heterozygotes would be created—the S-factor-containing complex coming through the eggs, the other complex succeeding in the sperm. In the case of *biennis I*, the original crosses were between a *biennis*-like population, at least some members of which carried an S-factor, and a *strigosa*-like population in which some individuals, at least, were free of incompatibility genes.

Later (1958), Steiner and Schultz found that this situation was not confined to *biennis I*. Several races of *strigosa*, *biennis II* and *biennis III* as well as one *parviflora* race were shown to have alpha complexes containing S-factors.

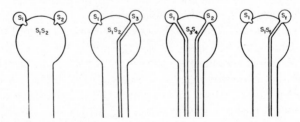

Fig. 13.1. The action of S-factors. A pollen grain with a given S-factor cannot germinate on a stigma whose cells possess the same S-factor. A grain lacking a S-factor can germinate on any stigma. Reprinted from Proceedings of the American Philosophical Society by permission.

A complicating factor, however, was discovered, in that some races were found to have alpha complexes with pollen lethals of a different kind, which inactivated their pollen completely, so that pollen carrying these alphas was unable to germinate on the stigmas of any race. Such true pollen lethals would, of course, mask the presence of S-factors if present.

In spite of this complication, however, it was still shown that several races belonging to the groups just mentioned possess S-factor-carrying alpha complexes, and it is probable that still other races, in which true pollen lethals are present, also have hidden S-factors.

From this work it is evident that S-factors were at one time wide-spread among the ancestors of the present-day forms of the subgenus *Oenothera*, no doubt in a balanced condition, and still exist in an un-balanced condition to an extent as yet undetermined, serving as pollen lethals which are now balanced, not with other S-factors, but with competitive situations that give the effect of egg lethals.

CHAPTER 14

Plastid Behavior

The role that plastids play in the determination and transmission of hereditary traits is well illustrated by the classical example of *Oenothera*. From the beginning of genetical work on this genus, the appearance of albinism or varying degrees of chlorosis, either of whole plants or of sectors, has been a matter of special interest. de Vries called attention to this situation in *O. lamarckiana* in his "Mutationstheorie" (1903, **I** p. 612, **II** p. 355). He attempted to learn something about the inheritance of chlorosis by crossing flowers from green and yellow sectors of the same or different individuals, and by growing selfed seeds from green and yellow sectors. In general he found a tendency for seeds from yellow sectors to produce more yellowish progeny than seeds from green sectors, but the results were not clear-cut and at that time he carried the experiments no farther than the F_1. In his "Gruppenweise Artbildung" he noted cases where the progeny of reciprocal crosses differed in respect to plastid behavior. For instance (p. 131) the cross *lamarckiana* × *hookeri* gave in F_1 the twin hybrids *laeta* (green) and *velutina* (yellowish and weak). The laetas, when selfed, yielded what he took to be green laetas and yellowish velutinas. In the reciprocal cross, *hookeri* × *lamarckiana*, the F_1 included laetas and velutinas but both classes were green. On selfing, the laetas produced laetas and velutinas, all green; the velutinas produced all green velutinas. Chlorosis, therefore, showed up only in the *velutina* progeny of the cross with *lamarckiana* as mother. This situation is summarized in Table 14.I in which de Vries' *laeta* and *velutina* are interpreted in terms of Renner's complexes.

Renner (1924) also found chlorotic or albino plants and corroborated in his own work the results that de Vries had obtained. He noted, however, that chlorotic hybrid seedlings often show green flecks, streaks or sectors, and their green reciprocals may show yellow spots or sectors. He coupled this observation with those of Ishikawa (1918) who showed in his study of fertilization in *Oenothera* that considerable cytoplasm is often introduced into the zygote from the pollen tube, which contains many leucoplasts.* As a result, Renner came to two important conclusions: that plastids can be derived from both parents in *Oenothera*, and that different classes of plastids exist in different races of the genus, which

---

* It is not yet certain, however, whether cytoplasm of the pollen tube in contrast to that of the sperm cell, fuses with the cytoplasm of the egg.

react differently in association with given genome combinations. In the case of the *lamarckiana–hookeri* crosses, for example, *lamarckiana* plastids find it difficult to retain their chlorophyll in the presence of the complex-combinations *velans*·^{h}*hookeri* or ^{h}*hookeri*·^{h}*hookeri*, whereas

TABLE 14.I

Plastid behavior in crosses between *lamarckiana* and *hookeri*

	F_1		F_2	
lamarckiana × hookeri	*laeta* (gaudens·^{h}hookeri)	(green)	*laeta* (gaudens·^{h}hookeri) *velutina* (^{h}hookeri·^{h}hookeri) *velutina*	(green) (yellowish) (yellowish)
	velutina	(yellowish)		
hookeri × lamarckiana	(velans·^{h}hookeri) *laeta* (^{h}hookeri·gaudens)	(green)	(velans·^{h}hookeri) (^{h}hookeri·^{h}hookeri) *laeta* (^{h}hookeri·gaudens) *velutina* (^{h}hookeri·^{h}hookeri) *velutina* (^{h}hookeri·velans) (^{h}hookeri·^{h}hookeri)	(green) (green) (green)
	velutina (^{h}hookeri·velans)	(green)		

hookeri plastids do not experience this difficulty. If, therefore, in the *velutina* resulting from the cross *lamarckiana × hookeri*, some hookeri plastids are introduced from the pollen tube into a zygote, there may result islands of cells in which *hookeri* rather than *lamarckiana* plastids are present and these spots will become green. Conversely, in the *velutina* derived from the cross *hookeri × lamarckiana*, the introduction of *lamarckiana* plastids from the male parent into the zygote will result in flecks or sectors of yellow.

Renner found similar behavior in crosses between other races and on the basis of extensive study reached the following additional conclusions (1924):

1. The plastids of the *biennis* of de Vries differ from those of *lamarckiana*, *muricata*, *hookeri* and *cruciata*, but are similar or identical with those of *suaveolens*. *Muricata* plastids differ from those of *lamarckiana* and *hookeri*, *lamarckiana* plastids differ from those of *hookeri* and *cruciata*. In other words, in these six species, there are at least four different types of plastid, the *biennis-suaveolens* type, and the types found in *lamarckiana*, *muricata* and *hookeri*. He did not determine whether those of *muricata* and *cruciata* belong to the same type. Later work indicates that they do.

2. These differences have arisen through mutations which have permanently modified the characteristics of the plastids. These were true mutations, not pseudomutations.

Renner's conclusions regarding the entrance of plastid primordia from the sperm into the zygote were in agreement with those of Baur (1909) who suggested that in variegated *Pelargonium* the young zygote has two kinds of plastid, green and white, which later segregate into cells with purely green and purely white plastids, mixed cells being in some manner suppressed. Although there was at that time no evidence that plastids were ever derived from the father, Baur stated that the presence of two kinds of plastid in the zygote would be easily understood if it were found that plastids are transmitted into the egg with the sperm.

Correns (1922) and Noack (1925, p. 82), on the contrary, opposed the idea of two kinds of plastid in the zygote. Correns argued that, if this were true, one should find areas of differentiated cells in which the two kinds of plastid were mixed; he found it difficult to imagine the complete separation of green and white plastids derived from cells in which they were originally mixed.

Noack reasoned that all plastids in a plant are intrinsically alike and the failure of plastids in certain sectors or layers to become green is the result of some metabolic disturbance in those portions of the plant.

One of Renner's students, Krumholz (1925) made a careful embryological study of reciprocal crosses between *biennis* and *hookeri*, and between *suaveolens* and *hookeri*, as well as backcrosses and selfed F_2s. When *hookeri* was used as male parent, the F_1s (*rubens*·^{h}*hookeri*, or *flavens*·^{h}*hookeri*) were green with occasional yellow areas; when *hookeri* was used as female, the same complex-combinations were yellow, at least in early stages, with occasional green spots. He found that mixed cells, containing both green and defective plastids, were present only in early stages; later the two classes of plastid became segregated into different cells. This finding has been corroborated by Stubbe (1955, 1957, 1959) and Schötz (1954, 1958a, b) who have studied the matter intensively. As a result of early segregation of the two classes of plastid, various types of chimera are produced. With respect to the mechanism by which this segregation is accomplished, Michaelis (1955) pointed out that complete segregation of two kinds of plastid from a mixed cell into different daughter cells can be expected only if the initial number of plastids in the mixed cell is small. The early segregation of different types found in *Oenothera* suggests that the zygote has few plastid primordia to start with, including those that are brought in from the male parent.

Although the differences found in the plastids of the various races must have arisen originally by mutation, Renner emphasized in numerous papers that the changed behavior that plastids show when brought into association with unsuitable gene combinations is not the result of fresh mutation. When, for instance, *lamarckiana* plastids are in association with the complex-combination *velans·*ᵸ*hookeri*, they are deleteriously affected. If brought back into a favorable genic milieu, however, they again behave normally—their essential nature has not been changed. Another example cited by Renner (1937a) is the cross *cruciata*** (*pingens·flectens*) × *biennis* (*albicans·rubens*). This gives *pingens·rubens* with green *cruciata* plastids but with yellow sectors that contain *biennis* plastids. Although *biennis* plastids are unable to function with this genic combination, they can develop normally in the combinations *pingens· velans* and *rubens·velans*. When, therefore, flowers on a yellow sector of *pingens·rubens* were pollinated by *albicans·velans*, the progeny was partly *pingens·velans* and partly *rubens·velans*, both with mostly *biennis* plastids. These were green in both combinations. *Biennis* plastids which found themselves unable to develop chlorophyll in *pingens·rubens* were now able to function normally—they had not changed as a result of an unfortunate association, they still retained their original qualities. Renner uncovered many instances of this sort of behavior, involving various classes of plastid. In no case were plastids caused to mutate because of their association with foreign genomes. The failure on the part of plastids to turn green was the result of disharmony between a given class of plastids and a particular combination of complexes.

Such lack of harmony is rarely observed in nature, since most races are self-pollinated, and this reduces greatly the chance of a class of plastids coming into association with foreign complexes. When races are artificially crossed, however, cases of plastid-genome disharmony are frequently encountered.

Renner's work showing that several genetically different classes of plastid exist in *Oenothera* was extended by some of his students, especially by Stubbe (1959, 1960, 1963a, b) who has analyzed not only a number of European races but also a large number of complex-combinations (over 80) derived from some 14 North American races, in order to determine the number of different kinds of plastid present in the population and the kinds of complex that exist from the standpoint of their effects on the various plastid classes. He has found five distinct classes of plastid, designated by Roman numerals I—V, and three classes of genome designated as A, B and C. The interactions between these classes

---

* Renner later adopted the designation *O. atrovirens* for this form.

of plastid and genome are shown graphically in his diagram which is here reproduced as Fig. 14.1. Table 14.II shows Stubbe's classification of the plastomic and genomic composition of principal European races and

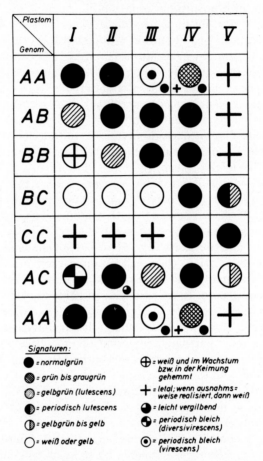

FIG. 14.1. Chart of Stubbe (1959). Five classes of plastid, based on their reaction to various genome combinations are listed above. Three classes of genome, from the standpoint of their effect on plastids, are recognized, various combinations of which appear at the left. Interactions of plastids to genome combinations are depicted graphically. Reprinted from *Z. Vererbungslehre* by permission.

also of the groupings of North American oenotheras which have been recognized in our laboratory and which will be discussed in detail later on.

We have made many hundreds of crosses between races belonging to these various groupings, and between North American races and those

TABLE 14.II

Classes of plastome and genome found in the various groupings of North American members of the subgenus *Oenothera*, and in certain races found in Europe, according to Stubbe (1960, 1963a)

Race or group of races	Class of plastome	Classes of genomes or complexes
NORTH AMERICAN GROUPS		
hookeri	I	A_2A_2 (homozygote)
elata	?	AA?
strigosa	I	A_1A_1 (alpha = A, beta = A)
biennis I	III	BA (alpha = B, beta = A)
biennis II	II	AB (alpha = A, beta = B)
biennis III	III	BB (alpha = B, beta = B)
grandiflora (true)	III?	BB (homozygote)
parviflora I	IV	BC (alpha = B, beta = C)
parviflora II	IV	AC (alpha = A, beta = C)
argillicola	V	CC (homozygote)
EUROPEAN RACES		
lamarckiana	III	AB (*velans* = A, *gaudens* = B)
biennis of de Vries	II	AB (*albicans* = A, *rubens* = B)
suaveolens	II	AB (*albicans* = A, *flavens* = B)
grandiflora of de V.	II	AB (*truncans* = A, *neoacuens* = B)
purpurata	II	AA (homozygote)
ammophila	IV	AC (*rigens* = A, *percurvans* = C)
atrovirens	IV	BC (*pingens* = B, *flectens* = C)
muricata	IV	AC (*rigens* = A, *curvans* = C)
parviflora	IV	BC (*augens* = B, *subcurvans* = C)
silesiaca	IV	BC (*subpingens* = B, *subcurvans* = C)
Bauri	I	AA (*laxans* = A, *undans* = A)

found in Europe, and while there are different levels of intensity of effect shown in different hybrid combinations, as indicated also in Stubbe's diagram, our findings conform in most respects to expectations based on his scheme (Cleland, 1962).*

The question may be raised at this point as to whether the incompatibilities resulting in defective plastid behavior are due to a conflict between genome and plastome, or whether it might not be the plasmon—the general cytoplasm—or some other element than the plastids in the cytoplasm that fails to cooperate properly with the nucleus. Renner made a number of crosses to check on this matter which suggested that it is the plastome itself, and not some other factor in the cytoplasm that fails to coordinate with particular genome combinations. When plastids

* But see p. 285.

are transmitted to a zygote through the sperm they come to lie in a cytoplasm that is primarily maternal in origin. Their behavior in the presence of a given complex-combination can then be tested in two different cytoplasms. For instance (Renner, 1936), the cross *suaveolens* × *muricata** produces as one of its twin hybrids *flavens–curvans*. These plants are yellowish or colorless. *Suaveolens* plastids are in this case unable to turn green in their own cytoplasm. In the reciprocal cross it is also possible to obtain *curvans·flavens* plants as metacline hybrids. In this case the plastids are predominantly from *muricata* and these are green in their own cytoplasm. In the first cross, however, green flecks or sectors are often present. These are areas where *muricata* plastids have come from the father. They are able to turn green in a foreign as well as in their own cytoplasm. Conversely, in the latter cross, islands or sectors of yellow cells appear. These contain *suaveolens* plastids, brought in with the sperm. Thus, *suaveolens* plastids cannot turn green in either their own or in *muricata* plasma when *flavens* and *curvans* are present. *Muricata* plastids, on the contrary, are green in both their own and in foreign cytoplasm in the presence of the same genes. The same situation is found in cases where *lamarckiana*, *hookeri*, and *biennis* are crossed with *muricata*. The plastids of these three races are yellow in both reciprocals when *curvans* is present, i.e., whether they are in their own or foreign cytoplasm. *Muricata* plastids, on the other hand, are green in both their own and foreign cytoplasms. It is therefore the plastome itself, and not the plasmon, that determines whether a given class of plastid can function in the presence of a given gene combination. The situation with respect to the examples given above is outlined in Table 14.III.

The existence of different classes of plastid has been further demonstrated by Stubbe (1957, 1958), who analyzed early stages in the embryology of hybrid plants at a stage when cells with mixtures of different kinds of plastid still exist. He found that distinct types of plastid could be easily recognized under the microscope, and that they retained their distinctive characters while they were present together in the same cell. In one case, he was able to obtain a hybrid with three different types of plastid, all recognizable within the same cell. This result was obtained as follows: *albicans·gaudens* with green *lamarckiana* plastids was pollinated by *albicans·*ʰ*hookeri* with mutated white *suaveolens* plastids. A few progeny (*albicans·*ʰ*hookeri*) were green-white chimeras, with a mixture of green *lamarckiana* and white *suaveolens* sectors in the floral region. Flowers in which both green and white sectors were present were pollinated by *albicans Grado·gaudens* (with yellow-green mutated

---

* Renner later adopted the designation *O. syrticola* for this form.

TABLE 14.III

Plastid behavior in certain *Oenothera* crosses in relation to plasmone and genome

Cross	Plasmone	Source of plastids	Color of plastids
suaveolens × *muricata* (flavens·curvans)	*suaveolens*	*suaveolens*	Yellow
	suaveolens	*muricata*	Green
muricata × *suaveolens* (curvans·flavens)	*muricata*	*muricata*	Green
	muricata	*suaveolens*	Yellow
lamarckiana × *muricata* (gaudens·curvans)	*lamarckiana*	*lamarckiana*	Yellow
	lamarckiana	*muricata*	Green
muricata × *lamarckiana* (curvans·gaudens)	*muricata*	*muricata*	Green
	muricata	*lamarckiana*	Yellow
hookeri × *muricata* (ʰhookeri·curvans)	*hookeri*	*hookeri*	Yellow
	hookeri	*muricata*	Green
muricata × *hookeri* (curvans·ʰhookeri)	*muricata*	*muricata*	Green
	muricata	*hookeri*	Yellow
biennis × *muricata* (rubens·curvans)	*biennis*	*biennis*	Yellow
	biennis	*muricata*	Green
muricata × *biennis* (curvans·rubens)	*muricata*	*muricata*	Green
	muricata	*hookeri*	Yellow

suaveolens plastids). Some of the progeny were three-colored with dark green, yellow-green and white sectors. The three kinds of plastid existed together in the same cells in early stages of such plants and were recognizable by their color and also by their size and internal structure. The fact that all three could exist side by side in the same cytoplasm without losing their distinctive structural and chemical characteristics indicates that these characters are inherent and not determined by the rest of the cytoplasm.

There are a few cases known where disharmony between certain genome-combinations and particular plastid types can be ascribed, so far as the genome is concerned, to specific gene loci. For instance, Renner (1943b) found that *lor*, a semi-lethal locus in *flectens* which produces narrow thread-shaped leaves in homozygous condition, if brought into association with *lamarckiana* plastids, prevents the latter from becoming green. If *lor lor* is associated in a zygote with *Fl Fl* (for bent stem tips) in the presence of foreign plastids, seed germination is prevented. Another case is reported by Stubbe (1955) in a strain of *suaveolens* known as "Weissherz". Here a recessive gene, present in homozygous condition, causes the base of the young rosette leaves to become chlorotic. This condition disappears with age so that mature "Weissherz" plants become entirely normal in appearance. This is the only case in *Oenothera* where within a race a single locus is known normally to produce a chlorophyll defect.*

Although association of certain classes of plastid with incompatible genomes does not result in mutation of the affected plastids, true plastid mutations are not unknown in *Oenothera*. Renner described in 1936 several such cases that had appeared in his cultures. In these cases the mutant plastids were incapable of becoming green no matter what genes were associated with them. That the plastids had themselves suffered a mutation was shown by the fact that the changes were cytoplasmically inherited: green × white produced green progeny (with occasional white areas derived from the pollen), white × green produced white seedlings (with green flecks often present). Whether these mutations arose spontaneously or were gene-induced in the first place cannot be determined with certainty, but the rarity of such mutations and the fact that the particular combinations of genomes that were present when the mutations occurred do not produce recurrent mutations, point to the conclusion that the mutations occurred spontaneously in the plastids themselves. Renner found plastid mutations in a frequency of about one in 2000 plants (1936, p. 280). Stubbe found 11 cases of spontaneous plastid mutation among 3530 plants (1953), a frequency of about three per thousand.

In 1946 another plastid mutation was found in a line of *hookeri* obtained in the F_2 of the cross *biennis* × *hookeri*, and carrying, therefore, *biennis* plastids (Schötz, 1954, 1968). A single individual of this *hookeri* had white sectors, with mutated *biennis* plastids; and crosses made to other races, using flowers on white sectors of this plant, showed the white sectors to have mutant plastids incapable of becoming green in any of the complex-combinations tested. This was an especially interesting case because of the use to which Schötz put it.

Following up earlier findings of Renner, Schötz reported (1954) in regard to competition between different kinds of plastid where these were brought together into a single plant by crossing. In these crosses he used flowers of white sectors of *O. hookeri* carrying mutated *biennis* plastids, and crossed these with flowers from a variety of green *Oenothera* races. In the hybrids he determined the "variegation value", i.e, the relative proportion of white and green tissue in the cotyledons and first two true leaves. If pollen was derived from a white sector of *hookeri*, and races such as *blandina* and *hookeri* (with normal plastids) were used as

---

* Stubbe (private communication) informs me that H. Kutzelnigg has obtained a series of *hookeri* lines with simple Mendelian defects.

egg parent, the hybrids were wholly green, i.e, the colorless mutated *biennis* plastids were wholly suppressed (variegation-value = 0).

At the other extreme, when the same pollen was used on certain other green races, as much as 20% or more of the area had plastids that were derived from the pollen and were colorless (e.g., *pingens·rubens cruciata* had a variegation-value of 20·671). The races fell into the following sequence based on the ability of their plastids to compete successfully with the white mutant type: *blandina*, and *hookeri* with *hookeri* plastids→ *bauri* → *lamarckiana* → *rubricaulis* → *suaveolens* → *biennis* → *syrticola* → *parviflora* → *rubricuspis* → *ammophila* → *atrovirens* → *pingens·rubens* → *pingens·rubens cruciata*.

If flowers of the white sectors of *hookeri* furnished the eggs, the hybrids were to a large extent white, but green plastids were present in the seedlings in varying degrees, depending upon the ability of the paternal plastids to survive the competition. The sequence of races was somewhat different from that shown by the reciprocal crosses: *lamarckiana* plastids showed the strongest ability to survive, about 20% of the leaf tissue being green. Then followed, in descending order of ability to compete: *blandina—rubricaulis—hookeri* (with *hookeri* plastids)—*suaveolens*, and certain forms whose hybrids never showed green tissue—*pingens· rubens, parviflora, syrticola, ammophila*, and *atrovirens*.

Schötz interpreted differences in competitive ability to be the result of different speeds of multiplication of the plastids themselves. As the zygote develops into the embryo and seedling, the more rapidly multiplying class of plastids gradually gains the upper hand. The relative expansion of green and white tissues depends on the relative speeds of multiplication of the two plastid classes.

Some cells, in this process, come to have only one class of plastid— green or white. In the course of time, the proportion of mixed cells becomes smaller and they finally disappear. At the end of the sorting out period, the size of the white and green areas depends upon the initial mixture of the two types of plastid in the zygote and their speed of multiplication. The sorting out period ends by the time that the first pair of foliage leaves have been developed, since Schötz found in general no mixed cells after this time.

If one always uses the same colorless plastid class, the average number of plastids in the egg, and reciprocally in the sperm, will remain constant, so far as this parent is concerned. Furthermore, from a microscopic study of mature embryo sacs, Schötz found that in different green races the number of plastids in the egg cells was essentially the same. The possibility that egg cells of different green species have different abilities to resist the intrusion of male plastids or cytoplasm can be ruled out.

As a result, the initial mix of plastids in the zygotes of different crosses between green races and the *hookeri* with mutated *biennis* plastids is the same and the colorless *biennis* plastids become a standard against which the division rate of the various green plastids can be measured. The variegation value at the end of the sorting out period (Schötz, 1954) then becomes a measure of the relative speed of division of the various kinds of plastid.

After the plastids have been fully segregated, differences in speed of division of the plastids no longer influence the number of cells formed, for cells with the faster dividing plastids tend to adjust their cell division rate to that of the cells which contain the more slowly dividing plastids. The proportion of green and white tissue, therefore, tends to remain constant, at the same level as that found in the young seedling. Certain processes at the growing point, such as the building of periclinal chimeras, may bring about alteration in the distribution of the differently colored tissues that have nothing to do with the different speeds of division of the plastids.

Schötz has furthermore shown (1954, 1958a) that, at least in many cases, the different speeds of multiplication depend upon the plastids themselves, rather than on the genes present. For example, *pingens·*[h]*hookeri* was obtained with three different plastid combinations as follows:

Pollen from white sectors of the strain of *hookeri* that carried mutated *biennis* plastids was used on

(a) *pingens·rubens* with *atrovirens* plastids
(b) the same with *lamarckiana* plastids
(c) sectors of the same containing *biennis* plastids (which tend to be colorless in *pingens·rubens*).

Three types of plastid were therefore brought into competition with mutated *biennis* plastids, all in the presence of the same genomic combination. They differed noticeably in their ability to compete with the mutant plastids: *lamarckiana* plastids were the strongest competitors (i.e., they multiplied the most rapidly), the *atrovirens* plastids were weakest, normal *biennis* plastids were in between.

There was evidence in some cases, however, that the genome could have some influence on the speed of multiplication: in some crosses where twin hybrids were produced, rates of multiplication differed slightly in the twins. For example, when *lamarckiana* was crossed with white-sectored *hookeri*, using flowers from white sectors of the latter, plastids differed somewhat in relative speed of multiplication in the twins *velans·*[h]*hookeri* and *gaudens·*[h]*hookeri*. The same was true of *rubens·*[h]*hookeri* and *tingens*·ʰ*hookeri*, derived from *rubricaulis* × white-sectored *hookeri* (again using flowers from white sectors).

In 1958, Schötz extended his findings on the relative speed of division of the plastids in different races. With respect to the twelve races discussed in 1954 he found that after 10 years, these races still showed the same relative speeds of chloroplast multiplication. To these were added additional races, making a total of 27. The chloroplasts of these races were tested against two mutant plastids, the previously used mutated *biennis* plastids present in a strain of *hookeri*, and mutated *lamarckiana* plastids present in a strain of *blandina*. The 27 races fell into three groups from the standpoint of the ability of their plastids to compete with the mutant plastids. Group 1 possessed "strong" plastids, i.e., those that when furnished by the eggs multiplied so rapidly that they entirely or almost entirely supplanted the mutant plastids brought in by the pollen. Group 2 had plastids of intermediate strength, and Group 3 contained "weak" plastids, i.e., those that permitted the mutant plastids to form sizeable white sectors. Schötz found an excellent correlation between his findings and those of Stubbe (1959, 1960) who found in nature five classes of plastid from the standpoint of their reaction to the presence of different categories of genome. Races with Stubbe's plastome classes I and III have "strong" plastids, those with plastome II have plastids of intermediate strength, and those with plastome IV (and V, e.g., *argillicola*) have weak plastids. These results are summarized in Table 14.IV, compiled from data in Schötz's 1968 paper.

Schötz and his student Heiser (Schötz and Heiser, 1969) have recently measured the relative speed of plastid division in 20 *Oenothera* races by determining the ratios of green to colorless plastids in mixed cells of hybrids between green races as female, and *hookeri* carrying mutant *biennis* plastids, or *blandina* with mutant *lamarckiana* plastids. Their results confirmed the earlier finding of Schötz with respect to the relative competitive strengths of the various races.

Schötz found another interesting fact. When green races are crossed as female with the two mutant strains, they show higher "variegation-value", i.e., a higher degree of white sectoring, with the *blandina* strain which carries mutant *lamarckiana* plastids than with the *hookeri* strain which has mutant *biennis* plastids. In other words, the mutated *lamarckiana* chloroplasts in *blandina* are more successful in competition than the mutated *biennis* plastids in *hookeri*. This is in line with the fact that normal green *lamarckiana* plastids (Group 1) multiply faster than normal *biennis* plastids (Group 2). Mutation of *lamarckiana* and *biennis* plastids, therefore, has not appreciably altered the relative speed of multiplication of these plastids.

Table 14.IV

Percentages of variegated plants, and variegation-values, in hybrid cultures resulting from crosses of certain green *Oenothera* species with *O. hookeri* (with mutated *biennis* plastids) and *O. blandina* (with mutated *lamarckiana* plastids) (Schötz, 1954, 1968)

Group	Maternal parent	Paternal parent	Mean percentages of variegated plants	Mean variegation value	Paternal parent	Mean percentages of variegated plants	Mean variegation value	Plastom class in mother
1	*hookeri*		0·0	0·00		8·2	0·21	I
	blandina		0·0	0·00		14·1	0·54	III
	bauri		7·4	0·10		43·8	2·97	I
	lamarckiana		3·1	0·17		26·4	1·47	III
	cockerelli	*hookeri*	0·0	0·00	*blanda*	9·9	0·27	I
	franciscana	with	0·8	0·02	with	19·9	0·68	I
	mollis	white	1·1	0·02	white	31·5	1·05	I
	deserens	*biennis*	1·3	0·05	*lamarckiana*	21·4	0·73	III
	chicaginensis	plastids	5·0	0·23	plastids	24·4	2·39	III
2	*rubricaulis*		23·2	1·39		41·8	4·93	II
	suaveolens		26·7	1·86		49·9	7·14	II
	biennis		22·2	1·51		53·8	7·31	II
	purpurata		14·9	0·77		45·0	3·53	II
	conferta		16·1	0·77		34·2	4·42	II
	nuda		12·4	0·81		54·6	5·60	II
	holscheri		31·1	1·87		57·5	11·54	II
	grandiflora		28·3	1·88		53·5	10·82	II
	coronifera		38·0	2·09		57·7	5·35	II
3	*syrticola*		48·7	7·75		63·3	10·28	IV
	parviflora		47·6	5·62		63·2	8·5	IV
	rubricuspis		52·3	9·96		68·7	17·16	IV
	atrovirens		70·1	15·72		72·6	23·93	IV
	ammophila		43·9	8·91		58·1	12·16	IV
	germanica		52·9	11·64		74·5	12·71	IV
	silesiaca		53·2	9·65		67·2	19·46	IV

Recent work on the plastids of *Oenothera*, especially by Schötz and his students, and by Stubbe, has proceeded in four directions: (1) a detailed study of greening and bleaching behavior of plastids in cases of plastome–genome disharmony; (2) quantitative chemical analyses of the pigment content of plastids belonging to different races at different stages in the development of plants with such disharmony; (3) structural studies of plastids of various sorts and under various conditions, using the light and electron microscope; (4) attempts to determine the biochemical nature of plastid mutations.

(1) Schötz (1958b) undertook an analysis of the pattern according to which plastid defects appear. Whether the defects are the result of plastid mutation or the presence of incompatibilities, they do not consist necessarily in the loss of ability to manufacture chlorophyll, but may rather result from the excessive photo-oxidation of pigments that have been formed. Plants in which a disharmony between genome and plastome is present tend to pass through three stages during the development of the seedling, first a greening stage, then a bleaching stage and later a second greening stage (in some cases there may even be a second bleaching period). Different complex-combinations differ in respect to the time of initiation of these stages and their duration—in some, bleaching does not begin until 1 or 2 weeks after germination; in others, it may begin before germination. The duration of the bleaching period varies from form to form. It may last only a few days, or several weeks; in the latter case the only plants that are able to survive are those that receive from the sperm enough plastids capable of remaining green and of competing in speed of multiplication. The second greening period is completely suppressed in some combinations, while in others the plants recover more or less completely and flower normally. Some hybrids that remain bleached and succumb under conditions of full sunlight are able to survive if grown in dim light. Schötz recognized six "bleaching types" which he named for certain complex-combinations that typify them (1958b):

(a) In the *albicans·velans* type, with *lamarckiana* plastids, the seedlings are green at first, but soon the new cells at the base of the cotyledons fail to become green as they are formed, so that the proximal half of the mature cotyledons becomes yellow, while the distal half is green. The bleaching period lasts until eleven or twelve true leaves have been formed, these leaves having a faint tinge of green, sufficient to keep them alive, after which newly formed cells are again able to develop a full complement of chlorophyll. All leaf tissues formed subsequently have the normal green color.

(b) In the *laxans·dilatans* type, with *bauri* plastids, the seedlings are yellowish green for the first 6–10 days, then the basal meristem of the cotyledons produces yellow cells for a period of 4–5 days, after which green cells are again formed. As a result, mature cotyledons have a yellow band across the middle, separating the original yellow-green region from the later-formed green region. The yellow-green region may gradually become chlorotic; this is a secondary effect involving degeneration of originally green plastids as opposed to the primary effect seen in the earlier bleached cells, which were yellow from the start.

(c) In the ^h*hookeri·curvans* type, with *hookeri* plastids, the seedlings are wholly green for about two weeks, then the green areas almost completely bleach, and at the same time new cells as they develop remain yellow. If kept in dim light, the seedlings may survive until about the fifth week when leaf meristems again begin to produce green cells, and from that time on all newly formed chloroplast-bearing tissues are green, but plants must be grown in dim light.

(d) In the *pingens·rubens* type, with *biennis* plastids, the seedlings are at first green, but the cotyledons begin to fade after five days and usually die. In dim light they may be carried on until the first foliage leaves appear. These are green, but somewhat deficient in chlorophyll, so that the plants continue to be viable only in weak light.

(e) In the ^h*hookeri·flectens* type, with *hookeri* plastids, the seedlings are colorless unless green *atrovirens* plastids from the pollen are present in sufficient amount to maintain life. This condition lasts for about 3 weeks. The first foliage leaves are green-edged and successive leaves thereafter become fully green but the plants are viable only in dim light.

(f) In the *rubens·flectens* type, with *biennis* plastids, the seedlings are yellow and die unless the pollen contributes a sufficient proportion of green *atrovirens* plastids. In this type there is no greening of the *biennis* plastids in subsequent stages.

Each of the above classes exemplifies a type of behavior that has been found in other crosses as well as in those mentioned, although minor differences in level of reaction and in timing are found in different combinations. There is quite a variation, therefore, in the type and strength of the effect when disharmonious plastid–genome combinations are formed.

It should be reiterated that incompatibilities between plastids and genome-combinations, such as have been described, are a rarity in nature. Most oenotheras are self-pollinating and rarely cross with other races. The chances that disharmonious combinations will arise, therefore, are minimal. Within a successful line of descent, of course, plastids and genes must be compatible, otherwise the line would die out very quickly.

(2) Schötz began in 1962 a quantitative analysis of the pigment content

of plastids under compatible and incompatible conditions, and at different periods during the growing season. In this work he has had the collaboration of Senser (Schötz and Senser, 1964, 1965; Senser and Schötz, 1964) and Bathelt (Schötz and Bathelt, 1964a, b, 1965). These workers have determined in normal green tissues the relative amounts of chlorophyll a and b, and have identified and measured the amounts of the various carotinoids present. They find quite a range of variation in total and relative amounts of the various pigments in different races and hybrids, but these differences bear little relation to the racial groupings or to the genetic classes of plastid and of genome that Stubbe has found to exist in the population. Schötz and associates have also found that races differ in respect of the seasonal fluctuations of pigment content, but again these variations show little relation to the different categories of plastid and genome. In some races there is a steady decrease in the amount of pigment throughout the growing season, in others there is a decrease followed by an increase, in still others the reverse is true, and finally there are races in which the amounts remain essentially constant throughout the season.

Studies have also been made of pigment content in plastids that are deleteriously affected by the associated genome-combination. In these, pigment content is greatly reduced, for example to as little as 3% of normal in dim light, or 1% in bright light in *pingens·flectens* and *rubens·flectens* (both with *biennis* plastids) (Schötz and Bathelt 1965). In *albicans·velans* (with *lamarckiana* plastids), the loss in pigment content is less (90% reduction in bright light).

Reduction in amount of pigment in bright light was found by Schötz and Bathelt (1964b, 1965) to occur in normal green plastids as well as in chlorotic ones, but the pattern of reduction was not always the same. In the *albicans·velans* type, chlorophyll b is reduced faster than a in green plastids, but more slowly in chlorotic plastids. In both, carotin is reduced more rapidly than xanthophyll. The same is true in the *pingens·rubens* type. In this combination, total chlorophyll content is reduced more rapidly than total carotinoid content, even to the point where there is less chlorophyll present than carotinoid. In the second greening period in this type, the chlorophyll content builds up to 100% of normal in dim light but only to 70–80% in an ordinary greenhouse and to 60% or less in the field. It is for this reason that these plants, even though turning green after the bleaching period, continue to be viable only in weak light.

The situation in *"Weissherz"* is somewhat different from that in disharmonious plastid–genome combinations. In this race, the regions that become yellow show, in comparison with green areas, a much reduced ratio of chlorophyll b to chlorophyll a, carotin to xanthophyll,

and of total chlorophyll to total carotinoids. The ratios in normal green
and temporarily chlorotic areas in this race were determined as follows:
(Schötz, 1962b)

In green areas:

$\dfrac{\text{Chlorophyll } a}{\text{Chlorophyll } b} = 3 \cdot 02$ $\dfrac{\text{xanthophyll}}{\text{carotin}} = 1 \cdot 97$ $\dfrac{\text{Chlorophylls } a \text{ and } b}{\text{carotinoids}} = 3 \cdot 95$

In chlorotic areas:

$\dfrac{\text{Chlorophyll } a}{\text{Chlorophyll } b} = 5 \cdot 79$ $\dfrac{\text{xanthophyll}}{\text{carotin}} = 5 \cdot 37$ $\dfrac{\text{chlorophylls } a \text{ and } b}{\text{carotinoids}} = 1 \cdot 19$

When the chlorotic region in "*Weissherz*" becomes green again, the
proportions return essentially to those typical of green areas that have
remained green continuously.

(3) Stubbe and von Wettstein (1955), as well as Schötz and co-workers
(Senser, 1962; Senser and Schötz, 1964; Schötz and Senser, 1964;
Schötz, 1965a, b; Schötz and Diers, 1965, 1968; Schötz, Diers and
Bathelt, 1968; Schötz, Senser and Bathelt, 1966; Diers and Schötz, 1965,
1966, 1968, 1969; Diers Schötz and Bathelt, 1968) have made extensive
studies of the structure of *Oenothera* plastids, using both the light and
electron microscope, and including both normal green plastids and those
that either develop from proplastids during the bleaching period, or
degenerate from the green to the chlorotic condition. In general, plastids
that arise from proplastids during the bleaching period do not develop
typical grana, but undergo a process of vacuolization that may result
in bursting of the plastid, a process that begins at different stages of
development in different complex-combinations. Plastids that become
green and then degenerate, as in the *pingens·rubens* type, experience a
disorganization of the grana, the development of lipid material from the
grana and the appearance of vacuoles. Plastids multiply only when
immature, by a process of fission. In some cells during the bleaching
stage, (Senser and Schötz, 1964) plastids fail to divide but grow instead,
so that only 1–3 plastids are present in the cell and these attain a diameter
of 15–20 μm, as opposed to the normal 6–8 μm.

Diers, Schötz and Bathelt, in their recent papers, have placed especial
emphasis on the structure of thylakoids, comparing normal plastids with
those which show disharmony. The material used was principally *lamarckiana, hookeri* and the hybrid (*lamarckiana* × *hookeri*) *velans·*[h]*hookeri*
with *lamarckiana* plastids, in which disharmony exists between genome
and plastome. They have found a variety of disturbances in structure as a
result of disharmony, including very long thylakoids, unusually large
and crowded grana, and in some cases an extensive system of sinuous,

interconnecting thylakoids that resemble the systems found in some blue-green algae.

Schötz and Diers (1965) and Schötz, Diers and Rüffer (1971) have also described a process of budding in plastids of *O. (lamarckiana × hookeri) velans·*ʰ*hookeri* and of *O. hookeri* by light and electron microscopy. Invaginations of the inner plastid membrane are formed which cut off small portions of the stroma. These may later protrude into the surrounding protoplasm and ultimately separate from the plastid as independent bodies. There is no evidence as yet, however, that they become additional plastids or that they are transformed into mitochrondria. Their function is unknown.

(4) Stubbe and co-workers have studied the biochemical characteristics of several recently discovered plastid mutants, using C^{14} and P^{32} labelling. These include three *hookeri* plastid mutants (Iα, Iγ, Iδ) and two *suaveolens* mutants (IIα, IIγ) (*hookeri* has plastids of Class I, *suaveolens* has Class II). These mutants have all been found to have suffered a blockage in the electron transport system; they have all the enzymes of the photosynthetic carbon cycle. In four of the mutants, the block is in photosystem 2—the system controlled by chlorophyll *b*. In one mutant (IIα) the block is in photosystem 1, the system controlled by chlorophyll *a*. The specific points of blockage have not as yet been determined (Hallier, 1966, 1967, 1968; Fork and Heber, 1968; Fork *et al.*, 1969; Hallier *et al.*, 1968).

An electron microscopic study of four of these mutant plastid types by Dolzmann (1968) shows widespread disorganization of grana and thylakoid structure.

If there is one fact that is demonstrated more conclusively than another by the plastid situation in *Oenothera* it is the continuity of plastids from one generation to the next. Plastids do not arise *de novo* in each generation, but are inherited as truly as are the genes. This fact was strongly emphasized by Stubbe (1962) and Schötz (1962a) in criticizing the conclusion drawn by Mühlethaler and Bell (1962) that in each generation plastids arise *de novo* as buds from the nucleus. As Stubbe and Schötz pointed out, the situation in *Oenothera* clearly disproves such an hypothesis. In *Oenothera*, genetically distinct classes of plastid may exist side by side in the same plant, which are transmitted from one generation to the next, not only through the egg but also through the sperm. Where different kinds enter the zygote from the two parents, these retain their identity and their innate characteristics, whether the genome-combination with which they find themselves associated is favorable or not. It is inconceivable that a fertilized egg nucleus could bud off two or even three distinct kinds of plastid corresponding exactly to the kinds found in the two parents. *Oenothera* clearly demonstrates the continuity of the plastidome and disproves the idea of the *de novo* origin of plastids.

As pointed out by Stubbe (1963a, 1966) plastids should be regarded as containing multiple genetic factors, since a given mutation may affect one characteristic and not others, showing that the genetic units in the plastid are separable. Thus, plastids in *Oenothera* have at least the following effects:

(a) They differ in respect to their ability to get along with particular combinations of gene complexes.

(b) Different kinds of plastid differ in the speed with which they multiply, even when these different types are all compatible with a given genome-combination (Schötz, 1954).

(c) Plastids may have an effect on leaf morphology. For example, *flavens·flavens* with plastome II has pale green leaves of normal shape, but the same genic combination with plastome III has dark green leaves with the edges rolled up.

(d) Plastids have an effect on the shape of starch grains in the pollen, and also on the germinability of pollen.

In order to demonstrate the influence of plastids on other than strictly plastid characters, as opposed to the influence of the plasma or genome, Stubbe performed the following experiment (1960): he used strains of *suaveolens* in which the complex *flavens* is alethal. By appropriate crosses he obtained two strains of *suaveolens* (*albicans·flavens*), of which one had *parviflora* plastids that were green with this combination, the other had white sectors which contained mutated *suaveolens* plastids. He crossed these two strains and obtained some *flavens·flavens* plants that were green-white chimeras. He self-pollinated flowers on both green and white sectors, and cross pollinated the two sectors reciprocally, with the following results:

Green sectors, selfed, gave no seeds (*parviflora* plastids present in the pollen).

White sectors, selfed, gave normal seed-set (mutant *suaveolens* plastids in pollen).

Green × white sectors gave normal seed-set (mutant *suaveolens* plastids in pollen).

White × green sectors gave no seeds (*parviflora* plastids in pollen).

In every case, pollen with the mutant *suaveolens* plastids germinated and as a result seeds were produced, but pollen with *parviflora* plastids failed to function. This behavior was confirmed when pollen was germinated *in vitro*. Pollen containing the mutant *suaveolens* plastids germinated, but that with *parviflora* plastids failed to germinate or germinated rarely, producing short pollen tubes. Since the genes present were the same in all cases, and all had *suaveolens* plasma, the different behavior could only be ascribed to the plastids*. Furthermore, all nonfunctional pollen tube grains showed spherical starch grains in their plastids, whereas functional pollen grains had spindle-shaped starch.

Summarizing the plastid situation in *Oenothera*, we may emphasize the following points:

(1) There exist five different classes of chloroplasts in the North American members of the subgenus *Oenothera* each of which reacts in its own particular way to each of three classes of genome found in the population.

(2) Plastids can be transmitted to the zygote, not only through the egg, but also through the sperm; in each case, of course, in the form of proplastids.

(3) Where the plastids received from egg and sperm are genetically different, they are able to exist side by side in the cells of the young embryo, each kind retaining its own distinctive morphological and physiological characteristics, unmodified by the presence of the other.

(4) Mixed cells, in which the different classes of plastid reside side by side, are found only in early stages of the seedling. The plastids soon become segregated into different cells, giving to the plant a sectored or mottled appearance. The extent of sectoring or mottling depends on the extent to which proplastids have entered with the sperm, and especially upon the relative speeds with which the different classes of plastid multiply.

(5) Certain classes of plastid are unable to develop and function properly in the presence of certain genome-combinations. The exact way in which this disharmony expresses itself depends upon what combination of plastid and genome is present. In general, the incompatibility expresses itself by a failure of the plastids to develop normally and to retain the normal amount of pigment. In some combinations, the disharmony is exhibited only during a certain period of development. This period may begin in some cases before germination of the seed; at the other extreme it may not begin until the seedling is several weeks old. The duration of the chlorotic period is different in different combinations—in some cases it is only an evanescent phenomenon; at the other extreme, the condition is permanent. At the close of the chlorotic period, the plants tend to become green, though in some of these cases, they remain sensitive to bright light. In some combinations the chlorotic areas are com-

---

* Recent work by Göpel (1970) suggests that the sporophyte may also exert some influence on pollen behavior. Pollen carrying *flavens* or *gaudens*, coupled with *parviflora* plastids (Plastom IV) is usually incapable of germination. Certain factors, however, among them the genotype of the sporophyte, make occasional germination possible.

posed of newly formed cells only; in other cases, cells that have been green become chlorotic.

(6) *Oenothera* presents cogent evidence for the relative independence of the plastome and genome. The genome produces the chemical and physical medium in which the plastidome must survive, the plastidome provides the initial source of energy without which the genome could not function. While the genome no doubt is able on occasion to induce mutation in plastids, the evidence in *Oenothera* is strong that plastids can mutate independently of the genome, and several such mutations have been found. Ordinarily the plastid maintains its distinctive qualities, no matter what its genomic milieu. If the milieu is proper, the plastid develops and functions normally; if it is improper, the plastid is unable to develop normally but it has not altered its intrinsic characteristics thereby, and if transferred to a suitable environment shows its normal behavior once more.

(7) Plastids contain more than one genetic determiner. They are able to produce morphological effects seemingly unrelated to the plastids or to plastid function.

From these facts, Stubbe draws the conclusion that plastids play a wider role in the structure, development and activity of the plant than is ordinarily recognized. The possibility of separating out the activities of genome, plasmone and plastome which *Oenothera* rather uniquely affords, makes it possible to shed light on the position of the plastids in the overall economy of the plant and to emphasize the degree of independence which they show, comprising as they do a separate system coordinate with the genome and under normal circumstances cooperating with it, though in many cases unable to cooperate with foreign genomes. The plastids in *Oenothera* exhibit continuity from one generation to another, specificity of behavior, and genetic activity. They are therefore without doubt bearers of genetic information.

PART IV

EVOLUTIONARY CONSIDERATIONS

CHAPTER 15

Nature of de Vries' Mutants

Oenothera has been a focus of evolutionary interest from the time it was first studied until the present. It first attracted the notice of de Vries because of the light that he thought it shed on the problem of the origin of species: it is again a center of evolutionary interest because of the fact that it offers a unique opportunity to trace in some detail the steps in the evolution of a genus, and to distinguish the more important factors that have made these steps possible.

We will first discuss de Vries' findings and evolutionary concepts in the light of present cytogenetic knowledge. We will then attempt to present the story of the evolution of the North American members of the subgenus *Oenothera* as it is known at the present time.

We have recounted above de Vries' discovery of *O. lamarckiana* and its mutants, and his efforts to explain both ordinary hereditary behavior and the appearance of aberrants in terms of his theory of Intracellular Pangenesis, an attempt which culminated in the formulation of his celebrated Mutation Theory of Evolution.

As might be expected, considerable opposition developed to his interpretation of the so-called mutations which he found in *Oenothera*. For example, Heribert-Nilsson (1912) concluded that the "mutations", which he had also found in the Swedish race of *O. lamarckiana*, were merely new combinations of factors already present in the parent race, and were not, therefore, the beginnings of new species or varieties.

Lotsy (1914) concluded that the "mutants" of de Vries resulted from the fact that *lamarckiana* was a "mixture of different types", and not a pure species. His definition of a pure species, to be sure, was a very restricted one—"a species is the total of all homozygous individuals of the same genetic constitution" (p. 76). His general position with regard to mutation was that "to prove the existence of mutation, the purity of the type from which the proof starts must be beyond the possibility of doubt", and he concluded "that *O. lamarckiana* is not a pure species, but a mixture of different types" (p. 79).

Davis (1911, 1916b) and Honing (1911) took the position that de Vries' "mutations" were segregations resulting from the fact that *lamarckiana* was of hybrid origin and therefore highly heterozygous. de Vries vigorously denied that *lamarckiana* had a hybrid origin.

de Vries' concepts were also questioned by van Overeem (1922) who was for some years his student and assistant, but who completed his doctorate at the University of Zürich. van Overeem concluded from a study of various forms that Davis was correct in considering *lamarckiana* to be a hybrid, one of whose parents was a form of *biennis*.

On the other side, there were workers who supported de Vries' mutational concepts, especially Gates (e.g., 1911b, 1914a, de Vries and Gates, 1928) and Bartlett (1915c). Gates was for many years a stout supporter of de Vries' hypothesis regarding the origin of his mutants. Bartlett (see p. 35) believed that his "mass mutations" were a result of mutation in the egg cells and were not due to segregation in a highly heterozygous species. de Vries himself considered Bartlett's mass mutations an outstanding demonstration of the correctness of his mutation theory.

MacDougal and co-workers (1905, 1907) also supported de Vries' conclusions. They grew extensive cultures of de Vries' races, and found most of the de Vriesian mutants in their own cultures. They also studied material from North America and other parts of the world. They found no reason to criticize de Vries' conclusions. One of MacDougal's co-workers was Shull who later made an extensive study of mutation in *Oenothera*. Shull's work, however, dealt mostly with point mutations, which did not make it necessary for him to enter the conflict between de Vries and his critics.

The rise of modern cytogenetics after 1900 made it difficult for workers, even though they followed de Vries in ascribing aberrations to mutation, to accept an interpretation based on intracellular pangenesis, which made no provision for paired alleles and the orderly segregation of alleles in germ cell formation. de Vries, however, although he undoubtedly came to have some understanding of the mechanism of meiosis and the relation of the genes to the chromosomes, placed little emphasis on this relationship, and continued to think in terms of pangenesis, to which he made reference as late as 1925. He did not alter his ideas regarding the origin of mutations, and whenever he classified his mutants from the standpoint of how they originated he continued to use the categories of progressive, retrogressive and degressive mutations, a concept that he never gave up.

Renner's breeding experiments, as well as his own, however, led de Vries to place increased emphasis upon breeding behavior in categorizing his mutants. In line with his classification of races into isogamous and heterogamous forms (see Chapter 2) he found that his mutants could be similarly distinguished. Some mutants, such as *blandina* and *nanella*, transmit their mutant characters through both sperm and egg and so are isogamous in their behavior. Others are able to transmit their mutant characters only through the egg. Their inability to transmit these characters through the pollen he ascribed to the presence of "androlethal factors" (1923b, 1925b). These forms were therefore, heterogamous in breeding behavior, since their progenies differed when derived through sperm and egg.

Among the heterogamous mutants, de Vries recognized two main classes, which he called dimorphic (1916a) and sesquiplex (1923b) mutants. Dimorphic mutants, when selfed, yield two classes of offspring, one class being typical *lamarckiana*, the other repeating the mutant form. This is the result, according to de Vries, of the presence, in addition to normal *velutina* and *laeta* eggs and sperm, of mutated *velutina* or *laeta* eggs. Self-fertilization of normal eggs gives rise to *lamarckiana*, of mutated eggs to the mutant. Dimorphics include such forms as *scintillans*, *cana*, *pallescens*, *lactuca* and *liquida*, all with 15 chromosomes. Sesquiplex mutants on the other hand breed true. They have two kinds of egg, one carrying the mutant genome, the other carrying, in the case of mutants from *lamarckiana*, a typical *lamarckiana* genome, either *laeta* or *velutina*. They produce only one kind of sperm, which carries the same unmodified *lamarckiana* genome found in the eggs. The mutant genome is eliminated in the sperm by a so-called "androlethal factor" (1923b). On self-fertilization, the sperm are able to fertilize those eggs that contain the mutant genome, but not those with the unmodified *lamarckiana* genome, because of the presence of the same (zygotic) lethal factor. As a result, the mutant breeds true, but with 50% bad seeds. Over the years, many new mutants have been found belonging to each of the above categories. In general, dimorphic mutants possess one or more extra chromosomes: sesquiplex mutants may be either diploid or trisomic.

de Vries recognized other categories of mutations in addition to the above. One group consisted of plants that were indistinguishable from *lamarckiana* phenotypically, or nearly so, but differed from it in breeding behavior. One such form was *laevifolia*, first observed in 1886 at Hilversum, later grown from rosettes collected at Hilversum in 1905. This form differed from the original *lamarckiana* in throwing non-brittle *erythrina* mutants instead of brittle *rubrinervis*. A second form was *simplex* (de Vries, 1919c, 1923a) which differed from *lamarckiana* primarily in lacking a functional *velutina* genome; in the eggs it had a normal *laeta* plus an alethal *laeta*-like genome, and in the sperm only *laeta*. A third form was *scindens* (de Vries, 1925c), indistinguishable from *lamarckiana* in appearance, but with fewer empty seeds, and giving on selfing a *velutina*-like segregate—*tarda*. The *velutina* gamete had in this case lost its lethal. A fourth strain, *ingeminans*, (de Vries, 1929) differed from *lamarckiana* in the fact that it produced almost no empty seeds. Like *lamarckiana*, but unlike *scindens*, it bred true. In outcrosses, it gave both velutinas and laetas when used as female, and exclusively laetas when used as male. de Vries considered it to have only *laeta* pollen, but two kinds of egg— *ingeminans* (a modified *velutina*) and *laeta**. It behaved, therefore, as a sesquiplex mutant. None of these four strains has been examined cytologically. They probably had 14 chromosomes.

There was still another class of mutants which de Vries (1917b) called "half-mutant". These carried two kinds of egg and the same two kinds of sperm. One kind of genome was alethal, the other a typical unmodified *lamarckiana* genome. Since its modified genome was alethal, this could exist in homozygous condition, so that each generation in selfed line included about a third of "full mutants" which contained the modified alethal genome in double dose. Several half-mutants were found in the course of time, all with 14 chromosomes.

Finally, to these various categories of mutants should be added *gigas* (a tetraploid) and *semigigas* (a triploid).

Many of de Vries' mutants appeared over and over again in succeeding generations. Others appeared rarely, some only once. Many mutants gave rise to other aberrants, which frequently seemed identical or very similar to forms produced by *lamarckiana* itself, but in other cases were apparently new. Some mutants have produced no aberrants (e.g. the homozygous segregants *deserens*, *decipens*, etc.). Others have produced many more than *lamarckiana* itself, both in number and in kind (e.g. *semigigas*). Some mutants have appeared in the progenies of several of the other mutants, some have come from only one.

Among the heterogamous mutants, certain ones stood out in de Vries' mind as especially important, either because of the frequency of their occurrence, or because they served as relatively prolific sources of other mutations. These mutants, all of which had an extra chromosome, de Vries termed "primary mutations" (de Vries and Boedijn, 1923). All of these were dimorphic in their breeding behavior, giving rise on selfing to individuals identical with themselves and also to the original *lamarckiana*. They transmitted their mutant traits only through the egg. These primaries were *lata*, *scintillans*, *spathulata*, *cana*, *liquida* and *pallescens*. For a time, *pulla* was classed as a primary (de Vries, 1924b), but later it was considered to belong to the *pallescens* group (de Vries, 1929).

From some of these, or from other sources, a number of additional forms arose, which were called secondary mutants, and were for the most

---

* de Vries explained the small percentage of empty seeds (3–7%) as due to strong preferential fertilization of the *ingeminans* eggs by the *laeta* pollen.

part true-breeding, i.e. sesquiplex mutants. On the basis of their resemblance to certain primaries, the secondaries were placed in groups, each group clustered about a primary.

As time went on, de Vries became increasingly conscious of chromosomes, and some of his students, especially Boedijn and Stomps, began to turn their attention to cytological studies. The fact that dimorphic and sesquiplex mutants seemed to fall into seven groups (if one included *pulla*), each composed of a primary with associated secondaries, coupled with the presence of 14 chromosomes in *O. lamarckiana*, suggested to de Vries and Boedijn (1923, 1924) that each of the groups might be related to a separate chromosome. Assuming the presence of seven pairs of chromosomes in *lamarckiana* (which later proved to be erroneous), and assuming also that each mutant is dependent on factors located in a single chromosome, they assigned each of their groups to a separate chromosome pair.* One group, which was especially large, was called the "central group" and to this they assigned the factors that differentiate between *laeta* and *velutina* with their lethals, as well as the factors for all of the 14-chromosome mutants. This group was considered to correspond to Shull's first linkage group. The other six groups were much smaller, each containing one primary together with its secondaries, each situated in a "lateral" chromosome. Each chromosome of *lamarckiana* upon which a cluster of mutants was based was thought to contain one or more groups of factors called "physiological chromomeres", (de Vries, 1925a) which were in a labile condition and therefore capable of mutating to produce the mutants based on the chromosome. The groupings as published by de Vries and Boedijn (1923) are shown in Table 15.I.

Three of these groups (the *central*, *lata* and *scintillans* groups) were larger than the other four. de Vries and Boedijn claimed that this finding was in line with the figures published by Cleland (1922) for *O. franciscana* which they thought showed three larger and four smaller pairs of chromosomes, although Cleland himself was unable to find significant size differences in *franciscana*. Later (1925), Boedijn studied *O. lamarckiana* and purported to find four larger and three smaller pairs (although *lamarckiana* has only one pair), and so assigned the *cana* group to one of the larger pairs.

de Vries and Boedijn did not have the benefit of information regarding chromosome behavior in *Oenothera* when they wrote these papers. Their 1923 paper was published the same year that Cleland described for the first time the existence of large closed circles; and Boedijn's paper on

---

* Triploid and tetraploid mutants were considered to involve all the chromosomes and were not related to any one chromosome.

TABLE 15.I

Distribution of mutant characters among the chromosomes of *Oenothera lamarckiana* according to de Vries and Boedijn (1923)

Central chromosome
   (14-chromosome mutants)

nanella
forms with homogeneous pollen and ovules
   blandina
   decipiens
   deserens
   tarda
   fragilis
certain sesquiplex mutants
   simplex
   secunda
   compacta
   elongata
   favilla
   linearis

Lateral chromosomes (15-chromosome mutants)	Primary mutant	Secondary mutants
lata chromosome	*lata*	*semi-lata*
		subovata
		sublinearis
		sesquiplex mutants
		albida
		flava
		delata
scintillans chromosome	*scintillans*	*diluta*
		militaris
		venusta
		sesquiplex mutants
		oblonga
		aurita
		auricula
		nitens
		distans
cana chromosome	*cana*	*candicans*
pallescens chromosome	*pallescens*	*lactuca*
liquida chromosome	*liquida*	
spathulata chromosome	*spathulata*	

lamarckiana, in which he thought he found only paired chromosomes, came in the same year that ⊙12 and one pair was shown to be the configuration of this race (Cleland, 1925). Boedijn did not accept the finding of a large circle in *lamarckiana*, and in 1928 published a second paper in which he claimed to have confirmed his earlier finding of seven independent pairs in this form, although he admitted (p. 27) that the pairs might be able to open later and become associated variously into chains. When de Vries later accepted the presence of ⊙12, 1 pair, in *lamarckiana* (1929) he related the central chromosome to the pair, and the lateral chromosomes to the ring of 12 (pp. 179, 189).

The presence of ⊙12, 1 pair, in *lamarckiana*, and the discovery that large circles have come into being through reciprocal translocation, removed the basis upon which rested de Vries' system of classification based on the chromosomes. Hoeppener and Renner (1929) pointed out that there are not merely seven kinds of chromosomes in *lamarckiana*, but rather thirteen. Each chromosome in the circle is different from each other circle chromosome with respect to its total genic content, so that there are 12 different chromosomes in the circle, to which must be added the paired chromosomes as a thirteenth kind. If individual mutations, then, are based upon single chromosomes, there should be, not seven groups of mutations, but 13.

Although de Vries' ideas in regard to the nature of his mutants underwent some modification during his lifetime, many of his original concepts were never given up, or only slightly modified. He always maintained that *lamarckiana* is passing through a mutable period during which hereditary changes occur with more than usual frequency. He ascribed all of his aberrants to the occurrence of premutations, by which certain pangenes (later, factors, but never genes) became labile and easily convertible into visible or recessive mutations. In his last paper (1935), published posthumously, he suggested that all premutations may have arisen within a very short interval of time, in a single line, perhaps in a single individual.

His concept of a labile factor changed over the years. At first he regarded it as indistinguishable in its phenotypic expression from an unmutated factor. It could arise from a dominant by a premutation that rendered it labile. When the labile factor mutated to the recessive condition, a retrogressive mutant was produced. It could also arise by a recessive factor becoming labile, in which case it became phenotypically a dominant, capable of becoming a stable dominant or reverting to the recessive condition again. When a recessive became labile, a degressive mutation had occurred. The trisomic mutants were interpreted as degressive mutants.

Later (1935) he visualized certain premutations as conversions of dominant factors to the recessive condition (p. 250). In the case of 14-chromosome mutants these could manifest themselves if they became homozygous. In trisomic mutants, however, they could not become homozygous because androlethal factors prevented their transmission through the sperm, and they owed their initial appearance rather to non-disjunction of the chromosomes in which they lay, so that they were able, in double dose, to outweigh the dominant antagonist. In ascribing the inability of the recessive mutant factor to be transmitted through the sperm, de Vries was unaware of the fact that $n + 1$ gametophytes are unable, in *Oenothera*, to survive in the pollen although they can survive in the ovules, and this is the reason why the mutant character of a trisomic is not transmitted through the pollen, but only through the eggs. de Vries placed the androlethal factors in the central chromosome. He was somewhat uncertain as to whether an androlethal factor was present for each of the primary characters in the lateral chromosomes or whether a single one was able to suppress all of them. He rather inclined in his final paper to the latter hypothesis.

He also ascribed the appearance of triploid mutants to factors situated in the central chromosome. Such forms arise too often, he thought, to be accounted for by the chance doubling of each chromosome. He did not visualize the possibility of the formation of restitution nuclei in meiosis, but thought of the doubling of each chromosome as genetically induced.

On the whole, it may be said that de Vries was little influenced throughout his career by developments in the field of cytogenetics, even by the discoveries made in the *Oenothera* field. He was influenced by Renner to the extent that he accepted the concept of balanced lethals, and he came to admit that empty seeds represent classes of zygotes homozygous for given lethals. He never accepted, however, Renner's findings in regard to the existence and nature of complexes and of complex-heterozygosity. To de Vries, *lamarckiana* and the other *Oenothera* races were essentially pure species, homozygous for most traits, heterozygous with respect to only a few characters which he believed were the result of mutations that had occurred within the species. The extensive heterozygosity found by Renner, suggesting hybrid origins, was not accepted by de Vries.

Similarly, de Vries was little influenced by developments in *Oenothera* cytology. Although he acknowledged in 1929 the presence of a circle of chromosomes in *lamarckiana*, the full significance of this type of behavior was not realized. He continued to think of *lamarckiana* as having only seven kinds of chromosomes. His so-called central chromosome was equated with the pair, the lateral chromosomes with the circle. All 14-chromosome mutants, the zygotic lethals, and the differences between *velutina* and *laeta* were assigned to the pair. It was not until after his death that Renner (1940b, 1942a) showed that some of de Vries' point mutants (e.g. *nanella*) and the zygotic lethals were based on genes in the circle.

de Vries obtained a great variety of forms from *semigigas*, ranging in chromosome number from 15 to 20 (see Dulfer, 1926, for chromosome numbers), but regarded them all as mutants, and attempted to assign each to one of the groupings associated with his primary trisomics. Had he been more conscious of the role of chromosomes, and studied meiosis in his *semigigas*, he might have discovered that the chromosomes become associated in a great variety of ways in this form, seldom forming large chains. As a result, a great variety of gametes, from the standpoint of genic content, is produced, many of which have a full set of genes plus additional genes, and the progeny is therefore quite heterogeneous. He missed the obvious fact, therefore, that the heterogeneity in the progeny of *semigigas* is the result of segregation of the many kinds of chromosome groups formed among the germ cells of this triploid.

It should be remembered, of course, that developments in the cytogenetics of *Oenothera* came when he was close to the end of his career. He was almost 80 years of age when the circle of twelve was found in *lamarckiana*, and the cytogenetic studies that demonstrated the relationship between chromosome linkage and genetic linkage, resulting in the location of individual genes in particular chromosomes, began to be published only four years prior to his death. He did not live to see the full demonstration of the parallel between chromosome behavior and genetic behavior.

The nature of de Vries' mutants is now for the most part well understood. They do not have the profound evolutionary significance that he thought they had. They are not examples of the way in which new species or varieties come into existence. They are of interest, however, in helping us understand the cytogenetic mechanisms operative in complex-heterozygotes.

We owe our present understanding of the characteristics of these so-called mutants, and the ways in which they have come into being, to the work of many persons, but especially to Renner (1943a, 1949), Shull (1921, 1925, 1928b), Catcheside (1933, 1936) and Emerson (1935, 1936). (Much of their work was done after the death of de Vries.)

Some aberrants have resulted from point mutations which have appeared either as the result of Mendelian segregation (*bullata, vetaurea,* etc.), or of crossing-over in forms with a large circle (*brevistylis sulfurea, nanella,* etc.). In the origin of others, translocations or other structural alterations have played a role (e.g. the half-mutants, to be discussed later). By far the largest number, however, have arisen as a result of alterations in chromosome number. *Semigigas* is a triploid and *gigas* a tetraploid, but the great bulk of hyperploid mutants have been trisomics.

Since trisomics constitute by far the largest and most important group of *Oenothera* mutants, it is not surprising that considerable attention has been given to their study. Different workers have developed different terminologies in classifying them; those used by de Vries (1929), Catcheside (1936) and Renner (1949) are set out in Table 15.II. With respect to the question as to the mechanism by which they originate, this problem has received especial attention from both Emerson and Catcheside, who have come independently to conclusions that differ only in detail. Both have found the following:

Dimorphic mutants that are based on circle chromosomes arise as a result of irregularities in the zigzag arrangement in which three adjacent chromosomes go to the same pole (Fig. 15.1). This is a phenomenon often seen. It will produce in each quartet of spores two that have eight chromosomes, each of which will include one entire genome plus one chromosome

TABLE 15.II

Different types of trisomic mutants. Terms were used originally for *lamarckiana* mutants but are in part applicable to those from other sources.

de Vries (1929)	Terminology of Catcheside (1936)	Renner (1949)	Characteristics
Dimorphic	Dimorphic	Isotrisomic	The extra is one of the chromosomes of the pair.[a] (A chromosome with a certain association of ends is therefore present three times.)
Dimorphic	Dimorphic	Anisotrisomic additive	The extra is one of the circle chromosomes (one chromosome present twice.)
Sesquiplex	Monomorphic	Anisotrisomic compensated	The extra ends are from two different circle chromosomes (no chromosome present in its entirety more than once).

[a] de Vries at one time assigned the trisomic *pulla* to the pair (his "central chromosome"). Later he changed his mind. All of his trisomic mutants were consequently assigned to circle chromosomes.

from the other genome. If in the case of *lamarckiana*, for example, a *velans* complex enters a germ cell along with a single *gaudens* chromosome, and the resultant eight-chromosome germ cell unites with a *gaudens* gamete, the resultant plant will be a trisomic containing *velans* and *gaudens* plus one *gaudens* chromosome as an extra. This trisomic will owe its peculiarities to the presence of a *gaudens* chromosome in duplicate. Since there are twelve chromosomes in the circle in *lamarckiana*, twelve possible kinds of n + 1 gamete can be produced in this way minus any that are impossible because of the presence of lethals. Half of these will

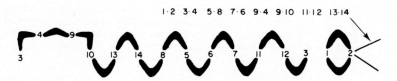

FIG. 15.1. The irregularity shown in the zigzag arrangement in *lamackiana* will produce an 8-chromosome gamete with a full *velans* complement plus 9·4 of *gaudens*.

have an extra *velans*, half an extra *gaudens* chromosome. A dimorphic derived from *lamarckiana* by self-pollination will produce four kinds of functional egg: *velans*, *gaudens*, *velans* + extra, *gaudens* + extra. Since pollen which carries an extra chromosome ordinarily fails to produce viable sperm, there will be only two kinds of sperm—*velans* and *gaudens*. Self-pollination will then produce two kinds of offspring—*velans·gaudens* (normal *lamarckiana*) or *velans·gaudens* plus the extra (the trisomic). This is then a dimorphic trisomic (Renner's "anisotrisomic additive" type). In addition to the dimorphics based on circle chromosomes, another is possible through non-disjunction involving the pair of chromosomes.

Sesquiplex (monomorphic) trisomics may be formed when two adjacent chromosomes in the circle go to the same instead of opposite poles, and somewhere else in the circle two other adjacent chromosomes go together to the same pole as the first two (Fig. 15.2). (If the two pairs of adjacent chromosomes go to opposite poles, the gametes resulting will have the normal number of seven chromosomes, but will be deficient for one segment while carrying another segment in duplicate. They will in consequence be inviable. (Fig. 15.3)). If the two sets of two go to the same pole (double non-disjunction in the same direction), there will be eight chromosomes at one pole, and six at the other. The eight-chromosome set will include a full complement of segments (some belonging to *velans*, some to *gaudens*) plus two additional segments, which will,

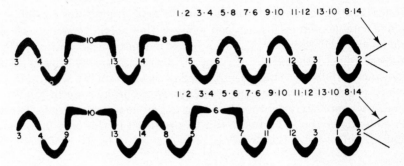

Fig. 15.2. The irregularities shown in *lamarckiana* will produce 8-chromosome gametes, viable in the eggs, that contain a full complement of ends plus two extra segments that belong to different chromosomes. The n + 1 gametes formed will contain a mixture of *velans* and *gaudens* segments, the number of segments derived from the two complexes depending upon the relative positions of the two irregularities in the chain.

however, belong to different chromosomes. Since all segments are represented in this eight-chromosome set, this will be viable in the egg; but it will not ordinarily be viable in the sperm, since extra-chromosome pollen as a rule is unable to produce viable sperm. This eight-chromosome egg will be fertilized by a normal seven-chromosome sperm. Whether it will succeed in self-fertilization with a *gaudens* sperm or a *velans* sperm will depend upon whether the eight-chromosome egg carries a *gaudens* or a *velans* zygotic lethal (since one of the extra chromosome ends may also carry one of the lethals, there is a chance that the eight-chromosome egg might have both in which case it could not produce offspring). While gametes resulting from this type of non-disjunction will always carry a mixture of *velans* and *gaudens* chromosomes, the exact mixture will depend upon where the two compensating pairs of adjacent chromosomes lie in the circle. They may be separated by one chromosome in one direction and seven in the other direction or by three and five, and these separations may be clockwise or anticlockwise with respect to a given pair of non-disjoining chromosomes. Altogether, after eliminating duplications, a total of 24 different eight-chromosome gametes can be formed by

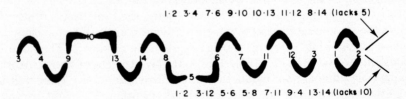

Fig. 15.3. When one pair of adjacent chromosomes goes to one pole, the other to the opposite pole, inviable gametes will result, each lacking a segment.

this mechanism from the circle of twelve in *lamarckiana*. These will differ in respect to the extra ends that are present and also in the mixture of *velans* and *gaudens* segments that they carry. The 24 different kinds of gamete are shown in Table 15.III, using the segmental arrangements that have been determined for *velans* and *gaudens*.*

$$velans = 1\cdot2 \quad 3\cdot4 \quad 5\cdot8 \quad 7\cdot6 \quad 9\cdot10 \quad 11\cdot12 \quad 13\cdot14$$
$$gaudens = 1\cdot2 \quad 3\cdot12 \quad 5\cdot6 \quad 7\cdot11 \quad 9\cdot4 \quad \quad 8\cdot14 \quad 13\cdot10$$

Trisomics produced as a result of this type of non-disjunction will breed true—they will not throw *lamarckiana* as do the dimorphic types. Suppose that the eight-chromosome genome depicted in Fig. 15.2 (above) united with normal *gaudens* to form the trisomic. During meiosis in this trisomic the chromosomes would assume the arrangement shown in Fig. 15.4 assuming that all pairing ends united (modified configurations such as a chain of 13 plus a pair, or a chain of 11 plus two pairs, etc., resulting from failure of complete pairing where three homologous ends are involved, would not change the result). Two kinds of gamete would be formed, assuming regular zigzag arrangement. One would be pure *gaudens*, the other an eight-chromosome gamete with, in this case, two *gaudens* chromosomes compensating for a missing *velans* chromosome. Both kinds of gamete would function in the egg, but only the *gaudens* would function in the sperm. Since *gaudens·gaudens* is lethal, only the eight-chromosome eggs could fuse with the *gaudens* sperm and the resultant progeny would repeat the trisomic, which therefore would breed true.

Catcheside and Emerson both agreed on the above facts, but they differed on a matter of minor importance, namely in their estimate of the total number of different trisomics that could be formed directly from *lamarckiana*. Emerson assumed the presence of a single zygotic lethal in each of the *lamarckiana* complexes. On this basis, many combinations of eight-chromosome gametes with *velans* or *gaudens* would be impossible, since both would carry the same lethal. On this supposition he calculated that the maximum number of trisomics of all kinds (dimorphic and monomorphic) that could be produced as a result of non-disjunction in the circle of twelve was 42. If we add the trisomic arising from non-disjunction of the paired chromosome, this makes a total of 43.

Catcheside on the other hand claimed that he had evidence suggesting that *velans* has no specific lethal. On this basis he calculated that non-disjunction in the circle of twelve could produce a total of 36 possible

* The system of numbering used here differs from that used by Catcheside and Renner in that the 11 and 12 ends are interchanged. Hence *gaudens* has 3·11 and 7·12 according to their system.

TABLE 15.III

The 24 different 8-chromosome gametes possible in *O. lamarckiana* as a result of double non-disjunction of circle chromosomes in the same direction. Duplicates eliminated

Type of non-disjunction	Resultant 8-chromosome gamete	Extra ends	Number of velans and gaudens circle chromosomes present in gamete
1·2 3·4-4·9 10·13-13·14 8·5 6·7 11·12 —　　　　　9·10　　14·8 5·6 7·11 12·3 1·2	→ 3·4 4·9 10·13 13·14 8·5 6·7 11·12 1·2	4, 13	5 velans
			2 gaudens
1·2 3·4-4·9 10·13 14·8-8·5 6·7 11·12 —　　　　　9·10 13·14 5·6 7·11 12·3 1·2	→ 3·4 4·9 10·13 14·8 8·5 6·7 11·12 1·2	4, 8	4 velans
			3 gaudens
1·2 3·4-4·9 10·13 14·8 5·6-6·7 11·12 —　　　　　9·10 13·14 8·5 7·11 12·3 1·2	→ 3·4 4·9 10·13 14·8 5·6 6·7 11·12 1·2	4, 6	3 velans
			4 gaudens
1·2 3·4 4·9 10·13 14·8 5·6 7·11-11·12 —　　　　9·10 13·14 8·5 6·7　　12·3 1·2	→ 3·4 4·9 10·13 14·8 5·6 7·11 11·12 1·2	4, 11	2 velans
			5 gaudens
1·2 4·9-9·10 13·14-14·8 5·6 7·11 12·3 —　　　　　10·13　　8·5 6·7 11·12 3·4 1·2	→ 4·9 9·10 13·14 14·8 5·6 7·11 12·3 1·2	9, 14	2 velans
			5 gaudens
1·2 4·9-9·10 13·14 8·5-5·6 7·11 12·3 —　　　　　10·13 14·8 6·7 11·12 3·4 1·2	→ 4·9 9·10 13·14 8·5 5·6 7·11 12·3 1·2	9, 5	3 velans
			4 gaudens
1·2 4·9-9·10 13·14 8·5 6·7-7·11 12·3 —　　　　　10·13 14·8 6·7 11·12 3·4 1—	→ 4·9 9·10 13·14 8·5 6·7 7·11 12·3 1·2	9, 7	4 velans
			3 gaudens

5 velans

2 gaudens

5 velans

2 gaudens

4 velans

3 gaudens

3 velans

4 gaudens

2 velans

5 gaudens

2 velans

5 gaudens

3 velans

4 gaudens

4 velans

3 gaudens

5 velans

2 gaudens

4 velans

3 gaudens

9, 12

10, 8

10, 6

10, 11

10, 3

13, 5

13, 7

13, 12

14, 6

14, 11

→ 4·9 9·10 13·14 8·5 6·7 11·12 12·3 1·2

→ 9·10 10·13 14·8 8·5 6·7 11·12 3·4 1·2

→ 9·10 10·13 14·8 5·6 6·7 11·12 3·4 1·2

→ 9·10 10·13 14·8 5·6 7·11 11·12 3·4 1·2

→ 9·10 10·13 14·8 5·6 7·11 12·3 3·4 1·2

→ 10·13 13·14 8·5 5·6 7·11 12·3 4·9 1·2

→ 10·13 13·14 8·5 6·7 7·11 12·3 4·9 1·2

→ 10·13 13·14 8·5 6·7 11·12 12·3 4·9 1·2

→ 13·14 14·8 5·6 6·7 11·12 3·4 9·10 1·2

→ 13·14 14·8 5·6 7·11 11·12 3·4 9·10 1·2

Table 15.III—*continued*

Type of non-disjunction	Resultant 8-chromosome gamete	Extra ends	Number of velans and gaudens circle chromosomes present in gamete
13·14-14·8 5·6 7·11 12·3-3·4 9·10 / 8·5 6·7 11·12 4·9 10·13 / 1·2 ; 1·2 5·6	→ 13·14 14·8 5·6 7·11 12·3 3·4 9·10 1·2	14, 3	3 velans
			4 gaudens
14·8-8·5 6·7-7·11 12·3 4·9 10·13 / 5·6 11·12 3·4 9·10 13·14 / 1·2 ; 1·2	→ 14·8 8·5 6·7 7·11 12·3 4·9 10·13 1·2	8, 7	2 velans
			5 gaudens
14·8-8·5 6·7 11·12-12·3 4·9 10·13 / 5·6 7·11 3·4 9·10 13·14 / 1·2 ; 1·2	→ 14·8 8·5 6·7 11·12 12·3 4·9 10·13 1·2	8, 12	3 velans
			4 gaudens
8·5-5·6 7·11-11·12 3·4 9·10 13·14 / 6·7 12·3 4·9 10·13 14·8 / 1·2 ; 1·2	→ 8·5 5·6 7·11 11·12 3·4 9·10 13·14 1·2	5, 11	5 velans
			2 gaudens
8·5-5·6 7·11 12·3-3·4 9·10 13·14 / 6·7 11·12 4·9 10·13 14·8 / 1·2 ; 1·2	→ 8·5 5·6 7·11 12·3 3·4 9·10 13·14 1·2	5, 3	4 velans
			3 gaudens
5·6-6·7 11·12-12·3 4·9 10·13 14·8 / 7·11 3·4 9·10 13·14 8·5 / 1·2 ; 1·2	→ 5·6 6·7 11·12 12·3 4·9 10·13 14·8 1·2	6, 12	2 velans
			5 gaudens
6·7-7·11 12·3-3·4 9·10 13·14 8·5 / 11·12 4·9 10·13 14·8 5·6 / 1·2 ; 1·2	→ 6·7 7·11 12·3 3·4 9·10 13·14 8·5 1·2	7, 3	5 velans
			2 gaudens

eight-chromosome gametes, each of which could combine with either *velans* or *gaudens* to produce a trisomic. This would add up to a total of 72 (dimorphic and monomorphic), plus the trisomic derived from non-disjunction of the chromosomes of the free pair.*

In any event, it is clear that a large number of different trisomics can be derived from a form with a large circle of chromosomes as a result of non-disjunction. It is not surprising, therefore, that most of the aberrants that have appeared in cultures of *lamarckiana* have proved to be trisomics.

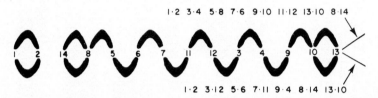

FIG. 15.4. Regular separation of adjacent chromosomes to opposite poles in a sesquiplex trisomic of *lamarckiana* will produce two kinds of gamete, one carrying an unmodified complex of *lamarckiana* (in this case *gaudens* (below)), the other an 8-chromosome gamete with two chromosomes of one complex compensating for a missing chromosome of the other (in this case, 13·10 and 8·14 of *gaudens* compensate for 13·14 of *velans*, with ends 8 and 10 present in duplicate)). Since pollen with extra chromosomes is inviable, only one class of pollen is present (in this case carrying *gaudens*). Since lethals render gaudens·gaudens inviable the only combination possible following selfing is between *gaudens* sperm and 8-chromosome eggs. Thus, the sesquiplex trisomic breeds true.

It has been found possible in a few cases to determine the extra chromosome or chromosome ends present in a given trisomic. Catcheside (1937b) found the extra in *lata* to be the 5·6 of *gaudens*. Other determinations have been made by Herzog (1940) and Renner (1940b, 1943a, 1949), some of them on de Vries' mutants, others on trisomics from other sources (Table 15.IV). It is interesting to note that one trisomic studied by Renner (1943a) involved the paired chromosomes of *lamarckiana*. This did not arise directly from *lamarckiana* but appeared in the progeny of the cross *O. lamarckiana subcruciata* × *biennis*. Among the *velans·rubens* individuals were four trisomic individuals, all of which had ⊙12 plus a group of three chromosomes. Since *rubens* has the same segmental arrangement as *gaudens*, *velans·rubens* would have the same circle of twelve, and the pair would be made up of the 1·2 chromosomes. Since eight-chromosome complexes do not come through the sperm, the non-disjunction in this case involved the 1·2 chromosomes in the *lamarckiana* (female) parent.

* Ford (1936) estimated the number of possible trisomics resulting from selfing of a plant with ⊙14 as 98. He ignored, however, the presence of lethals.

TABLE 15.IV

Trisomics whose extra chromosomes or chromosome ends have been identified

Trisomic	Source	Classification	Extra chromosome or chromosome ends	Reference
tripus	O. lamarckiana subcruciata × biennis	Dimorphic (isotrisomic)	1·2	Renner 1943a, 1949
cana	mutant of de Vries (= pallescens of de Vries) from lamarckiana and M-lamarckiana	Dimorphic	3·4 (of velans)	Herzog 1940
dependens		Dimorphic	3·11[a] (of gaudens)	Renner 1943a
incana	biennis	Dimorphic	9·4 (of rubens)	Renner 1949
lata	Mutant of de Vries	Dimorphic	5·6 (of gaudens)	Catcheside 1937b
macilenta	From M-lamarckiana via dependens and cana	Dimorphic	3·2 (of M-gaudens)[b]	Renner 1943a, 1949
scintillans	Mutant of de Vries	Dimorphic	13·10 (of gaudens)	Renner 1943a, 1949
candicans (mm cana)	From M-lamarckiana cana	Sesquiplex (monomorphic)	1 and 4 (of velans)	Herzog 1940, Renner 1949
albilaeta glossa	From albicans·velans scintillans × albicans·gaudens and other sources	Sesquiplex (monomorphic)	2 and 3 (of velans)	Renner 1943a, 1949
lonche	From (biennis × hookeri) albicans·[h]hookeri × s	Sesquiplex (monomorphic)	3 and 13 (of [h]hookeri)	Renner 1949
mira (MM dependens)	From M-lamarckiana and M-lamarckiana dependens	Sesquiplex (monomorphic)	3 and 11[a] (of gaudens) (3 derived originally from flectens)	Renner 1949

[a] Renner followed Catcheside's system in this paper. According to Cleland's system 12 should be substituted for 11.

[b] M-gaudens has 1·11 3·2 (1·12 3·2 of Cleland) instead of 1·2 3·11 (3·12). The 3·2 chromosome was derived from flectens of atrovirens. M-lamarckiana has ⊙14, instead of the usual ⊙12, 1 pair.

The trisomic then had three 1·2 chromosomes, two of them from *lamarcki-ana*, the third from *rubens*. The *lamarckiana* parent in this case had white midribs, so had the genes *rr*. The *rubens* parent had red midribs, and the 1·2 chromosome that it contributed happened to have *R* which dominated over the *rr*.* Renner called this trisomic *tripus* (1943b). It is the nearest approach to a truly isotrisomic plant that has yet been found. Incidentally, it is interesting that the extra segments have been deter-mined in five of the primary mutants of de Vries.

Although de Vries did not fully understand the reason for the presence of only one kind of sperm when two kinds of egg were present in ses-quiplex mutants, he correctly analyzed the behavior of sesquiplex mutants in concluding that they have but one kind of functional sperm, but two kinds of egg, one of which has an extra chromosome, the other being identical with the kind present in the sperm.

There is one type of mutant found by de Vries that is not yet fully understood, namely, the so-called half-mutant (see p. 194). It is not easy to visualize the process by which these 14-chromosome deviants with unbalanced lethals, which produce in each generation homozygous alethal segregants, have been formed. Several half-mutants have been examined with respect to chromosome configuration and a number of them have the same configuration—⊙6, 4 pairs. Furthermore, the segmental arrangements in three of these, *rubrinervis*, *erythrina* and *rubrisepala*, are identical. These three have a segmentally unaltered *velans* coupled with a lethal-free complex that has a mixture of *velans* and *gaudens* chromosomes (5·6 from *gaudens*, 3·4 9·10 11·12 from *velans*, 1·2 which is present in both *velans* and *gaudens*, and two interchange chromosomes, 7·14 and 13·8). This identity in segmental arrangement is striking, especially in view of the fact that these half-mutants have arisen independently on many different occasions. *Rubrinervis* is one of de Vries' earliest mutants and is reported (de Vries, 1919a) to have arisen 66 times from *lamarckiana* between 1890 and 1900. It has also appeared in the progeny of one of the trisomics (*oblonga*). *Erythrina* is another of the de Vriesian mutants. It arose originally in a line descended from rosettes collected at Hilversum in 1905 to which de Vries gave the name *O. similis* (1925c). It has arisen all-told at least nine times in five different families (de Vries, 1919b). None of these families has produced *rubrinervis*. *Rubrisepala* is still another half-mutant that has arisen several times in Heribert-Nilsson's Swedish *lamarckiana* (Håkansson, 1930a, p. 393). All of these forms, so far as they have been examined, have had segmentally identical complexes and the same chromosome con-figuration. They have undoubtedly arisen by the same mechanism on

* *R* is lethal in homozygous condition, so the *biennis* parent had to be *Rr*.

many different occasions. But what is this mechanism? Why has the same process occurred over and over again?

The only suggestions that have been made in the literature with regard to mechanisms will not explain why one particular process has occurred more frequently than others which are seemingly just as likely to occur. Darlington (1931) suggested that the alethal complexes of half-mutants arise through crossing over between homologous segments in otherwise non-homologous chromosomes. However, it would not be possible for the two exchange chromosomes in, for example, ^h*decipiens*, the alethal complex of *erythrina*, to be formed from the complexes of *lamarckiana* by a single interchange, although an appropriate sequence of two interchanges could produce this arrangement. This was pointed out by Catcheside in 1940.

Renner suggested (1943a) that interlocking of chromosomes might furnish the required mechanism for the production of the alethal complexes of half-mutants. It is possible that the chromosomes might lie in such a position prior to synapsis that when synapsis occurred, three chromosomes might find themselves interlocked (Fig. 15.5). As the chromosomes shortened, the stresses set up at the points of interlocking might bring about breakage and reunion, thus giving in effect a double interchange. In this way the segmental arrangement of the alethal complex of a half-mutant might be formed from the complexes of *lamarckiana* by a single event.

The problem still remains, however, as to why the same chromosomes should be involved time after time in the formation of the alethal complexes of half-mutants, especially when such a process involves what is

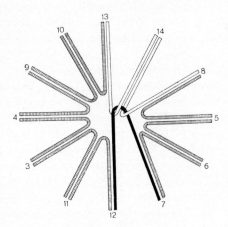

Fig. 15.5. Renner's diagram (1943a) illustrating a possible mechanism by which half-mutants can be produced by a single event. Explanation in the text.

in effect a double interchange. The lethals probably make some double interchanges impossible, and it may be that many theoretically possible exchanges would result in inviable germ cells. This may account for some of the empty pollen grains that characterize races with a large circle.

The same problem is encountered in the case of other half-mutants besides the three referred to. *Rubricalyx* is a half-mutant that has $\odot$ 8, 3 pairs (Cleland, 1925). Its alethal complex (*latifrons*) differs segmentally from ^{h}*decipiens*, etc., in having 7·10 9·13 8·14 instead of 7·14 9·10 13·8 (Catcheside, 1940). In this case also it is impossible to derive its segmental arrangement from the complexes of *lamarckiana* by a single interchange. The same is true of ^{h}*blandina*, which has 7·10 9·14 13·8 (Catcheside, 1940). This alethal complex belongs to a homozygous race that arose according to de Vries as a segregant from a half-mutant which was not recognized as such when it first appeared and which was later called *problandina* (de Vries, 1917a, 1923c). The half-mutant differed phenotypically from its segregant only in having about 25% empty seeds.

All of the above-mentioned mutants have, in addition to the modified alethal complex, an essentially unmodified *velans*. There is no theoretical reason why the alethal complex should not be able to combine with *gaudens* as easily as with *velans*, but because *gaudens* tends to dominate phenotypically, it is probable that such half-mutants, when they have appeared, have usually gone unnoticed. de Vries, however, found a *lamarckiana*-like plant that behaved like a half-mutant which he called *lucida*, and Renner obtained on two occasions the same half-mutant in progenies of his *O. M-lamarckiana* (1943a, pp. 59, 95). This form had the complexes *gaudens* and *subvelans* (= ^{h}*decipiens*), also called *minians*, and produced in the following generation a class of plants homozygous for the latter.

In the case of the so-called half-mutants, therefore, it must be admitted that the precise mechanism responsible for their production has not as yet been discovered.

In conclusion, it is now clear that some of de Vries' mutants were point mutations, the great majority were plants with altered chromosome numbers, and in a few cases, exchange of segments among the circle chromosomes has been involved. In no case can it be said that the mutants from *lamarckiana* represent in any sense incipient species or varieties, nor do they furnish a significant clue as to how new taxa originate in nature.

CHAPTER 16

Induced Mutations

Although recent work has shown that most of de Vries' "mutants" were not really the result of mutation in the strict sense, this does not mean that mutation does not occur in *Oenothera*. Point mutations have arisen in the cultures of several *Oenothera* workers, including those of de Vries, and the widespread variation observed in the *Oenothera* population is undoubtedly to be attributed in large part to such occurrences in Nature.

Attempts have been made by a number of workers to induce mutations in *Oenothera*, with some degree of success. The first of these was by Michaelis (1930) who placed plants of *hookeri* in the cold, and used pollen derived from cells undergoing meiosis during this period to pollinate untreated plants. He obtained trisomics as well as $2n - 1$ and $2n + 2$ plants, but no gene mutations. The first attempt to induce mutation by irradiation was made by Brittingham (1931). He used unfiltered radon tubes on buds of *lamarckiana* and *franciscana* and obtained many aberrant progeny which in most cases died in the rosette stage. In one case he succeeded in selfing an aberrant and obtained abnormal F_2 plants. He did not, however, study the chromosomes, or follow this work up. Another attempt at irradiation was made by Catcheside (1935, 1937a, 1939) who used X-rays on *blandina*. He irradiated pollen and placed it on untreated plants. In the F_1 he obtained some phenotypic aberrants with 50% bad pollen, some of them with ⊙4 (*blandina* has seven pairs), and also some plants with normal phenotype but with 50% bad pollen and in some cases changed chromosome configurations (five plants with ⊙4, one with ⊙6, one with two circles of four). One of the plants with ⊙4 showed the position affect described above (Chapter 11). Most of the interchange hybrids transmitted only non-interchange gametes to the F_2 showing that the interchange complements were defective. Several of the apparently normal F_1 plants gave rise to a proportion of morphologically crippled individuals in F_2.

In 1935, Rudloff and Stubbe reported on experiments in which they applied X-irradiated pollen of *hookeri* to untreated plants of the same race. They obtained many abnormal F_1 and F_2 plants which they described in detail and which in many cases they considered to be gene mutations, the first to be found in this strictly homozygous race. They made no cytological studies, but the absence of bad pollen suggested that the aberrants were not due to alteration in chromosomal structure or number.

Marquardt (1948) repeated this work on *hookeri* and obtained two plants which had ⊙4, ⊙4, 3 pairs. These plants had greatly reduced seed fertility, probably related to the fact that chiasmata frequently failed to form between circle chromosomes, resulting in abnormalities in disjunction.

Beginning in 1935, Oehlkers (1935b) and his students showed that other agents can induce chromosomal abnormalities. In that year, Oehlkers found that the ability of ring chromosomes to remain attached depended greatly upon external conditions. Using the hybrid *flavens·ʰhookeri* and its reciprocal (⊙4, 5 pairs) he found that under constant field conditions and constant temperature, the percentage of cases where chromosomes failed to remain attached was quite constant, but temperature changes, either to low or to high levels resulted in an increase in the number of such failures. Furthermore, the percentage differed from season to season or from field to field. Both temperature and nutritive conditions, therefore, affected the frequency of chiasma formation, and hence the ability of the circle to remain intact. He also found that the cytoplasm had some effect, since chiasma frequency differed in reciprocals, even when they were of the same genetic composition. The more extreme the environmental conditions, the greater the difference between reciprocals.

In later papers Oehlkers and his students showed that chiasma frequency is influenced also by osmotic relations and by the relative amounts of chlorophyll and carotinoids in the plastids (Oehlkers, 1936; Haselwarter, 1937; Kisch, 1937; Zürn, 1937a, b).

These experiments primarily affected chiasma frequency with the possible consequence of non-disjunction and trisomic formation. Beginning in 1943, however, Oehlkers (Oehlkers, 1943, 1949; Oehlkers and Linnert, 1949, 1951) found that certain chemicals were able to bring about breakage of the chromosomes, resulting in translocations and changed chromosome configurations. Again using *flavens·ʰhookeri* (⊙4), he found that KCl resulted in 2·5% of the pollen mother cells with altered configurations, and KNO$_3$ gave 4·5%. Certain organic compounds also were effective, but the most striking results were obtained with combinations of inorganic and organic compounds, especially ethyl urethane (1/20 M) plus KCl (1/200 M), which produced 38% of cells with altered chromosome configurations. He obtained the following new configurations:

Configurations representing but one translocation: ⊙6
                                                       ⊙4, ⊙4

Configurations representing two translocations: ⊙4, ⊙6, chain
                                                         of 8
                                                      ⊙4, ⊙4, chain
                                                         of 4

Configurations representing three or four translocations:

    chain of 10 (= ⊙10, 2 pairs)
    chains of 4 and 8 (= ⊙12, 1 pair, or ⊙4, ⊙8, 1 pair)

These results indicate that the chemicals used had a random and non-specific effect on the chromosomes. Oehlkers and Linnert (1951), however, found that cytoplasm also may influence the magnitude of the effect when urethane is used. The effect of urethane and KCl was twice as powerful when *flavens* and *hhookeri* were in *suaveolens* plasma as when they were in *hookeri* plasma (23% with altered configurations *vs* 10%).

In 1949 Oehlkers reported that other reagents were also effective. Certain narcotics such as acetophenone, glycol and acetanilid gave high percentages of translocation (28–36% of the cells). Alkaloids such as morphine (14–30%) and colchicine (21%) also had powerful effects. Only in the case of urethane, however, was a cytoplasmic effect obtained. In all cases, the effects seemed to be non-specific. They were also relatively independent of concentration and length of exposure. The effect seemed to be the indirect result of general cellular disturbances. The metabolic stage of the nucleus appeared to be the most susceptible, possibly including early prophase. In general the effects were similar to those obtained by Marquardt with X-rays. Some of the photographs of altered configurations that Oehlkers presented (1943) can be interpreted in more than one way. Nevertheless, it is clear that certain chemicals do have strong effects on chromosome structure and behavior, which vary greatly depending upon the chemical used. The chemical mechanism by which chromosomal alterations were induced was not investigated by Oehlkers. Freese (1967), however, has produced evidence in the case of urethane that this substance does not act directly in inducing alterations but is first converted enzymatically to hydroxyurethane, which by oxidation gives rise to H_2O_2 and radicals which alter the structure of DNA. In 1962, Oehlkers and Bergfeld-Gaertner reported on the effects of certain radioisotopes (P^{32} and S^{35}), both of which induced a significant amount of chromosome fragmentation, reciprocal translocation and alteration in chromosome number. Gene mutations affecting phenotypical characters were not mentioned.

In a discussion of induced mutations, mention should be made again of the work of Lewis (see p. 162) who tried to induce mutation of the S-locus, the self-incompatibility factor in *O. organensis*. It will be recalled that in no case did Lewis succeed in causing one S-factor to mutate into another functional allele. Almost half the mutations (as detected by the growth of pollen tubes in selfed styles and setting of seed) reverted in the next generation to their original condition. The rest resulted in permanent loss of function in the sperm without loss of stylar function, i.e., pollen containing the mutated allele could function in selfed styles, but styles carrying this allele prevented pollen with the unmutated allele from growing.

In conclusion, we may say that comparatively little attempt has been made to induce mutations in *Oenothera*, and what has been attempted has been more effective in reducing chiasma frequency and inducing translocations than in bringing about other forms of mutation.

CHAPTER 17

The Origin of *O. lamarckiana*

The study of evolution involves the problem of origins, and one of the moot questions for a number of years involved the origin of *lamarckiana* itself—the race upon which the burden of the Mutation Theory of Evolution largely rested. de Vries claimed that this was a pure species, derived from America. Davis was the leading opponent of this point of view, arguing that *lamarckiana* was of hybrid origin, and had originated in Europe by a cross between different races of American origin that had been introduced into Europe and had spread until they overlapped.

According to de Vries (1905a), the history of his race was as follows: His original material, which he found growing at Hilversum, was an escape from a nearby garden, where the owner had planted seeds obtained from the seedsman Ernst Benary of Erfurt. This material came originally from Carter and Company, London seedsmen, who claimed that it had been received from Texas. de Vries identified this material as *O. lamarckiana Ser.* Lamarck had grown material in the garden of the Muséum d'Histoire Naturelle in Paris which he identified as *Oenothera grandiflora*. Later, Seringe concluded that this particular material was not typical of *O. grandiflora* and renamed it *O. lamarckiana*. de Vries concluded that his material conformed to the description of the Paris specimen and adopted Seringe's name for his own material. He visited the Muséum d'Histoire Naturelle in 1895 and again in 1913, and examined Lamarck's specimens (1914a, b). He found two specimens of the renamed *O. lamarckiana* which he called specimens A and B (Figs 17.1, 17.2). The two specimens were loose on their sheets when de Vries examined them in 1895 and both bore the number 12, referring to no. 12, *O. grandiflora* in Lamarck's *Encyclopédie méthodique, Botanique*. Later they were fastened to new sheets and the numbers were lost. The two specimens were somewhat different, and it was a question as to which specimen should be regarded as the type of *O. lamarckiana Ser.* de Vries concluded that specimen A corresponded more closely to Lamarck's description in his *Encyclopédie*. Furthermore, the label on specimen B made reference to its fragrant flowers, which is a characteristic of *grandiflora* but not of *lamarckiana*. de Vries concluded, therefore, that specimen A represented the form of Lamarck's *grandiflora* which Seringe had renamed *O. lamarckiana*.

Not only was this true, but de Vries also concluded that specimen A

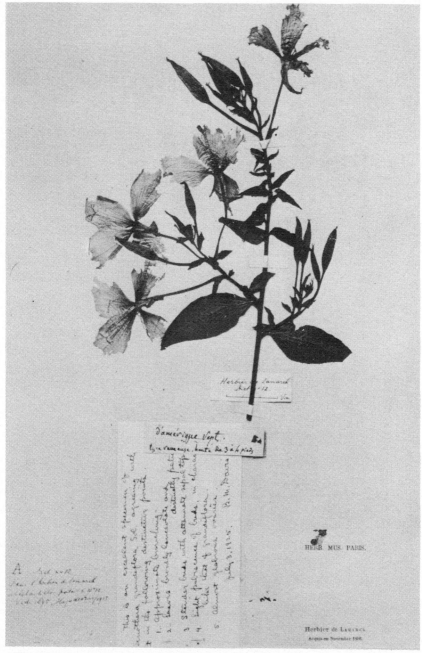

FIG. 17.1. Photograph of de Vries' specimen A, made by B. M. Davis in 1925. Compare with de Vries' photograph (de Vries, 1914b, plate XVII) in which a specimen from his own garden is placed beside the original specimen from Lamarck's herbarium for comparison. Proceedings of the American Philosophical Society by permission.

FIG. 17.2. de Vries' specimen B. Photograph by B. M. Davis in 1925. Proceedings of the American Philosophical Society by permission.

corresponded in every particular to his own experimental material. In 1913, he placed a tip from a pressed plant derived from his own garden on the sheet containing specimen A and then photographed the sheet (de Vries, 1914b, plate xvii), in order to show the close resemblance between his material and the preserved specimen.

Since Lamarck's description indicated that the material had come originally from North America, and since specimen A carried a notation in Lamarck's handwriting to this effect, de Vries concluded that his *lamarckiana* was a true species, introduced at some time from North America. At first he accepted Carter and Company's statement that seeds were derived from Texas.

de Vries' conclusion, however, was called in question by Davis (1912b). Believing that *lamarckiana* was a heterozygote and the so-called mutations nothing but segregants of the sort that a heterozygote might be expected to throw, Davis concluded that de Vries was incorrect in identifying his material with the Paris specimens. He accepted Miss Alice Eastwood's conclusion, after she had examined these specimens, that the sheet which de Vries designated as specimen B was the type of *O. lamarckiana*, and differed in no respect from *O. grandiflora* Solander, which meant that de Vries' material was not *lamarckiana*. Davis pointed to the fact that nothing resembling the *lamarckiana* of de Vries had been found in the Texas area, or anywhere else in North America, but that *grandiflora*, which had been discovered by William Bartram in Alabama, had been introduced about 1778 into Europe by Fothergill who had sponsored Bartram's expedition, so that it and *biennis* had had abundant opportunity to hybridize in Europe. He called attention to the fact that the sand banks along the coast north of Liverpool supported numerous colonies of large-flowered *Oenothera* and had done so as early as 1806 (*Smith's English Botany* **22**, 1534). In two papers (1911, 1912a) Davis called attention to certain hybrids between *biennis* and *grandiflora* that he had made, which he considered to resemble *lamarckiana*, and suggested that the latter had arisen as a result of a cross between *grandiflora* and *biennis*.

de Vries, of course, responded vigorously to the suggestion that *lamarckiana* was a hybrid and not a true species (1914b). He pointed to the fact that Davis (who up to that time had not personally seen the specimens in Paris) had accepted as the type of *lamarckiana* the wrong specimen, the one that he himself considered to represent *grandiflora* and not *lamarckiana*. He also visited, in company with H. H. Bartlett (see Fig. 17.3), the locality (Dixie Landing, Alabama) where Bartram had found *grandiflora* and where Davis had obtained his own material of this species (de Vries and Bartlett, 1912). He found the area to contain a mixture of *grandiflora*, *biennis* and intermediate forms and concluded that Davis, who considered *lamarckiana* to be hybrid in origin, could not be sure of the purity of his own material, which he was crossing with *biennis* in the hope of synthesizing a *lamarckiana*-like hybrid. He accepted, however, Davis' suggestion that his *lamarckiana* probably

Fig. 17. . Hugo de Vries and H. H. Bartlett, during a visit to Dixie Landing, Ala. in 1912. Kindness of Professor E. E. Steiner.

came from England. He agreed that Carter's statement as to its Texas origin might have been a mistake. He stated (1914b, p. 358)—"it is well known that such details are, as a rule, given more in the interest of advertising than in that of pure science. However, no horticulturist likes to offer for sale seeds with the announcement that the same form may be found as a wild flower in his own country."

In 1925 Davis had the opportunity of visiting Paris and spent several days at the Muséum d'Histoire Naturelle. He examined the sheets which contained de Vries' specimens A and B and came to the conclusion (1926, 1927) that they were both *O. grandiflora* Solander. A photograph of specimen A was published by de Vries (1914b, plate xvii) (see Fig.

17.1)* and one of specimen B is found in an article by Davis (1912b, plate 37) (see Fig. 17.2). Examination of these photographs tends to support Davis' conclusion regarding specimen B, but his contention that specimen A is also *grandiflora* is doubtful since the buds are broader and the sepal tips somewhat shorter than in the first specimen. Davis also agreed that this specimen was more puberulent than the other, both on the stems and the sepal tips, although he found hairs less abundant than in the *lamarckiana* of de Vries. Neither specimen, unfortunately, had mature capsules. Davis also examined other specimens in the Paris herbarium that de Vries had identified as *lamarckiana*—a specimen of the Abbé Pourret and one of Michaux among others (plates xviii, xix in de Vries 1914b; Figs 17.4, 17.5). The former he found to be *grandiflora*, the latter of uncertain affinity but not the *lamarckiana* of de Vries. He concluded in general that the Muséum d'Histoire Naturelle had no specimens of de Vries' *lamarckiana* of earlier date than 1860, that the flora in England prior to about 1860 was composed mainly of *biennis*, and that *lamarckiana* had not appeared until that date or thereabouts, the time when Carter and Company introduced it to the trade.

In order to test his hypothesis that *lamarckiana* had arisen by hybridization, Davis made crosses between species that he thought might be its parents, hoping to produce a hybrid that would have characteristics approaching those of *lamarckiana*. His first attempts, as we have said, involved *biennis* and *grandiflora* (1911, 1912a). His *biennis*, however, was not de Vries' Dutch *biennis* but came from two rosettes growing wild at Woods Hole, Massachusetts. His *grandiflora* came from the Dixie Landing area and proved to be rather heterogeneous. He selected certain strains of *grandiflora* and crossed them with his *biennis*. From the hybrids obtained he selected those nearest in appearance to *lamarckiana* and tested them from the standpoint of breeding behavior. He found that some tended to breed true, but with a certain proportion of aberrants which could be considered the equivalent of the "mutants" from *lamarckiana*. He concluded, therefore, that these results gave support to the hypothesis that *lamarckiana* arose through hybridization between a *biennis* and a *grandiflora*.

Davis soon found other hybrid material, however, that seemed to come closer to *lamarckiana* than anything he had previously produced (1916a, b; 1924). This came from a cross between the Dutch *biennis* and a *hookeri* from California, named by Bartlett *franciscana*. The cross *franciscana* B × *biennis* produced uniform F_1 progeny except for a

---

* Fig. 17.1 lacks the tip from de Vries' 1913 culture of *lamarckiana* shown in de Vries' plate. Our figure is from a photograph made by Davis in 1925.

FIG. 17.4. Photograph of specimen of the Abbé Pourret considered by de Vries (1914b, plate XVIII) to be *O. lamarckiana* Ser. Photograph by B. M. Davis in 1925. Proceedings of the American Philosophical Society by permission.

FIG. 17.5. A specimen from the Michaux herbarium, photographed by Davis in 1925. This specimen is included in plate XIX of de Vries (1914b) and considered by him to be *O. lamarckiana* Ser. Proceedings of the American Philosophical Society by permission.

varying tendency toward chlorosis. In the F_2 and subsequent generations, however, he obtained two main classes of plants, one *franciscana*-like, the other which he considered to be *lamarckiana*-like. There was considerable variation within these classes; one F_2 plant was especially close to *lamarckiana* phenotypically and became the progenitor of a selfed line to which Davis gave the name *neo-lamarckiana*. This line was carried for ten generations, each generation splitting into about half *franciscana*-like plants, a quarter *neo-lamarckiana*, and a quarter showing variation on these two themes. Davis concluded that *neo-lamarckiana* is like *lamarckiana* (1) in being an impure species of hybrid origin, showing how impure species may arise; (2) in maintaining a constant heterozygosity and the same genetic constitution; (3) in throwing in each generation variants of certain types, most of them in small numbers. It differed in its behavior from *lamarckiana* in throwing one class of segregants in large numbers (the *franciscana*-like form, appearing in up to 50% of each generation).

Phenotypically, *neo-lamarckiana* was not identical with *lamarckiana*, but resembled it closely. With many strains of *biennis* and of *hookeri*-like plants existing in nature, one would not expect to hit on the precise ones that crossed to yield *lamarckiana*. The resemblance, however, was quite close, as shown in parallel columns in his 1924 paper summarizing the results of these experiments.

By this work Davis believed that he had demonstrated the reasonableness of the hypothesis that *lamarckiana* is of hybrid origin, probably between a *biennis* and a *hookeri*, and consequently one must explain the so-called mutants of de Vries as segregants of one sort or another and not as evidence of rampant mutability giving rise to new forms.

This work was largely done before knowledge of Renner's genetic analyses became available, and before the discovery of chromosome rings. It is interesting, therefore, to realize that later work has proved Davis' ideas to be quite close to reality. We can now explain the behavior of his *neo-lamarckiana* in terms of Renner complexes. *Neo-lamarckiana* had the composition ^{h}*franciscana* $B \cdot rubens$: ^{h}*franciscana* is alethal and hence can appear in homozygous condition in each generation. The chromosome configuration of *neo-lamarckiana* was no doubt $\odot 10$, 2 pairs, that of the *franciscana* segregant 7 pairs. The presence of two pairs independent of the circle in *neo-lamarckiana* gave the opportunity for some independent segregation which was responsible for some of the variants obtained (the segregating pairs were $1 \cdot 2$ and $5 \cdot 6$). Finally, it now appears that Davis was not far off in choosing *biennis* and *hookeri* as the putative parents of *lamarckiana*. We now know that *rubens* is a complex of the beta *biennis II* (see Chapter 18) type, the *biennis II* races being characteristic of northeastern North America. While we have not found in nature a complex with the exact *rubens* segmental arrangement, we have found beta *biennis II* complexes only one interchange removed from it. The other complex, *velans*, suggests *hookeri* ancestry, as shown by its phenotypic effect, its segmental arrangement and by its reaction, in combination with other complexes, to the different classes of plastid. Its reaction to plastid classes might suggest that it has been derived from *strigosa*, subsp. *strigosa*, but the large, open-pollinated flowers, for which *lamarckiana* has become homozygous would seem to indicate a *hookeri*, rather than a *strigosa* origin.

What, then, shall we say regarding the probable origin of *lamarckiana*? First, no form remotely resembling *lamarckiana* has been found growing wild in North America. A few escapes have been found close to centers of population (*lamarckiana* is frequently grown as an ornamental), but there is no indication of its presence in truly wild situations, this in spite of the fact that oenotheras have been collected, often extensively, in most parts of the continent. Secondly, the presence in *lamarckiana* of one complex characteristic of the eastern seaboard of the United States and the other possibly characteristic of the west coast, coupled with the abundance for many years of *lamarckiana*-like forms on the west coast of England, north of Liverpool, suggests most strongly that it was on the ballast heaps of a European port—perhaps Liverpool—that the ancestors of *lamarckiana* met, crossed and produced *lamarckiana*. There can be little doubt that *lamarckiana* is of hybrid origin, that it was synthesized in Europe as the result of crosses between *biennis* and a *hookeri*-like form, and that de Vries' concept of its being a "pure species" was therefore erroneous.

O. lamarckiana has been a popular garden plant ever since its introduction into the trade by Carter and Company in 1860. It is now grown all over the world. I have picked it up in Japan and have obtained other races from Iran, and from Ft. Lewis, Washington. I have tested these races cytologically and genetically. They all have $\odot 12$, and their complexes have the typical segmental arrangements of *velans* and *gaudens*.

Incidentally, this tells us something about the frequency of successful reciprocal translocations in *Oenothera*. For many years *lamarckiana* has been grown in many parts of the world, yet its chromosome configuration has been maintained and its complexes have suffered no alteration in segmental arrangement. We will come back to this point later but it suggests that, although successful reciprocal translocations have been, relative to other organisms, enormously frequent, the actual frequency in terms of years and centuries has been low. In 40 years of growing oenotheras, often by the thousands, I have detected only one interchange in my garden (see p. 123).

It should now be added as a postscript that according to many taxonomists the *lamarckiana* studied by de Vries and Renner is a strain of *O. erythrosepala* Borb. (see Clapham, Tutin and Warburg, 1952; Munz, 1965), although Borbás (1903) in his flora of Hungary, stated that *O. lamarckiana* was unknown in that country, and a later flora of Hungary by Jávorka (1925) subordinates *erythrosepala* to *O. suaveolens*, under which he also includes *lamarckiana* (see Renner, 1956, p. 240, and Rossmann, 1963a, p. 452).

Since the name *O. lamarckiana* has become so firmly established in the genetical literature, however, we have chosen to retain its use in the present treatise. It would be a mistake to substitute for it the term *erythrosepala*, since de Vries, Renner and others have worked on only a few selected strains, and these should be clearly identified. There is no assurance that other strains of *erythrosepala* have the same complexes, the same chromosome configuration or the same segmental arrangements as those found in de Vries' and Renner's *lamarckiana*.* Whenever additional strains are found that do have the same configurations and arrangements, as I have found in the case of material from Teheran, Tokyo and Ft. Lewis (Washington), these should also be called *lamarckiana* since they are identical in their essential cytogenetic features with the classical strains to which this name has been applied.

---

* Material of *erythrosepala* recently sent me from Poland by Dr. Rostanski is identical phenotypically with *lamarckiana* and has the expected ◯12, 1 pair. I have not as yet had the opportunity of analyzing its complexes cytologically, but am confident that they will prove to have the required segmental arrangements.

CHAPTER 18

The Evolution of the North American Oenotheras (Subgenus *Oenothera*)

Although the reality of the evolutionary process is no longer questioned by biologists, the number of cases in which the story of evolution has been analyzed in detail is small. In some cases major evolutionary changes can be traced, as in the horse or elephant, but the factors that have caused these transformations are either unknown or matters of speculation. In the evening primrose, however, we have an organism in which both the evolutionary steps and their causal factors are becoming recognized.*

In order to understand how the North American members of the subgenus *Oenothera* have come into being, we must (1) first understand the taxonomic situation as it exists today. (2) We can then outline the technique by which it has been possible to analyze the various steps in the evolutionary development of the group. (3) We will then be in position to trace the history of the group. (4) Finally we will be able to discuss the factors that were most important in bringing about this evolution, and show how these have interacted to produce the results that we now observe.

1. POPULATION STRUCTURE IN THE SUBGENUS *Oenothera*

This is a group that has long presented difficulties to the systematist. On the one hand, the group presents an almost endless array of phenotypic variations. For example, we have grown hundreds of lines in our garden, derived from seed collected from many localities across the continent, and have never found strains from different localities, and rarely from nearby localities, that were identical in appearance. A single locality may contain several phenotypically diverse strains. These variations often grade into one another so gradually, however, that it is difficult or impossible to find clear-cut lines of separation, and so to be able to distinguish one taxon from another. Because of this fact, taxonomists have been inclined to recognize relatively few species, although realizing that the situation is much more complex than their classification shows. Typical of the reaction of taxonomists is the statement of

---

* For shorter reviews, see Cleland 1936, 1949b, 1957, 1958, 1960a, 1964.

Deam in his "Flora of Indiana" (1940, p. 705) who recognized four species that would be placed in the subgenus *Oenothera* but stated in connection with his *O. pycnocarpa*, "The status of this and the next three species is not yet definitely determined. Some authors regard them simply as varieties of *O. biennis* but I am regarding them as species as did the authors who described them. The plants are exceedingly variable and only an expert can name them with any degree of certainty. I have a large number of specimens which I am not including in this treatment because I cannot satisfactorily name them."

Also typical of the confusion that has existed among taxonomists is the statement contained in a footnote on page 1064 of Gray's Manual of Botany, 8th edition (1950): "A hopelessly confused and freely hybridizing group early introduced into Europe. . . The types of several species, described from European material, not wholly clarified; and further confusion added by the publication of many scores of 'mutants' or 'elementary species' as true 'species.' (Compare other strongly apomictic groups of similar behavior such as *Rubus*, subgenus *Eubatus* and European *Taraxacum officinale* and *Hieracium*)."

This statement not only confesses to conscious confusion but also displays some unconscious misunderstanding. It is not true, for instance, that the evening primroses as a group are "freely hybridizing"; as we shall see, they are for the most part self-pollinators and rarely hybridize, although hybridization has in the course of evolution played a major role in the creation of the groups now existent. Furthermore, the phrase "other strongly apomictic groups" implies that *Oenothera* is also apomictic, which is not the case. No apomixis has been discovered in this group.

It can be seen from such statements that the taxonomy of the group has been in a confused state. Only recently, with the help of the cytogenetic work to be described below, has it been possible to classify the North American oenotheras in such a way as to show their true phylogenetic relationships (Munz, 1965).

The population is therefore a varied one, the variations intergrading in such a manner as to make classification difficult. But although the population as a whole shows great variability, individual strains are, throughout most of the range, about as invariant as it is possible for strains to be. From the Rocky Mountains eastward, the population is made up almost entirely of a plethora of self-pollinating, true-breeding lines, each with a circle of 14 and balanced lethals. Because of the circle of 14, only two kinds of gamete are produced, genetically speaking. The lethals prevent the formation of homozygotes, and self-pollination prevents, or reduces to a very low level, outcrossing. As a result, each plant as a rule produces one and only one kind of offspring, all individuals produced being identical genetically with their parent, as well as with each other. There is therefore no variation whatever within a strain. The variation found in nature is due entirely to the very large number of strains that exist, each differing from the others, often in rather minor details of structure or behavior. There are probably thousands of such lines. Several lines may exist side by side but independently at a single site, each isolated reproductively from the others by reason of its self-pollinating habit. As a result, innumerable barriers are set up to gene flow. A mutation occurring in a given line is not likely to spread into other lines. If it is recessive, its existence may never become apparent, since it will be carried on in a permanent complex-heterozygote, masked by its dominant allele. Even if a mutation is dominant, it can still be confined to the line in which it originated.

It is true, however, that a small amount of outcrossing may occur, in spite of the self-pollinating habit. The anthers of a particular flower may produce only scanty pollen or none at all as a result of nutritional deficiency, or unfavorable environmental conditions. In such a case, there is a chance that pollen brought from other plants by bees, sphinx moths or hummingbirds may be able to function and to bring about fertilization of some or all of the eggs. Natural hybrids will thus be produced. Attempts to estimate the frequency of outcrossing have been made by Hoff (1962) who studied the progenies from 126 capsules from unguarded flowers of eleven complex-heterozygous races being grown in an experimental garden in close proximity to about 150 other races or hybrid combinations. These conditions afforded maximum opportunity for outcrossing, in strong contrast with natural habitats, where plants as a rule grow in rather small and isolated colonies, often rather homogeneous in composition. In one series of experiments, the percentage of plants that resulted from outcrossing ranged from 0·6 to 20·3 in different races. In another set of experiments, careful note was taken of the amount of pollen already present in a flower when it opened and this was correlated with the percentage of outcrosses present in the progeny of this flower. Flowers were classified, with respect to the amount of pollen which they produced, into four groups—with heavy, medium, light and sparse self-pollination. A high negative correlation was found between abundance of pollen produced by the flower and percentage of outcrosses. For example, flowers of *O. eriensis* with heavy yield of pollen produced 2·9% outcross progeny; those with medium yield gave 17·1%, those with sparse yield, 40%. Corresponding numbers for all plants of the three collections studied in this manner were: 2·19, 6·95, 16·07; 1·18, 2·55, 7·49; 0·56, 0·34, 2·13.

From these experiments it is clear that outcrossing can occur in habitually self-pollinating forms and that the amount will depend in each case, other things being equal, on the amount of pollen found in a flower's own stamens, and deposited on its own stigma. In the experimental garden, rich sources of foreign pollens are available, but in nature, this will very likely not be the case.*

If hybrids resulting from outcrossing possess ⊙14, and if the parental complexes have lethals, as would almost certainly be the case in any area east of the Rockies, a new self-pollinating, true-breeding line can be established. There is little doubt that the enormous number of isolated lines in nature is to a considerable extent the result of occasional or rare outcrossing between pre-existing lines.

If segmental arrangements were scattered at random throughout the population, the chances of hybrids thus produced having ⊙14 would be about 50% (Cleland and Blakeslee, 1931). Actually, however, neighboring lines are likely to be closely related, which means that their alpha complexes are apt to be identical or similar in segmental arrangement, the same with the beta complexes. Since, however, alpha and beta complexes in a race are so dissimilar segmentally that they give ⊙14 with each other, the chances of a particular beta giving ⊙14 with alphas that are similar or identical with its own associated alpha are probably much greater than 50%. This will be true especially if the races involved are strictly heterogamous. If, however, races are able to transmit alpha complexes through the sperm or betas through the eggs, the situation will rapidly change, depending on the extent to which this is true, since it will then be possible for two alphas or two betas to unite, as a result of which the chances of ⊙14 being formed are greatly reduced.

In any event, there are distinct possibilities that outcrossing may produce hybrids with some other configuration than ⊙14. These probably fail to survive in nature for reasons that we shall bring out later.

Summing up the situation that exists over most of the North American range of the subgenus *Oenothera*, i.e., the area from the Rocky Mountains eastward, the population consists of innumerable true-breeding and therefore permanent complex-heterozygotes, each isolated from the others by reason of its self-pollinating habit, but occasionally or rarely outcrossing, in which case hybrids are formed which tend to disappear if they have other than ⊙14, but which may become the progenitors of

---

* Dr. Hoff was able to pinpoint the origin of the pollen that resulted in some of his outcross progeny by analyzing the segmental arrangements of their complexes and relating these to forms known to be present in the garden. Incidentally, in most cases where several outcross progeny came from the seeds of a single capsule, they were identical, phenotypically and in chromosome configuration, showing that they had all resulted from a single pollination—a single visitation by a bee or moth.

new permanent and isolated complex-heterozygous lines if they have $\odot$14.

This is the situation in most of the range of the subgenus. It is, so far as is known, an essentially unique situation; the nearest approach to it is found in certain other subgenera of *Oenothera*, notably *Raimannia*, where some species have the same type of population structure. In the latter case, however, other species in the same geographical area are open-pollinated, with paired chromosomes. Only in the subgenus *Oeno-thera* does one encounter an entire population, covering millions of square miles, consisting almost exclusively of isolated, true-breeding, self-pollinating, circle-bearing complex-heterozygotes.

The situation in the more westerly part of the range, however, is quite different from this. The plants in Mexico and Central America, so far as known, and those resident in California and contiguous areas, have as a rule wholly paired chromosomes, do not possess lethals and are large-flowered and open-pollinated. Apart from *O. organensis*, which doubtfully belongs to the subgenus *Oenothera*, none of them is self-incompatible.

In the California area there is little heterogeneity in segmental arrangement. As one goes from the California area eastward, however, one finds increasing heterogeneity. In Arizona, New Mexico, Nevada and Utah, the members of the subgenus, although still large-flowered and open-pollinated, are apt to show small circles as part of their chromosome configurations. Interchanges have occurred so that a degree of variation in segmental arrangement has been introduced into the population. Since, however, lethals are rare or absent (they have doubtfully been found in a single race, from Albuquerque, New Mexico), plants do not breed true for their circles. Gametes which carry the same half of a circle can combine to give a plant with pairs instead of a circle. On the other hand, since pollination is open, pollen carrying one segmental arrangement is likely to reach a stigma of a plant whose eggs have a different arrangement, thus producing a plant with a small circle. The end result is a population whose individuals vary in configuration (7 pairs; $\odot$4; $\odot$4, $\odot$4; $\odot$6; etc.), but which maintains a balance between circle-free and circle-bearing individuals.

Population structure in the North American members of the subgenus, therefore, varies with geographical location. In California and Mexico plants are as a rule large-flowered, open-pollinated, lethal-free, self-compatible but normally outcrossing, and have mostly paired chromosomes. On the other hand, with very few exceptions to be mentioned shortly, all races in the eastern two-thirds of the continent, and even stretching to the Pacific Coast in the northwest, have a circle of 14, balanced lethals, small flowers and habitual self-pollination.

2. The Method Used in Tracing the Steps in the Evolution of the Group

Soon after the discovery of large rings of chromosomes in *Oenothera* an apparent correlation between chromosome configuration and the presence of homozygosity versus heterozygosity became recognized. When homozygous strains, such as the segregants from half-mutants (e.g., *decipiens, latifrons*) were examined cytologically (Cleland, 1923, 1925), they proved to have all-paired chromosomes. The same was true of natural complex-homozygotes, such as the *hookeri* of de Vries (Schwemmle, 1924). On the other hand, complex-heterozygotes always had most or all of their chromosomes in circles—⊙12 in *lamarckiana* (Cleland, 1925), ⊙6, ⊙8 in de Vries' *biennis* (Cleland, 1923), ⊙14 in most others. In other words, if a single complex was present in duplicate, the chromosomes were paired, if two different complexes were associated, large circles were present.

It was soon realized, furthermore (Cleland, 1931a), that identity in genetic composition was always associated with identity in chromosome configuration. Thus, the Swedish and Dutch *lamarckiana* both gave ⊙12, 1 pair. de Vries' *biennis* was studied by workers in Europe, the United States and Japan, and always showed ⊙6, ⊙8. The same situation applied to *muricata, suaveolens, chicaginensis, hookeri* and others. Furthermore, there were cases where different races have essentially identical complexes. For instance, *suaveolens* and *biennis* have *albicans* in common. The *rubens* of *biennis* and the *gaudens* of *lamarckiana* are very similar, even having the same zygotic lethal. In such cases, the two essentially identical complexes have identical segmental arrangements and give identical configurations in combination with a third complex.

An additional correlation was seen when Renner's genetical experiments were followed by cytological studies. Hoeppener and Renner (1928) published a diagram which purported to show the degree of genetical relationship existing between the complexes that Renner and his students had analyzed (Fig. 18.1). Each complex was represented by a circle. Overlapping circles represented closely related complexes. Those more distantly related were set apart in the diagram. Cleland made many crosses between these various complexes and studied the configurations of the hybrids obtained. He republished Hoeppener and Renner's diagram (Cleland, 1931a), adding to it the cytological findings. It was easily seen that complexes which are genetically closely related yield configurations when combined with each other in which most or all of their chromosomes are paired, whereas those which are more distantly related give configurations in which more of the chromosomes are involved in circles. A slight change in Hoeppener and Renner's diagram by which *rigens* is displaced slightly and no longer overlaps *velans* and *[h]hookeri* improves the correlation both cytologically and genetically (Fig. 18.2).

These early studies suggested that chromosome configuration may be used as an indication of the degree of relationship that exists between associated complexes. The closer the genetical relationship between complexes the more likely it is that they will have similar or identical segmental arrangements, and will therefore give mostly paired chromosomes when combined. Conversely, the more distant the relationship, the more unlike will their segmental arrangements be, and the greater

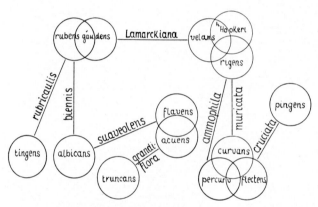

FIG. 18.1. Diagram of Hoeppener and Renner (1928) showing degree of relationship between various complexes based upon genetic analysis.

will be the likelihood that they will form large circles. Thus, chromosome configuration could be used as a tool with which to measure the degree of relationship between complexes.

When one stops to think of it, such a correlation is not surprising. It is generally expected that closely related forms will resemble each other more closely than more distantly related ones. Such similarities involve both morphological and biochemical characteristics. Similarity in segmental arrangement would also be expected in closely related complexes.

According to the Belling hypothesis of segmental interchange, all oenotheras may be assumed to have arisen from a common ancestor which had an initial segmental arrangement. As time went on, interchanges occurred which became incorporated into sectors of the population. The longer the time since the ancestors of two complexes began to

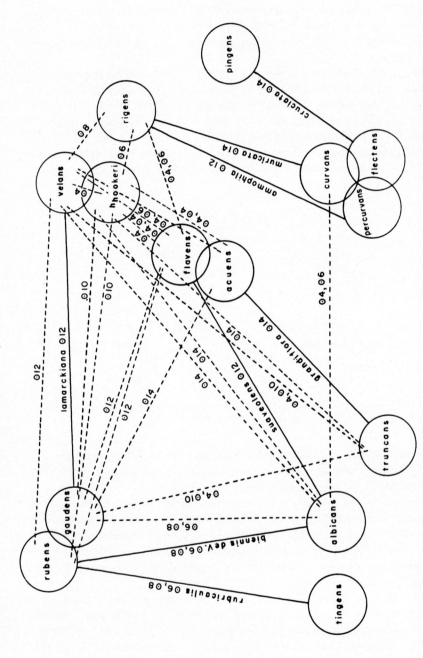

FIG. 18.2. Modification of Hoeppener and Renner's diagram (Cleland, 1931a) in which data with regard to chromosome configuration in various complex-combinations have been added.

diverge, the more time was available for interchanges to occur, and vice versa. If two complexes give $\odot$4, 5 pairs, with each other they differ by only a single interchange, and probably trace back to a recent common ancestor. The presence of a circle of 14, on the other hand, means that the associated complexes are only distantly related. How distant the relationship is we cannot tell for sure. A series of six interchanges properly distributed can build up a difference in segmental arrangement that will yield $\odot$14. If two associated complexes had each suffered three interchanges since the time when they first diverged from a common ancestor and these were different interchanges, they might yield $\odot$14 when combined. However, complexes which yield $\odot$14 may actually differ by more than six interchanges. No matter how many more interchanges take place, the maximum cytological effect will still be to give $\odot$14.

As a result of these considerations, we adopted a *working hypothesis to the effect that similarity in segmental arrangement is in general an indication of the degree of relationship between complexes.**

It was soon realized, however, that it was possible to go further than this. Not only could we gain some understanding of the *degree* of relationship between complexes; we should be able, by analyzing the segmental arrangements of individual complexes, to discover how many interchanges had occurred and in some cases specifically what interchanges had taken place during their development. Thus we could tell not only how closely complexes were related but also what type of relationship existed between them. Our technique was therefore to combine a complex whose affinities were unknown with a series of other complexes whose segmental arrangements had already been worked out, and determine the chromosome configuration in each hybrid combination. From this, by methods illustrated in Chapter 6, it was possible to determine the segmental arrangement of the new complex. One could then compare the arrangement of segments in this complex with that of other complexes whose arrangements were known, and by taking into consideration the geographical areas where these complexes were found and the phenotypes of the plants of which they were a part, it was possible to arrive at valid conclusions regarding the true affinities of the complexes involved.

---

* One must be aware, of course, of the possibility that exceptions to this rule may exist, for it is not inconceivable that similar or identical segmental arrangements could arise from diverse sources. Several examples are known among the derivatives of *lamarckiana*. For instance, *hblandina* appears to differ from the ancestral segmental arrangement (that of *hJohansen*) by only a single interchange. We know, however, that it is derived from *lamarckiana*, neither of whose complexes has the original arrangement, and by a process that had to involve a minimum of two interchanges. Such cases, however, are probably rare, and constitute exceptions that prove the rule.

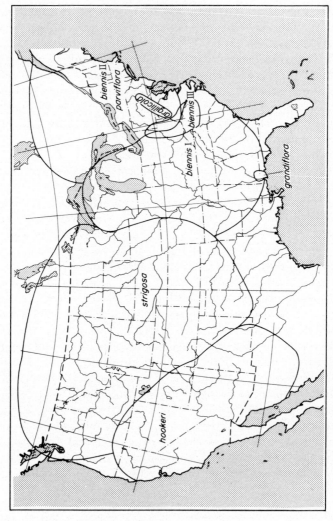

FIG. 18.3. Map showing geographical distribution of the major groups of the subgenus *Oenothera*.

Using this procedure, 438 complexes belonging to North American oenotheras have been analyzed completely and their relationships elucidated (see Appendix I for a list of these complexes and their segmental arrangements). As a result, we find that the races to which these complexes belong fall into ten groupings, each of which has a definite geographical range, distinctive phenotypic characteristics, and complexes whose segmental arrangements are similar or identical (Fig. 18.3). Stubbe has dubbed these groupings "superspecies," but Munz in his monograph (1965, p. 120 *et seq.*) has defined his species or subspecies in most cases as co-extensive with our groupings (Table 18.I). We consider each of these to represent a separate and distinct endpoint of evolutionary development.

TABLE 18.I

North American Oenotheras. Groupings of Cleland *et al.* compared with species classification of Munz (1965)

Cleland	Munz
elata	*elata*
hookeri	*hookeri*
	subsp. *hookeri*
	montereyensis
	Wolfii
	hirsutissima
	venusta
	ornata
	augustifolia
	grisea
	Hewettii
	Jamesii
	longissima
strigosa	*strigosa*
	subsp. *strigosa*
	canovirens
	cheradophila
biennis I	*biennis* subsp. *centralis*
biennis II	*biennis* subsp. *caeciarum*
biennis III	*biennis* subsp. *austromontana*
grandiflora	*grandiflora*
parviflora I	*parviflora* subsp. *augustissima*
parviflora II	*parviflora* subsp. *parviflora*
argillicola	*argillicola*
	var. *argillicola*
	pubescens

We shall now briefly characterize each group, as it is found today. After that we shall try to show how each group has evolved (see Fig. 18.4).

(a) *Hookeri.* This is a large and rather diverse group, but all have large, open-pollinated flowers (Figs 18.5, 18.6). Centered in California, it spreads northward as far as Washington (subsp. *ornata*), south into Mexico and east at least as far as Utah and New Mexico. Certain races found in Mexico and as far east as Texas and Oklahoma are separated by Munz into a separate species—*O. Jamesii.* In the California area, most forms have seven pairs. Lethals have not been found in this area. As mentioned above, however, small circles become more numerous in the eastern portion of the range and in one form (*Albuquerque B*), balanced lethals may be present in its circle of eight. Munz (1965) includes most of these forms within a single species, *O. hookeri* T and G., but recognizes nine different subspecies in line with the phenotypic diversity that exists in the group.

The commonest segmental arrangement in the *hookeri* group is the so-called *"Johansen"* arrangement, which we now consider the ancestral arrangement of the entire population of the subgenus *Oenothera.* It is 1·2 3·4 5·6 7·10 9·8 11·12 13·14. References to Fig 5.2 will show that these seven chromosomes are the most commonly found among the North American members of the subgenus, a result of the fact that in the course of evolution of most complexes not all chromosomes of the ancestral complex have been involved in interchange, so that most complexes still have one or more of the original chromosomes unaltered (see footnote on p. 295 for further discussion of this point).

(b) *Elata.* This group is found in Mexico and Central America. Only a few strains have been studied, most of them from Mexico, two from Guatemala. They have large, open-pollinated flowers, paired chromosomes, and no lethals. In phenotype they are intermediate between the hookeris on the west coast of the United States and the strigosas of the great plains (Figs. 18.7, 18.8). Segmentally, their complexes are on the average two interchanges removed from the commonest end arrangement found in the hookeris. It is not likely that one of the arrangements so far found in the elates is ancestral to the North American members of the subgenus *Oenothera* because of the fact that most of the *elata* chromosomes that depart from the [h]*Johansen* arrangement have end associations that are among the least often found among the complexes of the subgenus as a whole (compare *elata* complexes in Appendix I with those in other groups).

(c) *Strigosa.* This group extends from the Rocky Mountains eastward to the Mississippi River or slightly beyond, and in the northern part of

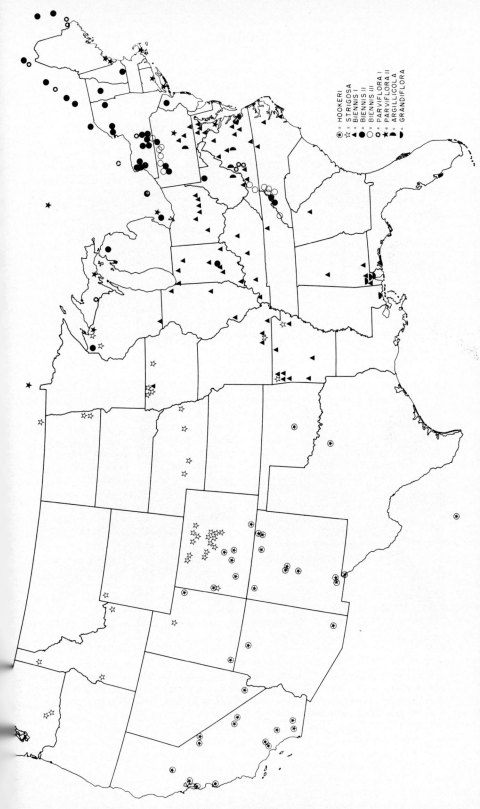

FIG. 18.4. Map showing locations from which strains have been obtained which have been analyzed cytogenetically by the author and co-workers.

Legend:
- = HOOKERI
- = STRIGOSA
- = BIENNIS I
- = BIENNIS II
- = BIENNIS III
- = PARVIFLORA I
- = PARVIFLORA II
- = ARGILLICOLA
- = GRANDIFLORA

FIG. 18.5. Rosettes of *O. hookeri*. (a) subsp. *hookeri* (strain *Johansen*); (b) subsp. *hirsutissima* (strain from Ramsey Canyon, Arizona); (c) subsp. *venusta* (strain from San Bernardino, Calif.); (d) Rosette of a strain from La Verken, Utah, closely related to the hookeris but classified by Munz as *O. longissima*.

FIG. 18.6. Flowering branches of *O. hookeri*. (a) subsp. *hirsutissima* (strain from Ramsey Canyon, Ariz.); (b) subsp. *Hewettii* (strain Las Cruces, N.M.); (c) subsp. *hookeri* (*franciscana* of de Vries); (d) *O. longissima* (from La Verken, Utah).

its range extends westward to the Pacific area. Its races all have ☉14, balanced lethals and relatively small self-pollinating flowers. The plants are quite strictly heterogamous, i.e., their alpha complexes are transmitted almost exclusively through the egg, the beta complexes through the sperm. So far as they have been analyzed, the alpha complexes show considerable similarity in segmental arrangement among them-

FIG. 18.7. Rosette of *O. elata* (strain from Cholula, Mexico).

selves, the betas show even more; but the alphas as a group are quite different from the betas as a whole, so much so that all alphas so far analyzed will give ☉14 with every beta complex that has been determined. Both alphas and betas differ from the original arrangement by an average of three interchanges.

From the standpoint of their effect on the phenotype, the alpha and beta *strigosa* complexes give very similar effects, producing woody stems and thick, relatively narrow leaves, both covered with appressed (strigose) hairs (Figs. 18.9, 18.10).

Munz' *O. strigosa* coincides with our group of the strigosas. He recognizes three subspecies: subsp. *strigosa*, found primarily in the Rocky Mountain area, and characterized by clearer green foliage, frequent presence of anthocyanin in leaves, buds and stems, and the presence of glandular, as well as longer spreading hairs on the buds; subsp. *canovirens*, widespread over the plains east of the Rockies with greyappressed long and short hairs on the buds; and *cheradophila*, found in the Pacific Northwest, with short appressed, but almost no long hairs.

(*d*) *Biennis I.* This and the next two groups to be mentioned are not

FIG. 18.8. Flowering spike of *O. elata* (strain from Zimapan, Mexico).

easily distinguished from one another in the field, and in this respect are unlike the groups mentioned above, which are rather easily recognized, although members of the subspecies *strigosa* often approach *hookeri* in appearance except for their smaller flowers. *Biennis I* is primarily a midwestern and southeastern group, ranging from the eastern edge of the prairie region to the Appalachians and, from Virginia southward, penetrating to the Atlantic coast. It extends from the Great Lakes on the north to the Gulf of Mexico on the south.

FIG. 18.9. Mature rosettes. (a) *O. strigosa* (strain from Palmer Lake, Colorado); (b) *O. biennis* subsp. *centralis* ("*biennis I*", strain from Galeton, Pennsylvania); (c) *O. biennis* subsp. *caeciarum* ("*biennis II*", strain from Wakefield, Mich.); (d) *O. biennis* subsp. *mustromontana* ("*biennis III*", strain from Newfound Gap, Tenn.).

These forms are similar to the strigosas in having ⊙14, balanced lethals and rather small self-pollinating flowers. They are, however, quite different in phenotype since they have broad, crinkly-wavy, thin leaves with relatively sparse pubescence and rather delicate, often brittle stems (Figs 18.9, 18.11). These characters are attributable to the alpha complex. The beta complex produces a *strigosa*-like phenotype which is masked in its effect by the presence of the alpha complex. The various races differ in the degree to which this masking occurs—in some, the *strigosa* characters are more readily seen, in others they are scarcely detectable even to an observer who is aware of the presence of a *strigosa*-like complex. In all races, however, the effect of the beta complex is so strongly masked that it was not realized by any worker that *biennis* carries a *strigosa* complex until this fact was established through cytogenetic analysis.

Biennis I races differ considerably in their degree of heterogametism —some are strictly heterogamous, others are essentially isogamous, still others show varying levels of intermediacy: in some, both complexes are transmitted through the egg, but only the beta through the sperm. In others, the reverse is true. In still others, the beta complex is carried by a few of the eggs, and/or the alpha by a small proportion of the sperms.

With respect to segmental arrangement, the situation in *biennis I* is unique in that the great majority of the alpha *biennis I* complexes have a common arrangement, which is but one interchange removed from the original. There are, of course, alphas that depart to some extent from this arrangement, but on the whole, the alpha *biennis I* complexes are the most homogeneous of any of the groups in the subgenus. The beta complexes show less homogeneity, although there are two arrangements, two interchanges apart, that are far more prevalent than any others. These two arrangements are present in about equal numbers, neither one confined to one part of the range, but both scattered from Arkansas to Virginia.

Biennis I coincides with *biennis* subsp. *centralis* of Munz (1965).

(e) *Biennis II.* This group is almost indistinguishable under field conditions from *biennis I*, although in the garden, where it is free from competition, its habit is generally lower and more spreading. In many races, bracts are caducous, so that ripening fruits develop on essentially leafless branches. *Biennis II* differs from *biennis I* in certain other ways: its geographical range includes eastern Canada and northeastern United States, extending westward into western New York and Ontario and southward along the Alleghenies to North Carolina; it also differs in the fact that the gametophytic lethals are reversed in comparison with *biennis I*. In *biennis II*, the *biennis* complex (the one producing broad,

Fig. 18.10a

Fig. 18.10. Flowering spikes of *O. strigosa* (a) subsp. *cheradophila* (strain from Granger, Wash.); (b) subsp. *canovirens* (strain from Omaha, Nebr.).

thin, crinkly and relatively hairless foliage, etc.) is the beta complex and is transmitted mostly through the sperm, the *strigosa*-like complex is the alpha.

With few exceptions, *biennis II* races have ⊙14. They have balanced lethals and rather small, self-pollinating flowers. They tend to be strictly heterogamous, although metaclines are occasionally produced when some races are outcrossed.

Biennis II coincides with *biennis* subsp. *caeciarum* of Munz (1965).

(*f*) *Biennis III*. This group is found in the areas where *biennis I* and *biennis II* overlap—from Virginia, North Carolina and eastern Tennessee

FIG. 18.10b

to northern Pennsylvania. Its races have ⊙14, balanced lethals and small, self-pollinating flowers. It has apparently arisen through crossing of *biennis I* as female with *biennis II* as male, since both of its complexes are of the *biennis* type, and unlike the other *biennis* groups, it has no *strigosa* in it. *Biennis III* then is pure *biennis*, but its alpha and beta complexes have developed by different routes.

Since *biennis III* has no *strigosa* in it, it is more easily distinguished phenotypically from the other *biennis* groups than they are from each other (Figs 18.9, 18.11). The absence of any trace of *strigosa* influence makes it quite easy to pick out the *biennis III* plants in an experimental

FIG. 18.11. Flowering branches of *biennis*-like plants. (a) A form from the pacific North-west which resembles *biennis* phenotypically but does not conform in segmental arrange-ment or in geographical position to this group. (From Camas, Wash.); (b) *O. biennis* subsp. *centralis* ("*biennis I*", strain from Indianapolis, Indiana); (c) *O. biennis* subsp. *caeciarum* ("*biennis II*", strain from Blowing Rock, No. Carolina); (d) *O. biennis* subsp. *austromontana* ("*biennis III*", strain from Coudersport, New York).

field. It is not so easy, however, in the wild, where plants do not show their foliage characters well because of the crowding.

Biennis III coincides with *biennis* subsp. *austromontana* of Munz (1965).

(*g*) *Grandiflora*. In the vicinity of Mobile, Alabama, a small population exists, mixed to a certain extent with *biennis I* plants, which has large, open-pollinated flowers, paired chromosomes, free of lethals. The segmental arrangement, so far as determined, is the ancestral arrangement. The plants are slow-growing annuals, and bloom only when the days are short so that when transferred to northern latitudes they rarely flower soon enough to ripen seeds.

(*h*) and (*i*) *Parviflora I* and *parviflora II* are two groups that occupy essentially the same range but have probably had independent origins. Both cover about the same territory as *biennis II*, both have shown ⊙14 (with the exception of one *parviflora II* form from Michigan, on the western edge of the range, which has ⊙4, ⊙10). They have balanced lethals and small self-pollinating flowers. They are strongly heterogamous. The beta complex in some cases comes through a small proportion of the eggs, but the alphas have never been known to come through the sperm (pollen grains that receive an alpha complex fill up with starch, and when nearly mature, die and shrivel down on the starch in characteristic fashion, forming Renner's so-called "inactives") (Fig. 3.6).

Parviflora I has alpha complexes of the *biennis* type. The alpha complexes of *parviflora II* are of the *strigosa* type. As a result, *parviflora I* races have clear green, relatively hairless or even glabrous foliage. *Parviflora II* is quite hairy, the hairs appressed and non-glandular in some, in others short glandular hairs are intermingled with longer strigose hairs (Figs 18.12, 18.13). The beta complexes of *parviflora I* and *II* are similar in phenotypic effect, and are quite distinctive among the members of the subgenus, giving to these groups their unique phenotype. They tend to produce narrow, often very narrow, leaves, stem tips that bend down and then up in characteristic fashion, and sepal tips that are subterminal. The latter two characters have never been separated genetically and are the result probably of closely linked genes. There are, however, according to Renner, a number of differing alleles in different races for these characters. Although these two traits are not found in the subgenus *Oenothera* except in the beta *parviflora* complexes and in the argilicolas, one of them is quite characteristic of the subgenus *Raimannia*, namely, the subterminal sepal tips; this suggests that the beta *parviflora* complexes are closely related to the raimannias, a conclusion that receives some support from the fact that Schwemmle (1927) was able to make a cross combining a beta *parviflora* with a *Raimannia* genome

(*Berteriana* × *muricata*), although as a rule *Oenothera* and *Raimannia* will not cross successfully.

Parviflora I corresponds to Munz' *O. parviflora* subsp. *angustissima*, *parviflora II* to his subspecies *parviflora*.

a   b

FIG. 18.12. Mature rosettes of *O. parviflora* (a) "*Parviflora I*" = subsp. *augustissima* (strain from St. Stephen, New Brunswick, Canada), (b) "Parviflora II" = subsp. *parviflora* (strain from Sunbury, Pennsylvania).

(*j*) *Argillicola*. This is a small population distributed along the Appalachians from Pennsylvania to Virginia, primarily on shale barrens. It has many primitive characters including large open-pollinated flowers, mostly paired chromosomes and apparent absence of lethals (Mickan in 1936 claimed that a race with ⊙4, 5 pairs, had balanced lethals but this was not confirmed and the strain was lost). It also has the bent stem tips and the subterminal sepal tips characteristic of the beta *parviflora* complexes. Stinson (1953), who has done the most work on the group, thinks that it may represent a relic of the ancient population which gave rise to the beta *parviflora* complexes. Its plastids, while able to function in the presence of a beta *parviflora* coupled with its own genome, differ from those of the parvifloras in that they experience difficulty in combinations with the genomes of other groups which are quite compatible with the *parviflora* plastids. Stubbe, therefore, places them in a separate class (class V) from the plastids of other groups. Its genomes, however, behave like the beta *parviflora* complexes in relation to the various classes of plastid. Stubbe, therefore, places them in category C along with the beta parvifloras.

The existence of these ten groupings of races has been demonstrated as the result of cytogenetic studies coupled with the facts of geographical distribution. It is therefore of interest to note that the most recent taxonomic treatment (Munz, 1965), based almost entirely on morphological criteria, agrees very closely with the results of our cytogenetic analyses (Table 18.I). The close correlation between cytological behavior, geographic distribution and morphological characters gives assurance that the classification arrived at conforms with the realities of phylogenetic relationship, and each group represents the end point of a distinct line of evolutionary development.

3. The Story of the Evolution of the Subgenus *Oenothera*

Our hypothesis begins, as stated earlier (p. 233), with the assumption that the series of events that has led to the present-day situation in this group would not have taken place had it not been that the ancestral stock from which these forms developed had a certain karyotype. The chromosomes of this primitive form were all alike in size and structure. They all had median centromeres, so that the arms of all chromosomes were equal. In addition, they had developed a concentration of heterochromatin on either side of the centromere and contiguous with it. It was this particular morphology that made possible the events that will now be described.

In most organisms, translocations have a net deleterious effect. This is because translocated chromosomes are likely to be of unequal size, either as a result of unequal exchanges, or because the chromosomes that have exchanged segments were of different sizes. In such cases, individuals that are heterozygous for a translocation will have chains or circles whose chromosomes will vary in size, the centromeres will be unevenly spaced around the circle, the forces responsible for disjunction will be unevenly distributed and consequently the regular separation of adjacent chromosomes will often not be achieved, resulting in deficiencies and consequent sterility (Fig. 7.2). In an open-pollinated strain, even the presence of translocation-homozygotes is a handicap, for whenever they cross with a non-translocation-homozygote, a heterozygote will be formed, with lowered fertility. As a result, in most cases of this sort, translocations tend to be selected against and to disappear.

FIG. 18.13a

FIG. 18.13. Flowering branches of *O. parviflora* (a) subsp. *parviflora* (*parviflora II*, from Sunbury, Pennsylvania); (b) subsp. *augustissima* (*parviflora I*, strain from Ithaca, New York).

The situation in *Oenothera*, however, was different. Heterochromatin is more easily broken than euchromatin, and heterochromatin in the *Oenothera* chromosome is concentrated close to the median centromere. Consequently, breaks were more likely to occur at the centromere region than elsewhere, and since all chromosomes were of the same size to begin with, and all arms of the same length, such translocations tended to be equal and the translocated chromosomes had the same morphology as the non-translocated ones. In a translocation-heterozygote, therefore, the spacing of centromeres in a circle was essentially equal, the forces

FIG. 18.13b

were evenly distributed, and as a result the regular zigzag arrangement with consequent disjunction of adjacent chromosomes was achieved in most cells. As a consequence, most gametes were complete and functional, and fertility was not noticeably lowered.* Translocations, then, were not

* As we have seen earlier (p. 114), essentially equal exchange can also occur between terminal pairing segments. This process probably takes place on occasions when chromosomes that are interlocked in the first meiotic prophase become disjoined during diakinesis. Such interchanges may also result in translocated chromatids with essentially unaltered morphology.

weeded out, and tended to accumulate in a population. As a population spread over widening areas, translocations now and then occurred, one translocation in one part of the range, another in another part. Some sectors of a population might be composed mainly of translocation homozygotes, others might contain a mixture of plants with translocated and non-translocated chromosomes, so that a proportion of the population possessed small circles. In the absence of lethals, such plants would not breed true for their circles but would produce homozygous segregants, which, however, would in turn cross with individuals possessing other segmental arrangements and so produce plants with circles, in the manner seen at the present time in the eastern portions of the *hookeri* range. In this way a balance was maintained between the all-pairing condition and the presence of small circles. In such a system, translocations tended not to be lost, but to be retained.

How then have the large circles come into being that characterize most of the subgenus? Have they developed by a gradual accumulation of interchanges, an exchange between a small circle and a pair enlarging the circle and this process continuing step by step until large circles were built up? We believe not. Large circles have come into existence suddenly, as the result of hybridization between different populations which have experienced different histories of interchange. These populations, before they overlapped and crossed, were probably in the state that is found at present in the eastern part of the *hookeri* range. They were open pollinated and showed a balance between individuals with small circles and individuals with paired chromosomes. In other words they had developed within the population, at least in most cases, a certain amount of heterogeneity in segmental arrangement, but this had not developed to the point where large circles had come into being.

Our evidence suggests that several populations have developed in successive periods of time and the present groupings have been initiated by hybridization between successive populations. This story will first be summarized in brief outline, and will then be discussed in more detail (see Fig. 18.14).

It appears that four populations have developed successively in the center of origin, which we conceive to have included Mexico and Central America. These have spread in successive waves across the North American continent. The first of these, Population 1, is represented at present by the argillicolas, a relict group on the shale barrens of the Appalachian Mountains. This population was later invaded by Population 2 which crossed with it, resulting in hybrids that became the progenitors of our *parviflora I* assemblage. Population 2, which possessed the mesophytic characters which we now find in *biennis*, was later invaded from the west by Population 3, a population possessing the relatively xerophytic features that now characterize the strigosas. This population crossed with Population 2 to give the beginnings of the present-day *biennis* groups. Two rather distinct populations of *biennis* developed, one in the midwest which we call *biennis I*, the other in the northeast which is

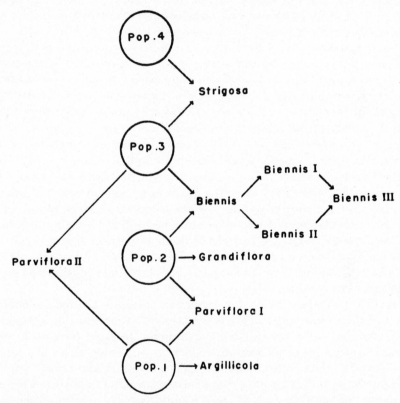

FIG. 18.14. Diagram to show the origin of present-day groupings of races from four ancestral populations. Reprinted from Proceedings of the American Philosophical Society by permission.

designated *biennis II*. At a later time, crosses between these two populations along a line where *biennis I* and *II* overlapped, gave rise to a third *biennis* population, *biennis III*.

Population 3, however, not only crossed with Population 2, but it also found members of Population 1 still present in the northeast and combined with them to produce the beginnings of our present-day *parviflora II* population.

A relict of Population 2 still survives around Mobile, Alabama, and is known as *grandiflora*.

The same process took place yet again when a fourth population developed and invaded the territory of Population 3, with which it crossed, giving rise to the progenitors of the present-day strigosas. Population 4 had a phenotype similar to Population 3, relatively zerophytic in character and adapted to the comparatively dry conditions of the prairies. No relicts of Populations 3 and 4 have been discovered.

Each population as it developed and spread experienced reciprocal translocations. Some translocations occurred early and became incorporated into large sectors of the population. Others occurred later and became characteristic only of the local areas where they first appeared. When two populations crossed, the probabilities were that any two individuals which hybridized would each have had a back history of interchange, but of different interchanges, so that their segmental arrangements would be quite unlike. Consequently the hybrid progeny which they produced would often have a circle of 14. We visualize, therefore, that the successive populations had small circles or none, and that large circles came into existence through hybridization between populations which had suffered different histories of interchange and hence had chromosome sets whose distributions of pairing segments were totally unlike. For reasons to be explained later (p. 268) hybrids with ⊙14 swamped out the original populations with their paired chromosomes or small circles.

A fifth population which developed and spread in the region mainly west of the Rocky Mountains has not been able to repeat this history of hybridization with pre-existing populations, perhaps because it has found no available population with which it could cross. This is the group of hookeris, which still retains the open pollination and mostly paired chromosomes that must have characterized the other populations before they met and crossed with different populations. There is some question as to whether the hookeris represent a later population than the others or whether it is early and primitive and owes its failure to overlap and cross with other populations to the fact that it developed in a more or less isolated area which other populations did not invade (see p. 294).

Finally, the elatas are still growing in the center of origin and have not spread out of this area to any extent. They have maintained themselves, therefore, uncontaminated by genes from other groups.

We shall now discuss in more detail this story of evolution and present the evidence upon which rests the hypothesis of overlapping and hybridization between successive populations, each with its own phenotype and distinctive history of interchange.

The following facts support the assumption that the center of origin of the subgenus *Oenothera* was Central America or Mexico: (1) The oenotheras in this area have paired chromosomes, freedom from lethals and large open-pollinated flowers which are no doubt primitive characters. (2) There is evidence that other organisms have spread northward into temperate North America from the tropical or subtropical areas to the south, evolving as they have spread. (3) The genus *Oenothera* is widely distributed throughout both North and South America with circles of 14 more prevalent in regions distant from Central America. The latter fact is true in North America for the subgenus *Oenothera*, and in both North and South America for *Raimannia*.

The first population that developed and spread from this region was one that evidently developed in an early period when the subgenera *Oenothera* and *Raimannia* were scarcely or only recently differentiated from a common source. This population, which we have designated Population 1, had very narrow, relatively hairless leaves, as well as bent stem tips and subterminal sepal tips, the latter character still being characteristic of many raimannias, but not found in the subgenus *Oenothera* except in those forms which have a Renner complex derived from Population 1. This population probably spread over the continent during a period when climatic conditions were more favorable than at present; but when the central part of the continent later became drier, it died out except in the eastern moister region. It has survived to the present day as a relict group on the shale barrens of the Appalachians, which now goes under the name of *argillicola*, and also as the group of the beta *parviflora* complexes.

A. The Origin of Parviflora I

The time came when a new and different population (Population 2) developed in the center of origin and spread over the North American continent. When it reached the area occupied by Population 1, hybrids were formed between the two populations. Because there had been different histories of interchange within the two, some hybrids had a circle of 14.* These were the progenitors of the group that we now call *parviflora I* (Munz' subspecies *angustissima*). Population 2 apparently developed during a period when the climate in the southwest was moist since its phenotype was distinctly mesophytic in character with broad

* It is conceivable that an occasional hybrid combination with ⊙12, or ⊙4, ⊙10, or ⊙6, ⊙8 was able to survive indefinitely (see p. 262). Since these configurations, however, are rarely found at the present time in Nature, it is probable that most of the hybrids ancestral to *parviflora I* had ⊙14. The same reasoning no doubt holds in the other cases to be discussed below.

thin leaves and scanty pubescence. Its characters were what we now recognize as the *biennis* characters, and it was this population that contributed the *biennis* phenotype to the members of the subgenus *Oenothera* (for initial studies, see Geckler, 1950).

Crossing between Populations 1 and 2 led to the combining in single plants of complexes with *parviflora* and *biennis* characters, and this combination is what distinguishes *parviflora I* from other groups of *Oenothera* today. The alpha complexes of *parviflora I* are of the *biennis* type, and when *parviflora I* is used as female in outcrosses the progeny are *biennis*-like in appearance. When *parviflora I* is used as male parent, the progeny, when successful, are of the *parviflora* type (they are often unsuccessful, due to plastid incompatibilities).

When Populations 1 and 2 first overlapped, both were probably open pollinated, with small circles or none and without balanced lethals. The question then arises as to how balanced lethals came into existence. As long as lethals were not present, a plant with a ⊙14 would not breed true for its circle and would split off individuals homozygous for one complex or the other, whose chromosomes would be paired. A clue to the origin of balanced lethals may have been furnished by Steiner (1956, 1961) and Steiner and Schultz (1958), as mentioned earlier in our discussion of incompatibility (Chapter 13). They have shown that a good many *biennis* races possess what was originally an S-factor (self-incompatibility factor) which now operates as a pollen or male gameto-phytic lethal. Population 2 presumably included some sectors that carried S-factors. If individuals from such sectors crossed with Population 1 individuals which did not have S-factors, giving progeny with ⊙14, these hybrids had an unbalanced S-factor, and the half of the pollen receiving this factor would be unable to germinate or grow tubes in selfed styles. Such a hybrid would thus behave as though it had a pollen lethal.

Admittedly, it is not possible to be as sure, in the case of the parvi-floras, that this was the source of the original male gametophytic lethals, as Steiner has found it to be in *biennis*, because all races of *parviflora I* so far examined (and the same applies, as we shall see, to *parviflora II*) have an additional pollen lethal which causes the death of all pollen grains containing the alpha complex while they are still immature. As mentioned above, such pollen develops spherical starch grains instead of the usual spindle-shaped grains, and the young microspore, when it dies, shrivels down on the mass of starch grains and becomes what Renner called an "inactive". Since grains that carry the alpha complex die before maturity, one cannot determine whether they also carry an S-factor.

On the other hand, most of the *biennis* population, as we shall see, have no additional pollen lethals to complicate the picture, and it is therefore possible to demonstrate experimentally that the pollen lethal is in fact an S-factor. The widespread presence of an S-factor in *biennis* coupled with the fact that the alpha *parviflora I* complexes are *biennis*-like in character, lends plausibility to the hypothesis that S-factors became the initial pollen lethals in the original hybrids between Populations 1 and 2.

The most we can say at present, therefore, is that one of the balanced lethals in the ancestral *parviflora I* population may have been present from the beginning in the form of an S-factor in those hybrids that survived.

The egg "lethal", as suggested above (p. 100), may not have been a lethal at all, in the strict sense of that term. The complex that failed to come through the egg may simply have been unable to compete successfully in embryo sac development with its opposing complex (the so-called "Renner effect"). If therefore, in the hybrids between Populations 1 and 2, the complex from the latter had an S-factor but was able to outgrow the other complex in the ovules, whereas the complex from the former was able to produce functional pollen but unable to compete in the production of the female gametophyte, a balanced lethal situation was established similar to that exhibited by present-day forms, and was present from the very beginning.

That the so-called female gametophytic lethals are really the result of the Renner effect, and are not lethals at all, is shown by the fact that many complexes, including some beta parvifloras, which ordinarily fail completely to be transmitted through the egg in their own races are able to compete to a greater or lesser extent when brought, through outcrossing, into combination with complexes from other races. Furthermore, there are many races in which the beta complex is able to compete successfully with its associated alpha complex in a small percentage of cases, showing that its ordinary failure is not the result of a lethal gene but of relative inability to compete with its opponent. Even the poorest team in the league wins some games.

It is of interest at this point to inquire regarding the degree of heterogeneity in segmental arrangement that exists in the complexes of *parviflora I*. The beta parvifloras are quite heterogeneous, and are quite removed from the ancestral arrangement of the subgenus *Oenothera*, ranging from an apparent minimum of three interchanges away from the original to six. In this they are quite different from the argillicolas which range from the original arrangement to one that is three interchanges removed from the original—in other words, the *argillicola* complexes are quite primitive from the standpoint of segmental arrangement (Stinson,

1953). If the argillicolas are a relic of the original population that is now represented principally by the beta *parviflora* complexes, one may wonder why the former are so close to the original segmentally, while the latter are so far removed.

It has been shown by Darlington (1931) that successful interchanges can take place without restriction between paired chromosomes or between two circles, or a circle and a pair, but are confined within circles to exchanges within a single complex or to exchanges of non-corresponding arms in opposing complexes (exchanges of corresponding arms of opposing complexes, e.g., two right-handed arms, result in inviable gametes). One would rather expect *a priori*, therefore, that somewhat more heterogeneity in segmental arrangement would develop in a population that does not have large circles than in one with ⊙14. Some facts seem to argue for this conclusion; some, however, appear to suggest the opposite—that heterogeneity in segmental arrangement is more the result of exchanges within large circles than between pairs or between pairs and circles.

In support of the latter hypothesis might be cited the fact that when exchanges of non-corresponding arms of opposing complexes take place in a form with ⊙14, the resultant interchange complexes may sometimes be quite unlike the parental complexes in segmental arrangement, the degree of resemblance depending upon the position in the circle of the exchanging chromosomes (Fig. 18.15). In a circle of 14, the two chromosomes in opposing complexes that exchange segments could be separated by two chromosomes in one direction and ten in the other, or by four and eight, or six and six chromosomes. In the first case each translocation complex would appear to differ by only a single interchange from one of the parental complexes. In the second case, however, a translocation complex would appear to be two interchanges removed from the complex which it more closely resembled, and in the third case, it would seem to be three interchanges removed from both parental complexes. In other words, it is possible for a single interchange between opposing complexes to produce complexes that seem to be as much as three interchanges removed from the parental complexes. It might be argued, therefore, that a very few successful translocations between opposing complexes in a circle of 14 could soon bring about a major shift in segmental arrangement. It should be added parenthetically that in these cases the translocation complexes would also have chromosomes or chromosome segments from both parental complexes.

There is, however, a fact on the other side that would seem to militate against the importance of this particular process as a major mechanism for achieving a high level of heterogeneity in segmental arrangement.

This is the fact that, while an exchange of non-corresponding arms of opposing complexes in a circle of 14 may produce gametes that appear to differ by as much as three interchanges from the parental complexes, such gametes, if they function in self-fertilization, will have to combine with one of the parental complexes and hence will in most cases produce progeny with an altered chromosome configuration, such as ⊙8, 3 pairs,

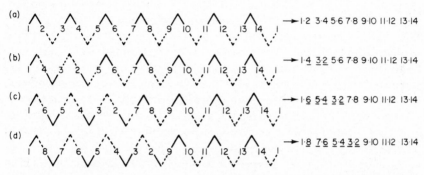

FIG. 18.15. Single exchanges of non-corresponding arms of opposing complexes within a circle of 14 may produce complexes that appear to differ from the parental complexes by more than a single interchange. (a) Separation of adjacent chromosomes in a circle of 14 composed of unmodified complexes, will yield unmodified complexes, one of which is shown; (b) Exchange of arms *2* and *4* will produce a complex that differs from one parental complex by a single interchange and includes three arms from the opposing complex; (c) Exchange of *2* and *6* will produce a complex that appears to differ from the complex which it more closely resembles by two interchanges, and includes 5 arms from one complex and nine from the other; (d) Exchange of *2* and *8* will produce a complex that appears to differ from both parental complexes by three interchanges and possesses seven arms from each.

in the case of an exchange between chromosomes that are separated in the circle by six chromosomes. The fact that races with other than ⊙14 almost never occur in nature suggests that a strain with such a configuration would be unlikely to survive indefinitely. The principal hope for survival of such an altered complex would be for it soon to become associated through outcrossing with a complex from some other race, with which it would form ⊙14. The chances of this occurring would seem to be minimal, though not necessarily nil; this is assuming, however, that self-pollination has become established and widespread. It is probable, on the other hand, that a considerable period of time actually elapsed between the initial formation of hybrids with circles of 14 and the advent of self-pollination. Centuries or even millenia may have passed during which hybrid strains with large circles were still large-flowered and open-pollinated, as is still the case with *lamarckiana* and de Vries' *grandiflora.* During this period, complexes resulting from exchanges of non-corresponding arms of opposing complexes could be carried by insects to other plants, and stand a chance of combining with other complexes with which they might give ⊙14.

There is one fact, however, that suggests that an important, perhaps the most important, source of heterogeneity in segmental arrangement has been exchange between chromosomes of the same complex within a large circle. Such an exchange can involve either corresponding or non-corresponding segments. If corresponding segments are exchanged in a plant with a circle of 14, gametes will be formed that will give altered configurations in the selfed progeny, but the number of possible configurations is strictly limited: if the chromosomes which exchange are separated in the circle by one chromosome in one direction and eleven in the other a gamete will be produced which will give ⊙12 with the opposing complex following self-pollination. If they are separated by three and nine chromosomes, the configuration produced will be ⊙4, ⊙10; if by five and seven, the configuration will be ⊙6, ⊙8. No other configurations are possible in the F_1 following self-fertilization, although in later generations from plants with ⊙4, ⊙10 or ⊙6, ⊙8, one circle may be lost if at least one half of this circle is alethal (it is not likely that one half of a circle of ten would be alethal, thus resulting in plants with ⊙4, 5 pairs). Plants with single circles of six, eight or ten are not likely to survive in nature.

The significant fact now appears. Among the 249 complex-heterozygous races from North America that we have examined in our laboratory, belonging to the strigosas, parvifloras and the three groups of *biennis*, only seven have had a configuration other than ⊙14, and these have all had one of the three configurations that could be derived from ⊙14 only by the exchange of corresponding segments within a single complex (three races showing ⊙12, four races ⊙4, ⊙10) (Table 18.2). In addition, a few individual plants with aberrant configurations have appeared in the experimental garden and all but two of these have had configurations derivable from the races to which they belonged by the same mechanism: in races with ⊙14, four plants have had ⊙12, one ⊙4, ⊙10 and one ⊙6, ⊙8; in a race with ⊙12, one had ⊙8. The two exceptions are a plant of the strain *Yakima* (⊙14) that had ⊙6, 4 pairs, and a plant of *Sunbury* (⊙14) that had ⊙10, 2 pairs. Since the parents of both of these individuals were found to have the typical configuration, the exceptions did not represent F_2 individuals in which a circle of eight or four had been lost, but were probably the result of exchanges of non-corresponding segments of opposing complexes which could give the configurations in question. The fact that every wild complex-heterozygous race with a configuration other than ⊙14 that we have observed has shown a configuration that can be derived from ⊙14 by the exchange of corresponding segments within a single complex, and by no other process, suggests that this mechanism has been of prime importance in bringing about heterogeneity in segmental arrangement. This conclusion is strengthened by the fact that all wild races so far found in Europe with configurations other than ⊙14, have had one of the three configurations obtainable only by this mechanism (Table 18.II).

If the exchange of corresponding segments within a single complex of a circle of 14 has been an important source of heterogeneity, is it not equally likely that exchanges of non-corresponding segments within a complex have also been of importance? An exchange of this type will produce gametes that will give ⊙14 following selfing, and will therefore result in progeny that should be viable, though with an altered segmental arrangement in one complex. Such an exchange, if it occurs between chromosomes in which odd- and even-numbered segments are associated, as we have postulated for the ancestral arrangement, will produce chromosomes, one of which will have two even-numbered ends, the other will have two odd-numbered ends. In fact, exchange of non-corresponding segments within a single complex is the only way by which even–even or odd–odd chromosomes can be formed within a large circle all of whose chromosomes initially had an odd–even association.

We may ask, therefore, whether it is possible to determine the relative importance of exchanges of non-corresponding arms within a single complex by an analysis of the frequency of odd–odd and even–even chromosomes. This appears to be difficult, for two reasons: (1) Exchanges that took place between pairs or small circles in the ancestral populations had an equal chance of producing odd–even chromosomes, or odd–odd and even–even chromosomes, so that we cannot be sure how many odd–odd or even–even chromosomes now extant were produced before or after circles of 14 were established. (2) It depends upon the arrangement of ends in the opposing complex whether non-corresponding arms in a complex give odd–even chromosomes when exchanged, or odd–odd and/or even–even chromosomes. For example, in a chain in which chromosomes have the ends–1·2–2·3–3·4–, non-corresponding arms of the first and third chromosomes will give odd–odd and even–even chromosomes if exchanged; whereas, if the initial arrangement is— 1·2–2·4–4·3–, the chromosomes formed will both have an odd–even association.

It should also be remembered on the one hand that the only way in which odd–odd and even–even chromosomes can be formed from odd–even ones within a complex is by the exchange of non-corresponding

TABLE 18.II

Occurrences of all configurations other than ⊙14 that have been found in races or single plants of complex-heterozygotes (normal configurations of races in parentheses)

	Configuration	Race	Single plant
NORTH AMERICA	⊙12, 1 pair	*biennis I* *chicaginensis* *reynoldsii* *biennis II* *Lanoraie*	*biennis I* *grandiflora* of de Vries (⊙14) *biennis II* *Corning II* (⊙14) *Moore* (⊙14) *biennis III* *Newfound Gap* (⊙14) *strigosa* *Hood River* (⊙14)
	⊙4, ⊙10	*biennis I* *Beaufort* *parviflora II* *Manistique* *Pine* (species?) *Sandia III*	*biennis I* *Warrenton* (⊙14)
	⊙6, ⊙8		*biennis I* *Iowa I* (⊙14)
	⊙10, 2 pairs		*parviflora II* *Sunbury* (⊙14)
	⊙8, 3 pairs		*biennis I* *reynoldsii* (⊙12)
	⊙6, 4 pairs		*strigosa* *Yakima* (⊙14)
EUROPE	⊙12, 1 pair	*ammophila* *coronifera* *lamarckiana* *suaveolens*	
	⊙4, ⊙10	*cantabrigiana* *parviflora* of Renner	
	⊙6, ⊙8	*biennis* of de Vries	
	⊙4, ⊙8		*suaveolens* (⊙12)

arms: on the other hand, even–even and odd–odd associations are lost whenever odd–odd and even–even chromosomes exchange segments, whether these segments be corresponding or non-corresponding. All other types of exchange maintain the status quo.

In a complex whose chromosomes all have an odd–even association, there is a 50% chance that an exchange of arms within the complex will produce an even–even and an odd–odd chromosome. At the other ex-

treme, when a complex has three even–even and three odd–odd chromosomes, 18 of the 42 possible exchanges between chromosomes of the
complex will eliminate an odd–odd and an even–even chromosome and
produce instead two odd–even chromosomes; and 12 of the exchanges
will result in the loss of one odd–odd or even–even chromosome. There is,
in the latter case, no possibility of an increase in the number of odd–odd
or even–even chromosomes. In general, it may be said that where few or
no chromosomes in a complex have odd–odd or even–even associations,
there is a high probability that interchanges within the complex will
produce chromosomes with such arrangements; with increasing numbers
of such chromosomes in the complex the chances will become progressively fewer for increase, and more numerous for decrease in their
number. A balance thus tends to become established.

A tabulation has been made of the frequency with which each of the
91 end-combinations has been found among the chromosomes of North
American complex-heterozygotes, omitting those chromosomes that
belong to the original arrangement. We have found 1036 odd–even, 314
even–even and 311 odd–odd chromosomes, i.e., about 37·6% of the
chromosomes found have even–even or odd–odd combinations. How
many of these were in existence prior to the establishment of circles of
14 we cannot say. From this discussion, therefore, it appears that we
cannot learn much regarding the importance of exchanges of non-
corresponding segments within a complex in building up heterogeneity
in segmental arrangement in the population. There is no reason, however.
to suppose that such exchanges are not as important as exchanges of
corresponding arms. They will give rise to progeny with a circle of 14
but with a different segmental arrangement within the complex. Such
a plant should be fully viable.

The fact that the relict races representing Populations 1 and 2 as well
as complex-homozygous species such as *hookeri* show little evidence of
translocation suggests that the bulk of the heterogeneity in segmental
arrangement found in *parviflora I* may have come from exchanges that
have taken place after hybridization had produced the progenitors of the
present-day parvifloras. Some of this probably occurred during the
period after inter-population hybridization had taken place but before
self-pollination became established. To a considerably greater extent,
however, it probably occurred after self-pollination was established,
primarily as a result of exchanges within single Renner complexes. It
is not at all unlikely that chromosomes bound together in chains are more
likely to be thrown into contact during the complicated contortions that
no doubt take place during meiosis than are paired chromosomes, and
so are more likely to exchange segments.

There is one other fact which argues against the relative frequency of exchanges of non-corresponding arms of associated complexes. This is the fact that, no matter how heterogeneous the segmental arrangements are in the various races of a species, the complexes remain quite typical of their particular group in their phenotypic effect. Alpha *parviflora I* complexes remain *biennis*-like in character, beta *parviflora I* complexes retain their typical beta *parviflora I* characteristics, as shown by out-crosses. If exchanges occurred frequently between opposing complexes, a blending would result, in the course of time, of the characters of the two very diverse types of complex, so that reciprocal hybrids of a given strain would become similar phenotypically. This is not the case. Reciprocals in all cases are very distinct. This applies not only to the *parviflora I* races, it applies also to *parviflora II*, *biennis I* and *biennis II*. The evidence is not so clear, as we shall see later, in the case of *biennis III* and the strigosas, since opposing complexes in these groups produce for other reasons very similar phenotypes.

Whatever the truth may be regarding the origin of the heterogeneity in segmental arrangement shown by the beta *parviflora I* complexes, the great variety that they show is probably a reflection of the length of time that this group of complexes has been in existence. This is the oldest group of complexes in the subgenus *Oenothera*.

We have seen that the segmental arrangements of the beta *parviflora I* complexes are quite varied. What about the arrangements among the alpha complexes? These also show considerable variety. On the other hand, they show in some cases a close affinity to the alpha complexes of the *biennis III* group, which group in turn is not far removed from the alpha *biennis I* complexes. Thus, the segmental arrangement of α *Clifton Forge*, a *parviflora I* complex, is identical with that of α *Smethport* and other *biennis III* complexes; α *Roanoke A* and α *Gala I* (both α *parviflora I*) are identical with α *Newfound Gap* and other *biennis III* complexes. α *St. George*, another α *parviflora I* complex, differs from α *Newfound Gap*, only in the fact that 5·6 has exchanged with 11·12 instead of 13·14, giving 5·12 11·6 instead of 5·14 13·6; and 3·4, in exchanging with 7·10, has produced 3·10 7·4 instead of 3·7 4·10.* Similarity in segmental

---

* We shall have occasion in the discussion which follows to refer frequently to specific interchanges that have occurred in the evolution of various complexes. In all cases, we shall assume, on the basis of the facts presented above, that these interchanges have occurred between chromosomes of single complexes in circle-bearing strains, or (which amounts to the same thing) between independent chromosome pairs or rings in members of original populations before inter-population crosses produced large circles, rather than between non-corresponding arms of opposing complexes. The chance that complexes resulting from the latter process would survive are so slight that this mechanism can be ignored.

arrangement coupled with similarity in their effect on the phenotype suggests strongly, therefore, that the alpha *parviflora I* complexes have come from ancestors close to those that later became progenitors of *biennis III*.

One other question remains—were the hybrids between Populations 1 and 2 formed by the crossing of Population 1 as male with Population 2 as female, or the reverse, or could the cross have succeeded either way? There is one fact which indicates that successful crosses were possible only between Population 1 as female and Population 2 as male, and not in the reverse direction. This has to do with the classes of plastids found in the *parviflora* and *biennis* groups. According to Stubbe (1959) the parvifloras have plastome class IV whereas *biennis I* and *III* have plastome III, and *biennis II* has plastome II. If the plastids in the original populations had the same properties as those found in present-day populations, the plastids of Population 2 would probably not have been able to turn green in a *biennis–parviflora* gene combination. The plastids of Population 1, however, would have had no difficulty. We can assume, therefore, that the crosses that gave rise to *parviflora I* were between Population 1 as female and Population 2 as male. Curiously enough, of course, the complex that came in from the male parent became the egg or alpha complex in the hybrid, and the one that came in from the female parent became the pollen or alpha complex. This presents no difficulty, however, since S-factors do not inactivate pollen but merely prevent its germination on selfed styles. The S-factor could therefore have entered the hybrid from either the male or the female side, but would function in the hybrid as a pollen lethal in either case.

The *parviflora I* group arose, therefore, by the cross Population 1 (*parviflora*-like) × Population 2 (*biennis*-like). It had plastids from Population 1 and S-factors from Population 2. It is probable that there was some heterogeneity in segmental arrangement in both populations when they overlapped, so that not all hybrids between them that had ⊙14 had the same segmental arrangement. No doubt many crosses occurred also that produced configurations other than ⊙14. These tended to lose out in the competition because they soon lost some of their hybrid vigor, and furthermore were unable to breed true. It is probable also that crosses occurred in the opposite direction, which resulted in offspring inviable because of plastid incompatibility. The hybrids that survived and gave rise to the modern *parviflora I* races had the fortunate combination of a ⊙14, plastids derived from Population 1, balanced lethals resulting from S-factors and distinct differences in competitive ability between the complexes in embryo sac development. Such hybrids probably arose many times in different localities and they probably had different segmental arrangements in these different areas, even though all of those which survived had $\odot$14. Some of the heterogeneity in segmental arrangement that we find in the *parviflora I* strains, both among the alpha and beta complexes, probably stems from the original hybrids, although further exchanges especially within complexes of plants with a circle of 14 probably added significantly to the variety.

The two populations that overlapped and crossed were probably large flowered and open pollinated, which made crossing easy. The argillicolas have retained this original condition, as have the grandifloras which are relicts of the original Population 2. At first, therefore, the hybrids with circles of 14 probably experienced a great deal of outcrossing, although any self-compatible plant that has many flowers on a given day will be self-pollinated as well as cross-pollinated when bees and moths go from one flower to another on the same plant. Each succeeding generation, therefore, would include the products both of cross-pollination and self-pollination. The latter process would produce but one kind of progeny because of the large circle coupled with balanced lethals. At first, the products of outcrossing might be varied, many plants representing crosses back to one parent or the other, or to individuals that differed segmentally from either parent, and these might represent a considerable proportion of the progeny. Since, however, a fair percentage of the progeny would arise in each generation from self-pollination effected through the agency of insects, and such plants would be identical with each other and with their parent genetically, colonies would in time develop in which an increasing proportion of the individuals would have $\odot$14, balanced lethals and identical genotypes. As the proportion of this type in the population increased, the chances would increase that outcrossing would give the same result as self-pollination since it would be between plants of identical genetic composition. This would decrease the proportion in each generation of individuals that were products of outcrossing to other and different forms, so that the true-breeding hybrid form would tend to swamp out other forms in the area where the hybrid first appeared. This, of course, does not take into account the effect of translocations, within complexes and between non-corresponding arms of opposing complexes, which might tend to increase segmental heterogeneity of the population after circles of 14 were established. Translocations occur, however, only rarely.

With respect to self-pollination, it might not have taken much of a change in some cases to convert flowers that were normally dependent on insects into habitually self-pollinating flowers. In many open-pollinated forms of the present day, the anthers come very close to the stigmas. In some, the stigma lobes bend downward and occasionally or frequently come in contact with the anthers. Some forms, such as the *hookeri* of de Vries, often become self-pollinators when grown under greenhouse conditions—the anthers grow proportionately longer in comparison with the style and overlap the stigmas. In some cases, therefore, a minor change in the relative length of style and stamens might have produced self-pollinating flowers.

Once this change was effected, there was no longer survival value in the presence of large flowers, and any mutation in the direction of smaller flowers would have no deleterious effect and might tend to be retained. In the case of the parvifloras, the flowers have actually become the smallest to be found among the North American members of the subgenus *Oenothera*.

B. *The Origin of* Parviflora II

Subsequent to the development and spread of Population 2, a third population (Population 3) developed and spread into North America. This population may have developed at a time when the climate was somewhat drier, since it was more xerophytic in character than Population 2. In its vegetative characters, it had what we now recognize as the *strigosa* phenotype: the leaves were narrower and thicker than those of Population 2, they were covered with hairs, the stems were stout and woody. It is probable that by the time this population developed, Population 2 had more or less died out in the areas that had become dry, since we find no trace of a *biennis*-like phenotype in the western part of the continent except in isolated areas in the mountains where moist conditions have continued to exist.

As Population 3 spread into the east, it came in contact with Population 2, and farther east encountered individuals of Population 1 that had hitherto escaped contamination with Population 2. History, therefore, repeated itself—crosses between Populations 2 and 3 took place here and there throughout the eastern half of the continent; and in the northeastern area crosses occurred between Populations 3 and 1. The resulting hybrids became the progenitors in the first case of the present-day *biennis* group, in the second case of the *parviflora II* population.

Parviflora II is a group that more or less coincides with *parviflora I* in its geographical distribution and, in common with the latter, it has beta complexes that produce narrow leaves, bent stem tips and subterminal sepal tips. Its alpha complexes, however, produce phenotypes of the *strigosa* type rather than the *biennis* type. The plants as a result have leaves that are thick and hairy, though usually quite narrow.

The story of the origin of *parviflora II* probably parallels that of *parviflora I*. Individuals of Population 1 as female were pollinated by individuals of Population 3. Some of the resultant hybrids received an S-factor from Population 3 and plastids from Population 1. The Population 1 complexes were capable of transmission through the sperm but failed in some cases to compete on equal terms in the production of eggs. In some present-day races of *parviflora II* the *parviflora* or beta complex succeeds in a small percentage of cases in the egg, in others it fails completely. The alpha (*strigosa*) complex never comes through the sperm. This was probably the result initially of S-factors. In most *parviflora II* forms, however, as is true in the case of *parviflora I*, an additional pollen lethal has been added to the genome, resulting in the formation of Renner's "inactives"—grains that die after they have reached the stage of starch formation. When such lethals are present, it is not possible to test for the presence of S-factors.

With respect to segmental arrangements, the situation so far as the beta complexes are concerned is similar to that in the *parviflora I* group. No two complexes have been found to have the same arrangement. In *parviflora II* as in *parviflora I*, however, the number of beta complexes in which segmental arrangements have been worked out in full is small, owing to the fact that only *biennis II* and *parviflora* have plastids that will function in the presence of their own alphas plus a beta *parviflora* complex. Most other hybrid combinations that contain beta parvifloras lack chlorophyll if they appear at all. The explanation for the diversity of beta *parviflora II* segmental arrangements is no doubt the same as for the beta *parviflora I* races.

As for the alpha *parviflora II* complexes, we find that there is one segmental arrangement that is apparently quite widespread, the one characteristic of *rigens* of de Vries' *muricata*: 1·2 3·4 5·6 7·11 9·10 8·14 13·12. This has been found in races in Pennsylvania, Connecticut, Massachusetts and Maine. Closely related arrangements have been found in Michigan.

From their segmental arrangements it is difficult to determine the relationships of the alpha *parviflora II* complexes to the other groups. Since Population 3 is considered to be the same population that gave rise to the beta *strigosa* complexes, and since the latter group, as we shall see, is almost wholly characterized by the presence of 1·4 3·2, it is surprising to find some alpha *parviflora II* complexes with 1·2 3·4, including the commonest arrangement so far found, that of *rigens*. The latter, however, is only two interchanges removed from the arrangement of α *Manistique*, which shows evidence of the original presence of 1·4 3·2, as do most of the other alpha *parviflora II* complexes which do not have the *rigens* arrangement. It is possible that the *rigens* arrangement has arisen by the reverse interchange 1·4 3·2 → 1·2 3·4 in a form that already had the other five chromosomes of the *rigens* formula and which gave rise to the α *Manistique* arrangement (see appendix 1) by the exchange 1·4 13·12 → 1·12 13·4.

One of the α *parviflora II* complexes, α *Ashland A*, has an arrangement close to that of one of the beta *strigosa* complexes—it is two interchanges removed from β *Nebraska*. It is three interchanges removed from β *Ashland C*, belonging to a *strigosa* race found in the same locality.

One other arrangement besides that present in *rigens* has been found in more than one alpha *parviflora* complex: 1·12 3·13 5·6 7·2 9·11 4·10 8·14 has been found in α *Tidestromii* from Maryland and *pubens* from Massachusetts, and the formula for *accelerans* differs from this arrangement only in having 1·11 9·12 instead of 1·12 9·11.

The number of alpha *parviflora II* complexes that have been analyzed is unfortunately so small that we cannot tell whether there are additional segmental arrangements that are much more widely found in the population than others.

C. *The Origin of* Biennis I

As we have indicated, Population 3 also overlapped and crossed with Population 2, producing the progenitors of the present-day *biennis*. Some crosses resulted in plants with ⊙14 and certain of these also received from one parent an S-factor that became a male gametophytic lethal. Whenever the S-factor-containing complex was able to dominate over the opposing complex in the development of the female gametophyte, a so-called balanced lethal situation was established. In other words, the same procedure was followed in the establishment of truebreeding heterozygotes that we have postulated in the formation of *parviflora I* and *parviflora II*.

Throughout much of the range occupied by Population 2, it was this population that probably furnished the S-factors. It is from the *biennis* races that Steiner and his students have obtained most of their evidence for the presence of S-factors that have become male gametophytic lethals. In hybrids resulting from such crosses the complex derived from Population 2 would act as the alpha or egg complex, the complex derived from Population 3 would be the beta or pollen complex. It is not improbable, however, that S-factors were also present in some parts of Population 3, so that (as happened in the origin of *parviflora II*) hybrids could have been formed in which the S-factor was brought in with the *strigosa* complex. In this case, if this complex was able to dominate over its opponent in embryo sac development, the resulting hybrids would have an alpha *strigosa* complex and a beta *biennis* complex. This may account for one of the striking facts about the *biennis* assemblage, namely that throughout the middle west, and stretching to the Atlantic coast in part of its range the races all have *biennis*-like alpha complexes, *strigosa*-like betas, whereas in the northeastern part of the United States and eastern Canada the reverse is true. The former group we have called *biennis I* (= Munz' *biennis* subsp. *centralis*), the latter we have called *biennis II* (Munz' subsp. *caeciarum*).*

The *biennis I* group is a widespread assemblage and is unique in the relative uniformity in segmental arrangement of its alpha complexes (Preer, 1950). The great majority of these complexes have an arrangement that is only one interchange removed from the original, namely: 1·2 3·4 5·14 7·10 9·8 11·12 13·6. Complexes with this arrangement, which we may call standard for alpha *biennis I* complexes, are found throughout the range of the group. To date 56 of the 78 alpha *biennis I* complexes that have been fully analyzed have been found to have this arrangement. Of the remaining 32, six are only one interchange removed from it, 13 are two interchanges, and three are three interchanges removed. Two alpha complexes (α *chicaginensis* = *excellens*, and α *Greenland II*) have the original arrangement for the subgenus *Oenothera*; and six complexes have arrangements that may have arisen directly from the original by single interchanges (α *Cambridge II*, α *New Concord I* and *II*, and α *Warsaw* have 9·13 8·14, and *Beaufort* and *acuens* have 1·4 3·2). Of the remaining, 13 complexes seem to have come from the standard alpha *biennis I* arrangement, since they all have 5·14 13·6 or one of these, or in two cases have suffered the exchange 5·14 13·6 → 5·13 6·14 (α *Athens A* and *C*). In one case (α *Iowa I*), the mode of origin is unclear. The complexes that differ from the standard alpha *biennis* arrangement are listed as follows (all chromosomes other than those mentioned are the same as in the standard):

One interchange removed from the usual alpha biennis I arrangement

αCitronelle	Alabama	7·9	8·10
αDyer	Indiana	1·13	2·6
αLa Crosse	Virginia	3·7	4·10
αTyronza	Arkansas	9·11	8·12
excellens	Illinois	5·6	13·14
αGreenland II	Arkansas	5·6	13·14

---

* It is significant that the alpha *parviflora II* and alpha *biennis II* complexes have essentially the same geographical range. They probably arose from the same sector of Population 3.

Two interchanges removed

αBestwater I	Arkansas	1·4	3·8	9·2	
αBestwater II	Arkansas	1·4	3·8	9·2	
αHarrisonburg	Virginia	1·4	3·2;	5·10	7·14
αReynoldsii	Tennessee	1·4	3·2;	5·10	7·14
αHebron I	Ohio	1·4	3·2;	7·6	13·10
αHebron III	Ohio	1·4	3·3;	7·6	13·10
αRichmond	Virginia	1·9	2·8;	3·10	7·4
αBeaufort	North Carolina	1·4	3·2;	5·6	13·14
acuens	Alabama	1·4	3·2;	5·6	13·14
αCambridge II	Ohio	5·6	9·13	8·14	
αNew Concord I	Ohio	5·6	9·13	8·14	
αNew Concord II	Ohio	5·6	9·13	8·14	
αWarsaw	Indiana	5·6	9·13	8·14	

Three interchanges removed

αAthens A	Georgia	1·4	3·10	7·2;	5·13	6·14
αAthens C	Georgia	1·4	3·10	7·2;	5·13	6·14
αIowa I	Iowa	3·8	5·4	9·13	6·14	

There is no geographical pattern to the variation from the standard segmental arrangement. It is probable that the various arrangements have arisen close to the localities where they have been found (Fig. 18.16).

The beta *biennis I* complexes show less uniformity in segmental arrangement than do the alpha complexes, although many cases of identity are found. Instead of there being a single dominant arrangement, there are two, found with nearly equal frequency, each distributed throughout the middle portion of the *biennis I* range. One of these arrangements, which we may call the β *Paducah* arrangement, has been found in 17 complexes; 18 complexes have the other arrangement, the *punctulans* arrangement.

The β *Paducah* and *punctulans* arrangements are apparently two interchanges apart. The β *Paducah* arrangement has 1·14 13·4, 3·2 5·9; the *punctulans* arrangement has 1·4 13·14, 3·9 5·2. Both complexes have 7·8 6·12 11·10.

A larger proportion of the beta complexes deviate from these norms than is the case among the alphas. Thirty-three of the 67 betas so far analyzed have been found to have arrangements other than the two prevalent ones.

As we shall see, the presence of 1·4 3·2 instead of the original 1·2 3·4 is characteristic of complexes derived from Population 3. It is therefore characteristic of the beta *biennis I* complexes, the alpha *biennis II*

FIG. 18.16. Map showing geographical locations of the alpha *biennis I* complexes whose segmental arrangements have been determined. Circles represent complexes with the standard alpha *biennis I* segmental arrangement. Stars indicate those complexes which depart to some extent from the standard.

complexes and some at least of the alpha *parviflora II* complexes. Population 3 must have suffered the interchange 1·2 3·4 → 1·4 3·2 very early in development so that these chromosomes have become characteristic of essentially all complexes of the beta *strigosa* type. It would seem that the arrangement from which most or all of the beta *biennis I* complexes have descended was 1·4 3·2 5·6 7·8 9·10 11·12 13·14. In the development of the beta *Paducah* arrangement one of three sequences of interchange took place involving 5·6 7·8 9·10 11·12. Either 5·6 and 9·10 became 5·9 6·10, followed by the exchange 6·10 11·12 → 6·12 11·10, or 9·10 and 11·12 exchanged to form 9·12 11·10, followed by the exchange 5·6 9·12 → 5·9 6·12, or 5·6 11·12 became 5·11 6·12, followed by 5·11 9·10 → 5·9 11·10. Following or preceding these interchanges the exchange 1·4 13·14 → 1·14 13·4 occurred.

The same sequence of exchanges took place in the formation of the *punctulans* arrangement, except that, instead of an exchange occurring between 1·4 and 13·14, one took place between 3·2 and 5·9, giving 3·9 5·2. Thus, the two groups had a common history except for one interchange, one group suffering an exchange between 1·4 and 13·14, the other between 3·2 and 5·9.

If the last interchange had been 3·2 6·12 → 3·6 2·12, the arrangement would have been that found in certain individuals of the *Evansville I* and *II* races.

The two major groups of β *biennis I* complexes occupy very much the same geographical area—a region stretching from Arkansas northeastward along the Ohio Valley and across the Appalachians into Maryland and Virginia. Neither arrangement has been found in the northern or the southern part of the range of *biennis I* (Fig. 18.17).

Many of the beta complexes that deviate from these two patterns are found in the Ozarks, or in the southeastern part of the range, although all parts of the range show some deviants.

The arrangements found among the 33 complexes that have departed from the beta *Paducah* or *punctulans* pattern are extremely varied, and superficially they show little resemblance to either of these patterns or to each other. Most of them seem to be separated from most of the others by at least four interchanges. If they could be combined (many cannot because they are pollen complexes of strictly heterogamous races) they would give large circles in a majority of the cases. Only here and there do we find complexes that could give small circles with one another, suggesting possible relationship. Most arrangements have been found in only a single complex; they have probably arisen recently and occupy a restricted geographical area.

Only two of these arrangements have been found in three or more complexes. One of these (the β *Rich Mountain* arrangement) has been found in four complexes found in Arkansas and Alabama. It has two chromosomes in common with *punctulans* but none with β *Paducah*. The other has been found in five beta *biennis I* complexes from Georgia and Virginia, including the *truncans* complex of de Vries' *grandiflora*, which is a *biennis I* race collected near Castleberry, Alabama. This arrangement

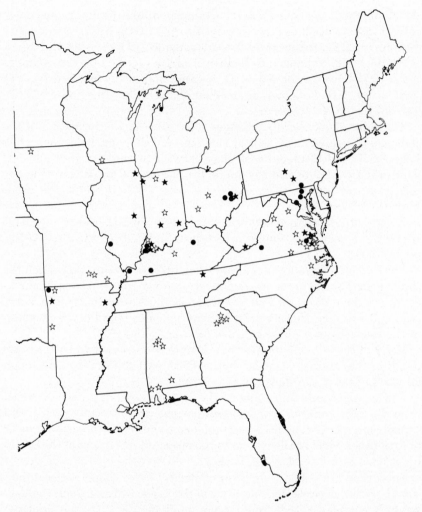

FIG. 18.17. Map showing geographical locations of the beta *biennis I* complexes whose segmental arrangements have been determined. Circles represent complexes with the beta *Paducah* arrangement; black stars indicate those with the *punctulans* arrangement, white stars those with other arrangements.

has also been found in *Charlotteville I*, a plant which was evidently a hybrid between *biennis I* and *argillicola*, in which it served as the alpha complex (in de Vries' *grandiflora*, *truncans* comes through both sperm and egg). Only one interchange removed from the *truncans* arrangement is the beta complex of *Magnolia*, from Kentucky; and a race from La Crosse, Virginia, has a beta complex that, while different from β *Magnolia*, is also one interchange removed from the *truncans* arrangement. It is probable that an intensive study of the southeastern states would show that the β *Rich Mountain* and the *truncans* arrangements are quite widespread in their respective areas.

Neither of these arrangements, however, is very close to the commonest arrangements found among the beta *biennis I* complexes. The *truncans* arrangement gives ⊙10, 2 pairs, with the *punctulans* arrangement; ⊙12, 1 pair, with that of β *Paducah*. This is as close as *truncans* comes to any arrangement so far found (except for β *Magnolia* and β *La Crosse*). One can almost wager that *truncans* will give ⊙14 when one combines it with an unknown complex. It has gotten as far away from the arrangements of other complexes as any that has been analyzed to date. The presence of two chromosomes in common with *punctulans*, however, neither of which is an original or relict chromosome, suggests that it has developed originally from the latter or some arrangement close to it.

In the other cluster of complexes with an identical segmental arrangement (the β *Rich Mountain* arrangement) we may have a situation that is intermediate between the ancestral and the *truncans* conditions. This arrangement has two chromosomes in common with *punctulans* (1·4 13·14) and two in common with *truncans* (3·7 and 8·12). It may stem back to a time when an ancestral form had 1·4 3·2 5·6 7·8 9·12 11·10 13·14. An exchange between 5·6 and 11·10 could have produced 5·11 6·10; two interchanges involving 3·2 7·8 9·12 could have yielded 3·7 9·2 8·12.

One cannot, of course, be certain, when one is dealing with as varied a population as this, that possible sequences such as these were the actual interchanges that took place. The 33 complexes with segmental arrangements different from the beta *Paducah* and the *punctulans* arrangements have yielded 25 different arrangements. Out of 91 possible kinds of chromosomes from the standpoint of end arrangements, 67 have been found among the beta *biennis I* complexes so far analyzed. These could have come about in a variety of ways. It is possible, however, to make the following general statements: (a) The beta *biennis I* complexes have been derived from the same original population (Population 3) that has produced the alpha *parviflora II*, alpha *biennis II* and beta *strigosa* complexes, all producing a *strigosa* phenotype. (b) Most beta *biennis I* complexes belong in a main stream of evolution, in which the exchange 1·2 3·4 → 1·4 3·2 was followed by 7·10 9·8 → 7·8 9·10 and this by two successive interchanges involving 5·6 9·10 11·12 to give 5·9 6·12 11·10. This led by a single interchange to the beta *Paducah* arrangement, and by a different single interchange to the *punctulans* arrangement. From one or the other of these arrangements, or possibly by independent routes, the various other arrangements arose. (c) The two chief arrangements have survived over a considerable part of the *biennis I* range, especially in the Ohio River Valley and in the region to the east including Virginia and Maryland. Deviant arrangements are especially characteristic of the mountainous areas of the Ozarks and Appalachians, where isolating factors have been more prevalent, and of the regions to the east of the Appalachians into which complexes have emerged after experiencing varying histories of interchange in isolated mountain valleys. At least one of these variants in segmental arrangement seems to have become quite widespread in the area to the east of the Appalachians—the *truncans* arrangement. The diversity in segmental arrangement is probably to be correlated with broken terrain more than with age of the population.

D. *The Origin of* Grandiflora

Population 3 has apparently failed to contaminate a small island of Population 2 plants existing around Mobile, Alabama, the so-called grandifloras. de Vries found on his visit to this area that there was a considerable admixture of *grandiflora* and the surrounding *biennis* forms and it would be interesting to carry out an intensive analysis of this area to determine the extent to which *grandiflora* is being transformed into *biennis*. Since *grandiflora* has no lethals it could theoretically combine either with the *biennis*-like alpha *biennis* complex or with the *strigosa*-like beta complex of *biennis I*. Since, however, it is open pollinated and the surrounding *biennis* forms are close pollinated, it is more likely that the crosses would involve *grandiflora* as female than the other way around. Such crosses would normally bring to the hybrid the beta complex of the *biennis* parent and give typical *biennis I* forms. As a matter of fact the few strains of *biennis I* that I have obtained and grown from this region are very similar to *grandiflora* phenotypically and probably have obtained their alpha complex from *grandiflora*. If this is the case, they should not have pollen lethals, since *grandiflora* is not known to harbor S-factors. They should therefore split off in each generation alpha · alpha plants, i.e., pure *grandiflora*, and it may well be that the *grandiflora* population itself is made up, at least in part, of such segregants. The breeding behavior of most of the *biennis* forms from that vicinity is unknown. It is of interest, however, that the *acuens* complex of de Vries' *grandiflora*, which is not a true *grandiflora*, but a southern type of *biennis I*, is alethal and appears in each generation in homozygous condition, the so-called *ochracea* of de Vries.

E. *The Origin of* Biennis II

Turning now to the *biennis II* population, this is an assemblage that occupies a different range from *biennis I* although the two overlap to some extent in the region stretching from Pennsylvania to the mountains of Tennessee and North Carolina (Geckler, 1950). As shown above, it would be difficult if not impossible to distinguish *biennis II* from *biennis I* in the field. In the experimental garden, however, where each plant has space to spread, differences between the two groups become in general recognizable. *Biennis II* plants tend to be lower, more spreading and the upper leaves which have served as bracts fall off in many forms. When grown as annuals (seeds sown in mid-winter in the greenhouse), they are somewhat more liable to produce rosettes that last throughout the growing season from which flowering shoots may develop relatively late. Once they begin to flower, however, they average more flowers per day per shoot and run through their flowering period relatively more rapidly than is true of *biennis I*. These features, however, are more characteristic of the strongly biennial long-day races found to the north than of those that reside in the southern portions of the range. Some *biennis II* plants resemble *biennis I* forms very closely.

The most striking differences between *biennis I* and *biennis II*, however, are not observable phenotypically. These are: (1) the *strigosa* and *biennis* complexes are reversed in position, i.e., the *strigosa* complex is transmitted through the egg, the *biennis* complex through the sperm; (2) the complexes are segmentally rather different from the complexes of *biennis I*. These differences, of course, were not recognized until *biennis I* and *II* had been investigated in some depth cytogenetically. All *biennis II* races have shown ⊙14, balanced lethals and self-pollination.

We are confronted at once, of course, with the question as to why the *biennis* races in the two geographical areas should show these differences. While we cannot be dogmatic on the basis of our present level of knowledge, it is possible, as mentioned above, that members of Population 2 in the northeastern part of the range that lacked S-factors were brought into contact with members of Population 3 that did have these factors, so that it was the *strigosa* rather than the *biennis* complex that contributed the S-factor. Steiner and Schultz (1958) list several *strigosa* and *biennis II* races for which they have evidence of the presence of S-factors, so that this is not an unreasonable hypothesis. In fact, it is quite possible that the *biennis II* population has stemmed from a single such cross, as evidenced by the close similarity or identity in segmental arrangement found among the alpha *biennis II* complexes, and to a slightly less extent among the beta *biennis II* complexes (see below).

The alpha *biennis II* complexes show their *strigosa* affinities in their segmental arrangements. There is one arrangement that is much more prevalent than any other in the material we have studied, an arrangement that is found throughout the range of the group—from Pennsylvania to New Brunswick, Canada. This originally had the 1·4 3·2 characteristic of the beta *strigosa* complexes. The 3·2 is still present but the 1·4 has been involved in a sequence of two interchanges involving 5·6 and 7·10, the result of which is to give 1·6 5·7 4·10. The entire formula is: 1·6 3·2 5·7 4·10 9·8 11·12 13·14.

This arrangement (which we generally refer to as the α *Waterbury* arrangement) has been found in 26 of the 39 alpha *biennis II* complexes so far analyzed completely. The other complexes have arrangements that are for the most part closely related to it. Three alpha complexes belonging to races on the western edge of the range have a segmental arrangement that is closer to the arrangement that is original to the North American oenotheras as a whole than it is to the α *Waterbury* arrangement. These are α *Morgantown* (West Virginia), α *St. Croix* (Wisconsin) and α *Williamsville* (western New York). They have 1.4 3·2 5·6 7·10 9·8 11·12 13·14 .This is probably the arrangement ancestral to all the beta *strigosa* complexes.

Other α *biennis II* complexes have arrangements similar to that of α *Waterbury* or to the α *Morgantown* arrangement. *Jugens* (= α *shulliana* from New Jersey) is one interchange removed from α *Waterbury*. α *Buck Creek* (North Carolina) is two interchanges removed from both the α *Waterbury* and the α *Morgantown* arrangements, α *Indian River* appears to be one interchange removed from both,* α *Micaville* is three interchanges removed from the α *Morgantown* arrangement, and α *Omaha* is three interchanges removed from both. Thus, the alpha complexes that do not have the prevalent arrangement have arrangements that are closely related to it. The alpha *biennis II* complexes constitute, therefore, a rather closely knit group segmentally, rather closely related in segmental arrangement to the ancestors of the beta strigosas. On the whole, they appear to be more primitive in segmental arrangement than the bulk of the beta *biennis I* complexes. This may be due to the fact that they are found in areas where the opportunities for geographical isolation are not as numerous as they are farther south. Barriers to geographical distribution are relatively minor in the northern plains states, the Ohio and St. Lawrence valleys. It is also possible, as just mentioned, that they are descended from a single hybridization between Populations 2 and 3.

The beta *biennis II* complexes, like the alpha *biennis I* complexes, are a fairly close-knit group. They show no evidence of the exchange 1·2 3·4 → 1·4 3·2. They are only moderately close to alpha *biennis I* in

---

* α *Indian River* may be intermediate between the original (α *Morgantown*) arrangement for the strigosas and the α *Waterbury* arrangement. The exchange 5·6 7·10→5·7 6·10 in the original gave the α *Indian River* arrangement, and the further exchange 1·4 6·10→1·6 4·10 gave the α *Waterbury* arrangement.

segmental arrangement, on an average a minimum of four interchanges removed. In one case it is possible to see how a beta *biennis II* complex could have arisen from the standard alpha *biennis I* arrangement. Beta *St. Anne* might have experienced the following interchanges: 1·2 13·6 → 1·13 2·6; 5·14 9·8 → 5·8 9·14; 7·10 11·12 → 7·11 10·12.

There are several arrangements that have been found to be rather prevalent among the beta *biennis II* complexes. In most cases, each of these arrangements has been found in widely separated localities, and is probably, therefore, quite widespread. Three of these arrangements are quite closely related. They have 1·2 5·6 7·10, and differ from the ancestral arrangement of the North American oenotheras by three interchanges involving 3·4 9·8 11·12 13·14. One frequently found arrangement (that of β *Buck Creek*), however, is less closely related to the original. It is a minimum of four interchanges removed from the common arrangement for the alpha *biennis I* complexes, is separated by four interchanges from two of the more frequent beta *biennis II* arrangements, and by five interchanges from the other commonly found arrangement.

Among the interchange chromosomes, 4·14 is the most frequently found in the beta *biennis II* complexes (in 18 out of the 35 of those whose arrangements have been determined fully). This suggests that the interchange forming this chromosome occurred early—possibly through a translocation involving 3·4 and 13·14, giving 3·13 4·14. The next commonest chromosomes are 3·12 (in 16 complexes) and 11·13 (in 13). These may have arisen quite early through the exchange 3·13 11·12 → 3·12 11·13.

While we cannot be sure of the exact sequence of interchange in most cases, it is clear that the segmental arrangements of the beta *biennis II* complexes are of the *biennis* rather than the *strigosa* type. Being only moderately related, however, to the commonest arrangement among the alpha *biennis I* complexes, they have probably developed by a different route from the one that produced the midwestern alpha *biennis I* complexes.

There are two races of *biennis II* that have decidedly anomalous geographical positions, outside the expected range. These are *Omaha* (Nebraska) and *Bloomington III* (Indiana). *Omaha* may owe its position to the accidental transfer of seed out of the normal range. On the other hand it could represent an isolated case where an S-factor-bearing member of Population 3 met a member of Population 2 which lacked it; or an S-factor might have become transferred from one complex to the other via crossing over or translocation involving non-corresponding segments of opposing complexes.

In the case of *Bloomington III*, the race was picked up in the wild several miles from our experimental field, and we were naturally suspicious that it might contain complexes that had escaped from our collection. It turns out that its alpha complex is identical segmentally with *albicans* so that one of its complexes has probably come from our garden. The other complex, however, has not been found previously: it is, however, a typical beta *biennis II* complex in segmental arrangement and is only one interchange removed from one of the commonest beta *biennis II* arrangements, that of *maculans* and β *Ithaca*.

A third anomalous race may be mentioned at this point. *Elgin* is a cruciate race from *Illinois*, mentioned in chapter 12, in which the alpha is the *strigosa* complex, the beta is the *biennis* complex. It would therefore seem to be a member of the *biennis II* group, especially since the segmental arrangement of β *Elgin* is identical with one of the common arrangements found among the beta *biennis II* complexes in New York and Pennsylvania (the β *Elma V* arrangement). The alpha complex, however, bears no resemblance to the alpha *biennis II* complexes in segmental arrangement, but rather shows affinity with the alpha *strigosa* complexes. *Elgin* is therefore not a typical *biennis II* race but is probably the result of a cross between a *strigosa* as female and a *biennis II* as male. Its geographical position suggests that this was possible, since the strigosas have been found as far east as Wisconsin, and *biennis II* has also been found in Wisconsin. *Elgin* grows, therefore, at the extreme eastern edge of the *strigosa* range and the extreme western edge of the *biennis II* range and it is quite probable that it owes its existence to a cross between the two species and thus has complexes descended from Populations 2 and 4, rather than from 2 and 3, as is true of the *biennis II* races.

F. The Origin of Biennis III

The third category of *biennis* forms is what we have termed *biennis III* (Preer, 1950). This is an assemblage situated along the line where the ranges of *biennis I* and *biennis II* meet or overlap. It extends from northern Pennsylvania to the mountains of Virginia, North Carolina and Tennessee. When grown in the garden it is relatively easily distinguished from *biennis I* and *II*. The latter groups are combinations of a *biennis* and a *strigosa* complex, and although the *biennis* complex dominates to such a degree that the presence of a *strigosa* complex was not suspected until cytogenetic work demonstrated its presence, the *strigosa* complex has some influence on the phenotype, thickening the leaves somewhat, increasing the amount of pubescence and often showing other effects that one would scarcely recognize if he were not looking for them. *Biennis III*, however, has no *strigosa* complex. It has arisen by crossing between *biennis I* and *biennis II*, alpha *biennis I* combining with beta *biennis II* complexes. It is therefore pure *biennis* in character. As a result, the leaves are thinner, more evenly and less coarsely crinkled, less pubescent than in the other *biennis* forms.

The number of *biennis III* forms so far studied is limited. Thirteen races have been found, one of them with a somewhat questionable history in the experimental garden (*White Top*). The alpha complex has been fully determined in twelve of these, the beta complex in eleven out of the thirteen.

As might be expected, the alpha *biennis III* complexes are close to the alpha *biennis I* complexes segmentally. Most of them have arrangements that are only two interchanges removed from the commonest *biennis I* arrangement, although the interchanges involved in the formation of the different alpha *biennis III* complexes have varied, so that the twelve complexes which have been analyzed include eight different segmental arrangements.

The beta *biennis III* complexes on the other hand show very close affinities to the beta *biennis II* complexes. The two most commonly found arrangements among the beta *biennis III* complexes are identical with two of the commonest arrangements among the betas in *biennis II*. Two other complexes are identical with another of the common arrangements in *biennis II*. Only one beta *biennis III* complex has been found that is not identical in segmental arrangement with one of the beta *biennis II* complexes so far analyzed. It is not too much to say, therefore, that the beta *biennis II* and beta *biennis III* complexes belong to the same group and are distinguished one from the other only by the company they keep, i.e., the complexes with which they are associated.

From the above it is quite clear that *biennis III* has been formed by hybridization between *biennis I* and *biennis II*. The geographical location of these forms in the region where the other two groups meet, the presence of two *biennis* complexes instead of a *biennis–strigosa* combination, and the similarity (essential identity in the case of the beta complexes) in segmental arrangement between *biennis III* complexes and the corresponding complexes in the other groups support this hypothesis. Another point should also be added. According to Stubbe (1959) *biennis III* has the same class of plastids that is found in *biennis I*, which is to be expected following the cross *biennis I* female × *biennis II* male.

The question may be raised at this point as to whether crosses between *biennis I* and *II* have occurred in the reverse direction also. We should expect *alpha biennis II · beta biennis I* combinations to be formed, as well as *alpha biennis I · beta biennis II*. These would be pure *strigosa* in phenotype. Actually, only a single such form has been found, near La Follette,

Tennessee. This race showed $\odot 4$, $\odot 6$ in the first generation but later lost its circle of four and was carried on for some generations with $\odot 6$, 4 pairs. It was later lost. The complexes were analyzed after its configuration had become $\odot 6$, 4 pairs, and it was found to have one complex identical in segmental arrangement with β *Paducah*, the other had 7·11 6·8 10·12 instead of the 7·8 6·12 11·10 present in the latter, and was not identical with any alpha *biennis II* complexes so far analyzed, though closely related to β *Paducah* in segmental arrangement (plants with $\odot 6$, 4 pairs, probably had a double dose of the β *Paducah* half of the circle of 4 that was present in the original hybrid).

Does the fact that we have found only one case where crossing between *biennis I* and *II* has led to a *strigosa·strigosa* combination mean that the cross is more likely to occur in one direction than in the other? We believe not. It so happens that *La Follette* had a configuration that would not favor its survival in Nature ($\odot 4$, $\odot 6$). There are, however, as many chances of $\odot 14$ being formed in hybrids between the alpha *biennis II* and beta *biennis I* complexes that we have analyzed, as in reciprocal crosses. The commonest arrangements found in *biennis I* and *biennis II* would not give $\odot 14$ in combination, whichever way the cross was made. The commonest arrangement among the alpha *biennis I* complexes would give $\odot 10$ with most of the commoner arrangements among the beta *biennis II*'s. The commonest α *biennis II* arrangement would give $\odot 6$, $\odot 6$ with *punctulans* and $\odot 4$, $\odot 8$ with β *Paducah*. There are complexes in the two populations, however, that will give $\odot 14$ with each other. Whichever way the cross is made, such a result can be obtained. One cannot argue, therefore, that the chances for hybrids with $\odot 14$ are fewer following crosses in one direction than in the other. Furthermore, there should be no plastid incompatibility, no matter which way the crosses occurred. It is probably a mere matter of chance that we have so far failed to find the products of *biennis II* × *biennis I* as frequently as the reverse crosses. We have only found thirteen cases of the latter. An intensive investigation of the region where such crossing could occur might turn up a number of cases of hybrids formed in both directions. We would be surprised, however, to find large numbers of *biennis III* forms (or their reciprocals), because these would undoubtedly represent the products of crosses between self-pollinating races, races which hybridize with other strains only when flowers for some reason fail to produce enough pollen to effect self-fertilization of all their eggs and when other races happen to grow in close proximity. Such outcrossing is rare, and when it occurs, another configuration than $\odot 14$ may be present in the hybrid which would place it under a handicap. Since, therefore, crossing between *biennis I* and *biennis II* is rare, even in a mixed colony, and since the rare hybrids produced may have other than $\odot 14$, the number of successful true-breeding hybrids produced must be rather low. It is surprising that we have found as many as thirteen races that have arisen in this way.

A final word about the plastid situation in the *biennis* groups. According to Stubbe (1959) *biennis I* and *III* have plastid class III, whereas *biennis II* has class II. This conclusion was based on a study of more than 80 complex-combinations involving 14 races. We have kept a record of plastid behavior in the more than 3000 complex-combinations that we have studied. So far as the *biennis* groups are concerned, however, our results do not seem to require the hypothesis that the plastids of *biennis II* are different from those in *biennis I* and *III*. It is true that *biennis I* or *III* × *parviflora I* or *II* results in chlorotic or chlorophyll-less plants, whereas *biennis II* × *parviflora* produces green progeny. These two situations, however, are not comparable, since the first cross combines a *biennis* with a beta *parviflora* complex, whereas the second combines a *strigosa* and a beta *parviflora* complex. When metaclines occur in crosses of the first type, they are green—they combine *strigosa* and beta *parviflora* complexes as do *biennis II* × *parviflora* crosses (e.g., (*Birch Tree II* × *Gala II*) β *Birch Tree II* · β *Gala II*). Similarly, when metaclines occur in crosses of the second type, they are deficient in chlorophyll when they appear (usually they fail to germinate). These combine *biennis* and beta *parviflora* complexes as do *biennis I* × *parviflora* crosses (e.g., (*shulliana* × *Nobska*) *maculans* · β *Nobska*). In other words, plastids of both *biennis I* and *biennis II* react in the same way to similar complex-combinations. The acid test will be performed when the plastids of several *biennis I* and *biennis II* races are brought into association with identical genome combinations.

While minor variations in behavior are found among the various races within a group, our experience in general indicates that the plastids of all *biennis* forms show very similar behavior.

To sum up the basic facts regarding the origin of *biennis*, we postulate that both *biennis I* and *biennis II* arose from crosses between Populations 2 and 3. In the case of *biennis I*, Population 2 furnished what became the female or alpha complex, with its *biennis* characters and its S-factors; Population 3 furnished the male or beta complex. *Biennis II* received its beta complex from Population 2 and its S-factors probably came from Population 3. The plastid situation seems to suggest, however, that both *biennis I* and *biennis II* arose by crosses between Population 2 as female and Population 3 as male. Both groups seem to have *biennis*-type plastids. If Population 3 had served as female, plastids of the *strigosa* type would probably be present in the resulting races of *biennis*.

According to Stubbe, however, *strigosa* has class I plastids, the same class that is found in the hookeris. We may postulate, therefore, that both *biennis I* and *biennis II* arose from crosses between Population 2 as female and Population 3 as male; they differ, however, in that *biennis I* received its S-factor from Population 2 and *biennis II* received its S-factor from Population 3.

Having described the origin of the *biennis* groups in this way, we must hasten to qualify the conclusions drawn in one respect. We have spoken as though all eggs in *biennis I* carry the alpha complex which produces *biennis* characters and all sperms carry the *strigosa*-like beta complex, the reverse being true of *biennis II*. This would seem to imply that all *biennis* forms are strictly heterogamous. This is not true, however, especially in the case of *biennis I*, many of which are only partially heterogamous, and a few are essentially isogamous. The transmission of beta complexes through the egg is easily explained in terms of the so-called "Renner effect". The beta complex may be strong enough to win out occasionally over the alpha when it happens to occupy the micropylar end of the quartet of megaspores—the favored end. When, on the other hand, an alpha complex is transmitted occasionally or frequently through the sperm, as is true in some *biennis I* forms, it would seem to indicate that this complex does not have a true S-factor. It is probable that not all *biennis* races have S-factors. Instead, in some cases where alpha complexes come through a proportion of the sperm, we may have, as originally suggested by Renner (1919a, b) and Heribert-Nilsson (1920a), a process of competition in the growth of the male gametophyte. Some classes of pollen may develop their tubes faster than others, so that by the time the slower class of pollen tube reaches the ovules, all eggs have been fertilized by sperm from the faster growing complex, thus simulating the effect of a pollen lethal (see p. 103).

Since an approach to isogamy is more common among *biennis I* forms than others, this is perhaps the place to re-emphasize the fact that most North American members of the subgenus *Oenothera* that have ⊙14 breed true whether they are nearly isogamous, partially heterogamous or strictly heterogamous. This probably means that many forms that are not wholly heterogamous have zygotic lethals which prevent the development of complex-homozygous zygotes. It is likely that when hybrids with circles of 14 were first formed they had no zygotic lethals, although, as we have seen, gametophytic lethals were probably present from the start. As time went on, mutations occurred and an occasional mutation produced a lethal gene. Or it may be that some translocations were not quite equal and as a result complexes were formed that had a deficiency of an essential locus, so that they could not exist in homozygous condition. Thus, zygotic lethals developed.

When balanced gametophytic lethals are present in a strictly hetero-gamous form it is not possible to detect the presence of zygotic lethals directly. We may infer, however, that they are present in many if not in all such forms, since essentially all strains which produce metacline hybrids when outcrossed, fail to throw homozygotes when selfed.

G. *The Origin of* Strigosa*

We may now carry our story one step farther. There came a time when a fourth population developed and spread into the area east of the Rockies. Population 4 had phenotypic characteristics very similar to those of Population 3, but different interchanges had taken place in the two populations. In the western part of the range of Population 3, where it had not come into contact with Population 2, it was still composed of open-pollinated forms and was therefore capable of crossing easily with the invading Population 4. Once again hybridization occurred here and there and an occasional hybrid had ⊙14. Population 4 included plants with S-factors which became pollen lethals in the hybrids. In some cases the S-factor-carrying complexes brought in from Population 4 were able to win out in the competition to produce embryo sacs. They then became the alpha complex, the complex from Population 3 became the beta complex. The resultant hybrids became the progenitors of the modern strigosas.

The plastids of the strigosas belong to the same class as those found in the hookeris (Class I—Stubbe 1959). If they were contributed by Population 4, then Population 4 served as the female in the crosses with Population 3 and furnished both plastids and S-factors to the crosses. If they were contributed by Population 3, then Population 3 furnished the plastids and Population 4 furnished the S-factors.

The strigosas are on the whole strictly heterogamous forms. Out of 42 races analyzed in our laboratory, only nine have shown any exceptions to the rule of strict heterogamy. Two races have produced metaclines only rarely, and only when used as male (*Minturn* and *Tinytown*) (Norby, unpublished); in two races such metaclines have appeared more frequently but still in small proportions (*Haskett, Iowa II*). Three races have produced metaclines only when used as female. These are *Cockerelli*, *Sutherland* and *Brookston*. Only two races have produced metaclines when used both as male and female, namely *Iowa II* and *West Estes Park* (Norby, unpublished). Thirty-three races of the 42 studied have shown no metaclines among hundreds of hybrid offspring.

---

* See Cleland 1954a, b.

With regard to segmental arrangement, the situation is somewhat more complicated among the alpha strigosas (derived from Population 4) than among the betas (Population 3). We will discuss the beta strigosas first.

There is much similarity among the beta *strigosa* complexes in the distribution of segments. Among the 42 complexes of this group that have been completely analyzed, only 14 arrangements have been found. All originally had 1·4 3·2 and all but five still have both of these chromosomes. Four of these exceptions have 3·2, the 1·4 having exchanged with 13·14 to give 1·13 4·14. One complex has 1·3 4·2 instead of 1·4 3·2 (*β Heber*). Two main groups of beta complexes are found, one which originally had 7·10 9·8 and in which 5·6 7·10 have exchanged to form 5·10 7·6 (16 complexes) or 5·7 6·10 (3 complexes); the other which came to have 7·8 9·10 and still retains these chromosomes, but in which 5·6 13·14 have exchanged to form 5·14 13·6 (17 complexes) or 5·13 6·14 (1 complex). Two of the former group (the one with 7·10 9·8 originally) have suffered a further exchange involving 7·6, which exchanged with 13·14 to give 7·14 13·6; two others (both from the Ozarks) have had a more complicated history in which the original 9·8 took part in a sequence of interchanges involving 5·6, 11·12 and 13·14. Several of the complexes that originally had 7·10 9·8 have experienced, in addition to the above, a further interchange involving in one case 11·12 and 13·14, in another case 9·8 and 13·14, or all three of these chromosomes in two cases.

It is easy to visualize the sequence in which these interchanges took place as Population 3 spread over the continent (see Fig. 18.18). We suggest that the exchange 1·2 3·4 → 1·4 3·2 took place very early, as the population was emerging from the center of distribution, since this exchange is characteristic of all the beta complexes of the strigosas as well as of the beta *biennis I*, alpha *biennis II* and alpha *parviflora II* complexes, all of which have had a common origin in Population 3.

One sector of the spreading population advanced northward, then spread both to the west and the east in higher latitudes. It suffered a second interchange which became characteristic of all races in these areas. This was the exchange 7·10 9·8 → 7·8 9·10.* Later on, a third exchange took place in the ancestry of most of the group. This involved 5·6 and 13·14 to give 5·14 13·6 which characterizes all but one of the beta *strigosa* complexes so far analyzed from the region stretching from the Pacific Northwest to the northern Great Plains and is also found in western and central Colorado. Two lines branched off from the mainstream, however. One of these (represented by *β Heber*) suffered an

---

* This was apparently the arrangement that became ancestral to the beta *biennis I* complexes.

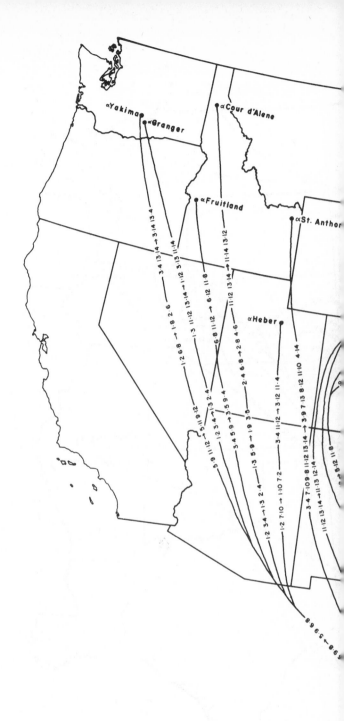

FIG. 18.19. Map showing the sequences of interchange that have c
major waves of migration, as did Population 3. The lines in this and t
they do indicate in a general way the relation of the various paths c

FI
have
of m
[f. p.

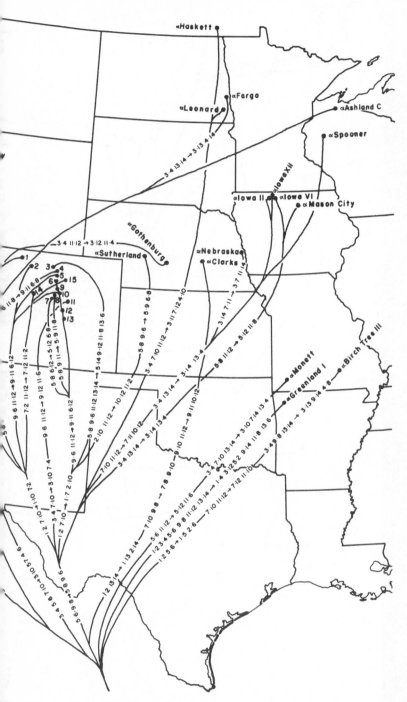

occurred in the evolution of the alpha *strigosa* complexes. Population 4 followed three
the previous figure are not meant to indicate the exact location of migratory routes, but
of migrations to one another.

[f. p. 289]

interchange from 1·4 3·2 to 1·3 2·4. Otherwise it is identical with the population just mentioned. The second line may or may not have branched off before the exchange took place that gave 5·14 13·6. This line, represented by β *Loveland*, experienced an exchange which gave 5·13 6·14 instead of 5·14 13·6. One complex has suffered two exchanges in addition to those which characterize most of the population in the northwest. β *Cœur d'Alene* has 5·9 10·14 instead of 5·14 9·10, and 7·12 11·8 instead of 7·8 11·12.

A second sector of the original population spread northward into the Rockies and across the central plains in the direction of Iowa and Wisconsin.* Very early in this migration the exchange 5·6 7·10 → 5·10 7·6 took place, which became a feature of most of the beta *strigosa* complexes found in these regions. In the case of β *Nebraska* no further exchanges took place, but in the line that led to β *Clarks*, β *Iowa II*, β *Mason City* and β *Spooner* the exchange 11·12 13·14 → 11·13 12·14 took place, and in the populations that invaded the Rocky Mountain area other exchanges occurred locally, 9·8 13·14 → 9·14 13·8; 1·4 13·14 → 1·13 4·14; 9·8 13·14 → 9·13 8·14, 7·6 13·14 → 13·6 7·14. Certain of the complexes in the upper Mississippi Valley are apparently derived from the line that acquired 9·14 13·8: in the case of β *Iowa VI*, 5·10 7·6 became 5·7 6·10 and in the case of β *Ashland C* and β *Iowa XII* a further exchange between 13·8 and 11·12 gave 11·13 8·12.

A third sector spread eastward into the Ozark region and here the interchanges seem to have been different and more complicated. Only two beta *strigosa* complexes have been analyzed from this area and they are identical in segmental arrangement. They retain the 1·4 3·2 and the original 7·10. There has been, however, a sequence of interchanges involving the other four chromosomes to give 5·11 9·6 8·14 13·12. It is not possible at present to determine what exchanges occurred, and in what order, to give this result.

In the case of the beta strigosas, therefore, we can often see what interchanges took place and in what order, and we can see how early interchanges in different areas became characteristic of large sectors of the population, and how subsequent interchanges occurred more locally as the population spread.

These exchanges may have occurred in Population 3 before it met Populations 1 and 2, and before it was in turn invaded by Population 4. This is almost certainly true of the interchanges that are widely distributed throughout the oenotheras. Interchanges that are of only local

---

* In addition to the beta *strigosa* complexes in these areas, the alpha *biennis II* complexes have probably developed from this sector.

distribution within the strigosas may have occurred after crosses between Populations 3 and 4 had produced plants with ⊙14.

The situation is somewhat more complicated when we turn to Population 4, from which the alpha *strigosa* complexes have evolved. Here there was considerably more variation in segmental arrangement than in Population 3. Once again we can see differences between those parts of the population that spread northwestward, southeastward into the Ozarks, or northward and northeastward into the Rocky Mountains and the Great Plains. There have been more interchanges, however, and the story is not quite as clear-cut as in the case of Population 3 (Fig. 18.19).

The complexes that reside in the central area, including Colorado and nearby regions, are on the whole closely related, differing for the most part by one to three interchanges from one another. Those in the Pacific Northwest show little relationship to the Colorado group segmentally. As in the case of Population 3, the complexes in the Ozarks show evidence of an early separation from the main stream of evolution of the population. They are quite unrelated to the rest of the alpha strigosas. The complexes of the central plains states are more closely allied to the Colorado group than to the others, as was also the case with most of the beta strigosas in this region.

All told, 39 alpha *strigosa* complexes have been completely analyzed. These include 27 different segmental arrangements.

None of the alpha *strigosa* complexes has the 1·4 3·2 that characterizes the beta strigosas. This exchange, which occurred at the very beginning of the evolution of Population 3, did not occur in Population 4, all of which have retained 1·2 and 3·4, or one or both of these have been involved in further exchanges.

In the case of the central group of complexes, extending from Colorado northeastward, there is one exchange that seems to have taken place very early in its evolution, namely 5·6 9·8 → 5·8 9·6. Many complexes of the group possess both of these chromosomes, while in others one or both have been modified by subsequent exchanges. Following this exchange, several pathways have been followed. Many complexes derived from the part of the population that spread northward west of the continental divide have experienced in their evolution the exchange 1·2 7·10 → 1·10 7·2. Following this, different interchanges have taken place in the development of different complexes: 3·4 11·12 → 3·12 11·4 in the case of α *Gothenburg*; 9·6 11·12 → 9·11 6·12 in α *Littleton I*, α *Longmont* and α *Yampa Valley*; 7·2 11·12 → 7·12 11·2 in α *Minturn* and α *Platteville*. It will be noted that some of the complexes in the northern plains area have probably come from this portion of Population 4, having crossed the divide in the Wyoming area where the passes are relatively low and broad. Thus, α *Loveland*, α *Fargo* and α *Ashland C* have probably arisen from this source, experiencing the exchange 5·8 11·12 → 5·12 11·8 on the way (In the case of α *Loveland* a further exchange occurred between 11·8 and 9·6 giving 9·11 6·8). This suggestion is based on the assumption that all complexes with 1·10 7·2 have come from the same source, which of course could be wrong. These chromosomes could have arisen independently east of the divide, the chief evidence for which is the fact that α *Littleton I* and α *Littleton II* differ only in the fact that one has 1·10 7·2, the other 1·7 2·10.

To the east of the divide, the original exchange that formed 5·8 9·6 was followed by 1·2 7·10 → 1·7 2·10. The further exchanges 9·6 11·12 → 9·11 6·12 led to such complexes as *curtans*, α *Forsberg*, α *Littleton II*, α *Skyview*, α *Palmer Lake* and α *North Colorado Springs*. An additional exchange 5·8 6·12 → 5·12 6·8 led to α *Turkey Creek Pass*, and 9·11 5·8 → 5·9 11·8 led to α *Tinytown*. A sequence of exchanges involving 5·8 9·6 11·12 13·14 produced the 5·14 9·12 11·8 13·6 of α *West Estes Park*, and the exchanges 2·10 11·12 → 10·12 11·2 and 5·8 9·6 → 5·9 6·8 produced α *Sutherland*.

Other complexes have arisen from lines derived from the ancestor in which 5·8 and 9·6 are found, but in which neither of the exchanges producing 1·10 7·2 or 1·7 2·10 occurred. The lines that led to α *Leonard* and α *Haskett* experienced a sequence of two interchanges involving 3·4 7·10 11·12 to give 3·11 7·12 4·10. Another line that spread northeastward suffered interchanges between 7·10 and 11·12 to give 7·11 10·12, and between 3·4 and 13·14 to yield 3·14 13·4, thus producing the arrangement found in α *Iowa II*, α *Iowa XII* and α *Spooner*. The additional exchange 7·11 3·14 → 3·7 11·14 produced the α *Nebraska* arrangement. Another sector of the population that advanced to the northeast suffered the exchanges 3·4 13·14 → 3·14 13·4 and 5·8 11·12 → 5·12 11·8, thus producing the α *Iowa VI* and α *Mason City* arrangements.

One complex in Colorado seems not to have been derived from the segment in which 5·6 9·8 became 5·8 9·6; in the development of α *Brookston*, 11·12 13·14 became 11·13 12·14 and a sequence of two exchanges involving 3·4 5·6 7·10 produced 3·10 5·7 4·6.

There is one complex found in the plains area that shows no evidence of the exchange that produced 5·8 9·6: in the development of α *Clarks*, 1·2 13·14 became 1·13 2·14, 7·10 9·8 became 7·8 9·10, and 9·10 11·12 became 9·11 10·12.

As in the case of Population 3, the sectors of the population that migrated into the Ozark area or into the far northwest seem to have been derived from sources other than that which produced the Colorado–northern plains complexes. Only three complexes have been analyzed from the Ozarks, and it is not possible to determine in detail all the inter-changes that have occurred in their evolution. α *Monett* has experienced the exchange 5·6 11·12 → 5·12 11·6, plus two exchanges involving 3·4 7·10 13·14 to give 3·10 7·14 13·4. During the evolution of α *Birch Tree III* the following interchanges have occurred: 1·2 5·6 → 1·5 2·6; 7·10 11·12 → 7·12 11·10; and a sequence of two exchanges involving 3·4 9·8 13·14 to give 3·13 9·14 4·8. The exact sequence of exchanges leading to the α *Greenland I* arrangement cannot be followed with assurance.

In the case of the far northwestern complexes, it appears that the exchange 5·6 9·8 → 5·9 6·8 occurred early in the ancestry of all but one of the complexes analyzed. This was followed, in the evolution of α *Yakima* by the exchanges 5·9 11·12 → 5·11 9·12, 1·2 6·8 → 1·8 2·6, and 3·4 13·14 → 3·14 13·4; in the development of α *Fruitland*, the exchanges were 3·4 5·9 → 3·5 9·4, and 6·8 11·12 → 6·12 11·8. Leading to the arrangement in α *Cœur d'Alene* the following interchanges took place subsequent to the original one—1·2 3·4 → 1·3 2·4, 1·3 5·9 → 1·9 3·5, 2·4 6·8 → 2·8 4·6, 11·12 13·14 → 11·14 13·12. The exchanges in the case of α *Heber* were 1·2 7·10 → 1·10 7·2 and 3·4 11·12 → 3·12 11·4. α *Granger* probably experienced in common with α *Cœur d'Alene* the exchange 1·2 3·4 → 1·3 2·4. This was followed by two interchanges involving 1·3 11·12 13·14 to give 1·12 3·13 11·14.

The only northwestern complex so far analyzed that did not stem from the sector of the population that had experienced the interchange 5·6 9·8 → 5·9 6·8 was α *St. Anthony*. This complex still retains 5·6 as well as 1·2 of the ancestral arrangement. The sequence of exchanges that has led to the α *St. Anthony* arrangement is too complicated to follow.

Summing up the story of the evolution of Population 4 as traceable by the analysis of segmental interchanges, we find, as in the case of Population 3, that three distinctive sectors developed, one advancing into Colorado and from there spreading northeastward into the upper plains states, one spreading northwestward, the third toward the lower Mississippi valley.

Most of the central group stem from a segment of the population that, early in its history, experienced the exchange 5·6 9·8 → 5·8 9·6. The portion of this segment that advanced west of the continental divide suffered the exchange 1·2 7·10 → 1·10 7·2. This segment spread through the lower passes of the northern Rockies and invaded the Dakotas and neighboring areas of Canada. In the segment that advanced east of the divide 1·2 7·10 also exchanged, but formed 1·7 2·10 instead of 1·10 7·2. These chromosomes became characteristic of many of the complexes found along the eastern edge of the Rockies and in some of the adjacent plains areas. Another segment spread into Nebraska, Iowa and Wisconsin.

This segment did not experience the exchange between 1·2 and 7·10: different complexes have suffered different exchanges following the initial one that gave 5·8 9·6: α *Brookston*, west of the divide, and α *Clarks* in Nebraska, apparently stem from segments of the population that did not experience the initial exchange to give 5·8 9·6.

In contrast to the central group of complexes, most of the northwestern group are apparently descended from a segment in which 5·6 9·8 became 5·9 6·8 (instead of 5·8 9·6). The complexes in the Ozark area have not shown evidence of a common segmental interchange.

Although the situation in the alpha *strigosa* group is more complicated than that in the betas, we can often see in both groups what interchanges took place, and in what order; we can note certain interchanges that occurred early and became characteristic of large sections of the population, and others that occurred later and are consequently found only in restricted areas. In the case of both Populations 3 and 4, we can visualize the way in which these populations spread, and can follow the sequence of interchanges as they have occurred in different regions.

As in the case of the earlier populations, the interchanges that are most widespread in the population probably occurred while the population was still open pollinated, with paired chromosomes or small circles. Interchanges found only locally may have occurred after circles of 14 had arisen by crossing of Populations 3 and 4.

H. *The Origin of* Elata *and* Hookeri

There remain for consideration two extant populations that do not have ⊙14, but instead have mostly paired chromosomes. These are the elatas and the hookeris. The elatas occupy an area close to what we have considered the center of distribution for the subgenus *Oenothera*. So far as we have been able to examine them (ten races, studied by Steiner, 1951, 1955), they do not have what we consider to be the ancestral arrangement of ends. Each race is homozygous for its own arrangement, and three arrangements have been found among the ten races. Plants found in Mexico have all shown 1·4 3·2. In addition, they differ from the original by one other interchange. In five races, 9·8 13·14 has become 9·14 13·8. In three races, 5·6 9·8 has become 5·9 6·8. The remaining two races are from Guatemala. These do not have 1·4 3·2. Instead, they originally had 1·2 3·4, but 3·4 has exchanged with 7·10, giving 3·10 7·4. In addition, in common with three races from Mexico, they have suffered the interchange 5·6 9·8 → 5·9 6·8. All elatas, therefore, that have been studied are two interchanges removed from the original.

There are probably other arrangements, of course, in the Central American population. The elatas are probably descendants of the oenotheras that stayed at home, a remnant of the pool from which successive migrations have spilled over into the regions to the north. Those that have been studied, however, have departed slightly from what we consider to have been the ancestral segmental arrangement. Like all other populations before inter-population hybridizations occurred, they are open pollinated, lethal free, with mostly paired chromosomes. Phenotypically, they lean in the direction of the strigosas in vegetative characters, toward the hookeris in flower size and in chromosome behavior.

Finally, the hookeris remain to be discussed. The main problem in connection with this group is whether to regard them as the oldest extant population, representing the ancestral type from which all the others have developed, or the most recent assemblage—the latest to migrate out of the center of distribution. If they are ancient, why have they remained uncontaminated by later migrations? Alternative answers to this question might be: (1) *hookeri* may have spread into its present range when the climate was moist. Drier conditions prevailing when subsequent populations arose and spread were such as to discourage migration into that area, so that the ancient hookeris have had no access to later populations; (2) on the other hand, there may actually have been contamination along the eastern border of the *hookeri* range and some of the strigosas may actually represent crosses between *hookeri* and members of Populations 3 or 4 rather than between the latter two. One of the subspecies of *strigosa* recognized by Munz (subsp. *strigosa*) often shows characters that are definitely *hookeri*-like—less greyish foliage, glandular hairs and more anthocyanin in stems, midribs and especially on bud cones. Except for flower size they could often be mistaken for one of the eastern forms of *hookeri*.

At one time I had under cultivation material collected from near Hesperus, Colorado. This material, derived from a single plant, proved to be a mixture of individuals, differing slightly in phenotype, but markedly in chromosome configuration. Different individuals had ⊙14, ⊙12 and ⊙4, ⊙4. All had smallish flowers (2–4 cm across). A plant with ⊙12 was selfed and yielded some plants with ⊙12 and others with seven pairs, showing that one of its complexes was alethal. When the original plant with ⊙12 was pollinated by the *hookeri* of de Vries, most of the progeny had ⊙6, 4 pairs, and closely resembled *hookeri*, even to having rather large flowers, for the most part open pollinated. One plant had ⊙4, ⊙8 with smaller self-pollinated flowers. The results suggest that one complex of *Hesperus* was a *hookeri*, the other a *strigosa* complex. Another cross, *Hesperus* (with ⊙12) × *Cockerelli* (a *strigosa*) gave plants with ⊙4, 5 pairs, with rather small self-pollinating flowers and *strigosa*-like phenotype. *Hesperus* may have been a recently formed hybrid between *strigosa* and *hookeri*; at least, its *hookeri*-like complex had no zygotic lethal. Its presence in southwestern Colorado, in an area in which both *strigosa* and *hookeri* are found, suggests that there may have been some crossing between the two species and that some hybrids have not as yet reached the stage where the complex of *hookeri* origin has acquired a zygotic lethal.

If crossing occurs between *hookeri* and *strigosa* at the present time, the chances are much greater that *hookeri* will act as the female parent than the reverse since the strigosas are self-pollinating, and pollen brought to them from *hookeri* would only rarely have a chance to function, whereas *strigosa* pollen brought to *hookeri* would almost certainly result in fertilization.

While it may not be possible to decide definitely whether the hookeris are old or very recent, the group seems to have retained its primitive characters, including large open-pollinated flowers, absence of lethals and paired chromosomes, or at the most small circles. In one other respect, *hookeri* has remained primitive—the commonest segmental arrangement among the hookeris is undoubtedly the arrangement ancestral to the entire North American population of the subgenus *Oenothera*. This can be inferred from the fact that the seven chromosomes that make up this genome have been found more frequently than any others throughout the entire range of the subgenus (Fig. 13.1).*

There is one consideration, however, that suggests the opposite conclusion, namely, that the hookeris are of recent origin. This is the fact that the hookeris and the strigosas have the same class of plastids

---

* Reference to Fig. 13.1 shows that the incidence of 1·4 and 3·2 among the chromosomes so far studied is almost as high as that of 1·2 and 3·4. Does this suggest that 1·4 and 3·2 could possibly have been the original, 1·2 and 3·4 the derived chromosomes? We believe not, for the following reasons: (1) the high frequency of 1·4 3·2 results from the great and extensive role that Population 3 has played in the evolution of *Oenothera*. Population 3 has contributed the alpha *parviflora II*, beta *biennis I*, alpha *biennis II* and beta *strigosa* complexes. Since 1·4 3·2 were formed very early in the development of Population 3, these chromosomes became characteristic of this entire population and of the four groups of complexes derived from it. Consequently, we have encountered these chromosomes many times in our survey. They trace back to the beginning of Population 3, but not necessarily to the ancestry of the subgenus as a whole. Chromosomes 1·2 3·4, on the other hand, are characteristic of the hookeris, the alpha strigosas, the alpha *biennis I*, beta *biennis II* and alpha *parviflora I* complexes, as well as of the relict species *argillicola* and *grandiflora*; (2) We find that 1·2 and 3·4 are commonly associated with the other five chromosomes that, by reason of their frequency, we consider ancestral. This is true especially in the hookeris, in alpha *biennis I* and beta *biennis II* complexes, and in the relict species. The 1·4 and 3·2 chromosomes are less frequently associated with these chromosomes. For these reasons we assume that 1·2 and 3·4 are the ancestral chromosomes rather than 1·4 and 3·2.

(Stubbe's class I). Population 4 was the latest of those that spread in successive waves over much of temperate North America, the Population 3 was the next to the last. The plastids of the strigosas have probably come from Population 4, from which the alpha *strigosa* complexes are descended, and the fact that this population and the hookeris have the same class of plastid would seem to argue for a recent origin of the latter.

4. Factors Responsible for the Evolution of *Oenothera*

We have now outlined the story of the evolution of the members of the subgenus *Oenothera* as it is revealed by the analysis of segmental arrangements, considered in relation to phenotypic characteristics and geographical distribution. The major factors responsible for this evolution have been discussed in some detail. In may be well, however, to summarize these briefly in order to have them clearly in mind. As we list them, it will be interesting to observe how a series of factors that are ordinarily deleterious in their effect when present alone can come to have positive survival value when combined in the right way.

(a) *Chromosome structure.* As we have pointed out earlier (Chap. 7), a necessary preclude to the evolution of the subgenus *Oenothera* was the presence, in the ancestors of this group, of chromosomes that were all equal in size, with median centromeres and arms of the same length, and with the centromere embedded in a region of heterochromatin, which is more easily broken than euchromatin.

(b) As a result, *translocations* were more likely to occur at the centromere than elsewhere, producing whole-arm, and therefore equal exchanges, so that translocated chromosomes possessed unaltered morphology, i.e., were all alike in size and structure, as they are at the present day. As a consequence, translocations did not have the adverse effect on fertility that they often have in other organisms, were not selected against, and hence tended to be retained and to accumulate in the population.

(c) *Hybridization.* A third factor of major importance has been hybridization. As we have seen, almost all forms from the Rockies eastward have arisen as a result of hybridization between unrelated populations that have had different histories of interchange. This means that the associated complexes in a plant are quite unrelated, in origin, in segmental arrangement and in genic composition. *Oenothera* races are therefore highly heterozygous complex-heterozygotes.

Hybridization is a factor that can have in other organisms a deleterious effect on fertility—hybrids are often more or less sterile because of structural hybridity. In the subgenus *Oenothera*, however, apart from reciprocal translocations, structural alterations have been conspicuous by their absence. So far as I am aware, not a single case of inversion has been reported in *Oenothera*, either spontaneous or induced. Nor have structural alterations brought about modifications in the structure of pairing ends to such an extent as to interfere with normal synapsis and chiasma formation. Almost any race, therefore, is capable of forming fertile progeny with any other—the only factor that operates against this is the presence of different classes of plastid in the population.

(*d*) *Lethals.* A fourth factor has been the development of lethals or of situations which simulate the presence of lethals. Lethals kill or sterilize and consequently are highly deleterious. In circle-bearing oenotheras, however, they have a compensating advantage which helps to overcome their bad effect. For one thing, a plant with ⊙14, because of its balanced lethals and large circle, breeds true. Not only does it breed true but, being highly heterozygous, it has a high level of hybrid vigor. This high level of heterozygosity and of hybrid vigor is maintained indefinitely by a single pair of lethals. To obtain as high a level of continuing heterozygosity in an all-pairing form, a race would have to possess balanced lethals in each pair. This would result in almost total sterility: one pair of balanced lethals would cut fertility to 50%, a second pair to 25%, etc. Seven independent pairs of lethals would cut fertility to 0·78%. With a circle of 14, however, a single pair will preserve the heterozygosity of all the chromosomes at an expense of only 50% loss in fertility.

(*e*) *Self-pollination.* Even this harmful effect of the lethals, however, is offset by another factor that has contributed greatly to the evolutionary development of the group. This is self-pollination. Ordinarily, self-pollination is deleterious because it results in a reduction of heterozygosity and consequent loss of hybrid vigor. The classical example is in maize where inbreds are small and puny in comparison with hybrids. In *Oenothera*, however, self-pollination cannot reduce heterozygosity, since the balanced lethals in a circle of 14 maintain maximum heterozygosity indefinitely. Self-pollination, therefore, loses its deleterious effect in *Oenothera*. On the other hand it becomes of positive benefit to the plant, for it ensures heavy pollination and consequent heavy seed set. In most oenotheras, the pollen bursts over the stigma 24 h or longer before the flower opens, the stigma soon becomes receptive, and the eggs are often fertilized before the flower opens. Thus, unless something has happened to prevent proper development of pollen, pollination is assured. In contrast, open-pollinated races must depend upon insects, and the prevalence of the latter on a given day often depends upon the weather on that day. It has been our experience in the experimental garden that self-pollinated races mature many more capsules than do the open-pollinated ones and have more seeds per capsule, although the opportunities for cross-pollination are much greater in such an environment than under normal field conditions.

In *Oenothera*, therefore, we have the interesting situation where several ordinarily deleterious characters add up in combination to a situation that has great survival value. Translocations, lethals and self-pollination, when coupled with the results of hybridization between diverse populations, have together produced a system with high survival value, in which plants are able to breed true to type, to profit from maximum hybrid vigor and to reproduce prolifically.

5. THE FUTURE

This unique and highly successful system of ensuring the maintenance of hybrid vigor is, however, not without its disadvantages. Present-day fitness has been achieved at the possible expense of future well-being. In some respects, circle-bearing oenotheras have run into an evolutionary blind alley. For one thing, they have created a system that prevents recessive mutations from being exposed so that Natural Selection can work on them. The self-pollinating habit which reduces outcrossing to a minimum tends to keep mutations within a given line of descent—gene flow is essentially eliminated by the innumerable reproductive barriers that exist within the population. Even if a mutation should be dominant, it would still be confined ordinarily to its own strain. Nor do the rare outcrosses do much to alleviate this situation. Only if a hybrid from such a cross has one or more pairs, or a small circle that is free of lethals, at least on one side, will there be a chance for recessive genes to become homozygous in following generations. The fact, however, that races with other than ⊙14 are almost totally absent from the area from the Rocky Mountains eastward indicates that forms with smaller circles and pairs stand little chance of indefinite survival, so that a form in which a recessive mutation appears in homozygous condition is not likely to survive, not because the mutation has been exposed but because of the reduction in hybrid vigor. If an outcross results in a hybrid with ⊙14, then a recessive mutant gene will find itself associated with a different allele, and while there is a possibility that it might be able in the new combination to express itself at least partially, or even to dominate, the chances of this are slight. Outcrosses resulting in hybrids with ⊙14 do, however, serve to distribute genes more widely. By such crosses, particular genes can come to reside in more than one inbreeding line.

On the other hand, a circle of 14 with balanced lethals is an excellent mechanism for preserving recessive mutations from annihilation. They will be carried along indefinitely and successive mutations may accumulate until a given complex is carrying quite a load of such mutations.

The chief possibility for future evolution would seem to be through the infrequent occurrence of outcrosses between races, resulting in the occasional production of new complex-combinations with ⊙14. Some of these combinations may have enhanced survival value in the environment in which they find themselves. At best, however, evolution in the future will probably be slow because of the barriers set up to gene flow and the rarity with which mutations are subject to the screening process of Natural Selection.

6. GEOLOGICAL CONSIDERATIONS

The question may be asked at this point regarding the length of time it has taken for the members of the subgenus *Oenothera* to have evolved. Are they a relatively old or new group? No fossil remains have been ascribed to this subgenus, but this is not surprising since its foliage characters are not especially distinctive, and fossil pollen, which would be easily recognized if found, is hardly to be expected, since the oenotheras are dry land plants and their pollen is heavy and not wind-blown. When one considers the complexity of the story as we have outlined it, and realizes that interchanges are, after all, rare occurrences in terms of years, even in *Oenothera*, it seems unlikely that the group as we know it today, with its multiplicity of segmental arrangements and its complicated population structure, could have developed within the short span of 10,000 years since the end of the Wisconsin glaciation. It is therefore more probable that the evolution of this group traces back into the Pleistocene.

How far back in the Pleistocene the subgenus goes is, of course, open to question. At first sight, one is struck by the fact that the number of successive *Oenothera* populations that have developed and spread over the continent is the same as the number of major periods of Pleistocene glaciation (the Nebraskan, Kansan, Illinoian and Wisconsin). If Populations 1–4 corresponded in their origin to these four glacial periods, however, the subgenus would trace its beginning back to a period a million or more years ago, which may seem excessive. It is perhaps unnecessary to postulate such a long history, since in some of the glacial periods major fluctuations have probably occurred; in the Wisconsin, for example, at least two periods of major glaciation separated by an interglacial period have been recognized, and there is some evidence that as many as six advances of the ice occurred during this period. It is not clear to what extent similar variations may have occurred during the other major glaciations. If the subgenus *Oenothera* goes back to about the beginning of the Wisconsin, this would allow for a lifetime of some 70,000 years, which could conceivably be a long enough period for it to have evolved to its present state. The subgenus could, of course, have originated during the middle of the Pleistocene rather than near its beginning or its end.

Whatever the truth may be in this regard, it is probable that periods of extreme glaciation were accompanied by increased rainfall in the now arid southwestern part of the continent, so that desert areas were more restricted than at present, and regions with mesophytic conditions were more extensive. Morrison (1965) has found from a study of the great lakes that once existed in the Great Basin area, especially Lake Lahontan and Lake Bonneville, that the alternate expansion and contraction of these bodies of water were the result largely of differences in rainfall, rather than glacial runoff. Pleistocene climate apparently fluctuated widely, both in temperature and in precipitation. In general, interglacial periods, at first relatively cool and dry, became increasingly warm and wet. Onset of a glacial period coincided with a drop in temperature, so that the climate in the Southwest became cool and wet (the so-called pluvial period). During the glacial period the climate remained cool, diminishing in moisture toward the end until, as the glaciers retreated and an interglacial period began, the climate had again become cool and dry. Roughly speaking, therefore, toward the end of an interglacial period, the climate in the Southwest was relatively warm and moist, during the glacial period cool and at first moist, later becoming drier. Moist conditions prevailed during the transition from interglacial to glacial periods, dry conditions characterized the transition from glacial to interglacial periods.

It is probable that the successive populations of *Oenothera* for whose existence we have evidence arose in the Central American–Mexican area during the relatively moist times which signalled the transition from interglacial to glacial conditions, and lasted into the glacial period. As the glaciers retreated these populations advanced northward, occupying the unglaciated areas, and later much of the region left exposed by the retreating ice. Specifically, Population 1 developed toward the end of a certain interglacial period or the beginning of a major glaciation. It migrated into the area south of the retreating glaciers and came to occupy much of the continent, later becoming exterminated in the Southwest and South by the aridity that developed during the succeeding interglacial period. Upon the advent of a second glaciation, it was wiped out in the north but survived south of the advancing ice sheet. Meanwhile, Population 2 developed during the moist period that preceded and accompanied the second glaciation, and migrated northward, meeting Population 1 where it still survived. Conditions at the center of origin during the development of Population 2 were probably quite moist, since the characters carried by this population were strongly mesophytic. As the second glaciation receded, both populations advanced northward, and during the period when they were sympatric crosses occurred that produced the progenitors of *parviflora I*. Furthermore, during the second interglacial period the increasing drought in the Southwest that coincided with the retreat of the glaciers wiped out most traces of Population 2 in that area.

Upon the return of moist climatic conditions accompanying the close of the second interglacial period and the onset of a third glacial advance, Population 3 developed, spread northward and eastward, met members of Populations 1 and 2 that still survived, and produced the progenitors of the *biennis* races as well as *parviflora II*. With the onset of the third interglacial period, these advanced behind the retreating glaciers along with members of the *parviflora I* population.

The conditions in the Southwest during the development of Population 3 were probably not as moist as they were during the development of Population 2, since Population 3 seems to have been better adapted to relatively dry conditions than the earlier populations, which enabled it to survive during the third interglacial period in the plains and mountainous areas, as the latter became freed of excessive glaciation.

During a fourth glacial period, Population 4 developed and migrated northward, encountering Population 3 which had survived in the Great Plains area, and crosses occurred that gave rise to the strigosas. Population 4 also had rather xerophytic characters and the strigosas which resulted from these crosses were well adapted to the relatively low level of summer precipitation that characterizes the Great Plains area where they are found. While the strigosas were being formed, the groups created by hybridization between the earlier populations survived south of the glaciers and again advanced northward with the disappearance of the ice.

Since it is unlikely that the extensive evolutionary process just outlined could have been accomplished in post-glacial times, it is more probable that the successive populations of *Oenothera* originated prior to the end of the Pleistocene. Whether the climatic fluctuations that occurred during the Wisconsin glaciation were of sufficient magnitude to have had the effects described may possibly be open to question. One or more of the earlier populations of *Oenothera* may have arisen in pre-Wisconsin times; one or more, or perhaps all of the major periods of

Pleistocene glaciation could have been involved. While the correspond-
ence between the number of *Oenothera* populations that developed and
the number of major Pleistocene glaciations may be nothing more than
a coincidence, it is at least highly probable that climatic fluctuations of
sufficient magnitude and in the requisite number occurred during the
Pleistocene to bring into existence the four populations which have given
rise through intercrossing to all of the species now found from the Rocky
Mountains eastward. I do not believe that the climatic conditions in
post-glacial times were such, or that the time was sufficient, to have
made possible the origin and crossing of these populations during this
period.

On the other hand, the fact that all races can cross with all others,
yielding fertile progeny (apart from the effect of plastid–genome incompatibility) would seem to argue against the suggestion that the story
as we have outlined it had its beginning as far back as the beginning of
the Pleistocene.

CHAPTER 19

The *Oenothera* Flora of Europe

As we have pointed out earlier, *Oenothera* is a western hemisphere genus, but it has reached Europe and other parts of the world at various times and in various ways.* Early botanical explorers sent seed back to their home countries which were grown in botanical gardens and in some cases escaped. But perhaps a more important mechanism was transmittal in ballast. Cargo ships to the western world often returned to Europe in ballast. The soil thus carried was deposited on ballast heaps near major ports. On these, seed contained in the soil germinated, and many American species were thus introduced into Europe. *Oenothera*, characteristic of disturbed areas at home, found both terrain and climate ideal. From these locations, evening primroses spread widely, and quickly became a dominant element in the flora along the coastal areas near such ports as Liverpool and Amsterdam.

Since ballast was derived from many parts of the New World, plants from different regions sprang up on the same ballast heaps. Hybridization occurred, and some of the strains now found in the Old World arose as new hybrids which did not exist in the western hemisphere. For example, we have seen that *O. lamarckiana* apparently arose by the combination of a beta *biennis II* complex from eastern North America with perhaps a *hookeri-* or a *strigosa*-like complex from the west.

Essentially the same process, therefore, has taken place in Europe as that which occurred in North America during the evolution of the subgenus *Oenothera*. Crossing between representatives of populations that had experienced divergent histories of interchange has given rise to new forms, many of which have possessed large circles and balanced lethals, and in many cases these are robust and have flourished in their new environment. Thus, Europe has become overrun by races, some of them of hybrid origin, their complexes derived from diverse sources. Intermingled with these are a number of American races that have managed to survive in competition with the newly formed hybrids. Among the American introductions is probably the race that de Vries called *O. muricata*, which is very similar in phenotype and the segmental arrangement of at least its alpha complex to a *parviflora II* race found in Pennsylvania (*Sunbury*). It is probable also that de Vries' *biennis* is an American race. Both of its complexes are typical of the *biennis II* group in

---

* Rostanski, however, argues (1968) that *biennis*, *rubricaulis* and *suaveolens* are indigenous to Europe or the Eurasian area. See p. 314.

eastern North America. Its exact counterpart has not been found in the latter area, but there are hundreds if not thousands of inbred lines within the *biennis II* group, only a few of which have been studied. *Biennis* is the most successful and widespread species of *Oenothera* in Europe and Renner's work shows that it has been involved in the production of several of the hybrid races that have become established in various parts of Europe.

Attempts to classify members of the subgenus *Oenothera* found in Europe go back beyond the creation of the binomial system of nomenclature. Linnaeus himself recognized only three species belonging to this subgenus, *O. biennis*, *O. muricata* and *O. parviflora*. In the list published by the German Botanical Society in 1940 (Mansfeld), this was reduced to two—*O. biennis* and *O. muricata*. Renner (1942b) disagreed with this treatment. Since he found a number of forms well established in the wild, he felt that these should be recognized in the taxonomic literature. As a result, he accepted a number of designations made by other workers, and proposed several new species in addition. In the case of certain well-established hybrid populations it was found possible to identify the species that had crossed to produce them.

Since Renner has made the only extensive study of the European *Oenothera* flora that is based on cytogenetic as well as phenotypic criteria, his conclusions possess a validity not found in purely taxonomic studies, and they are therefore presented herewith in condensed form, with a few comments.

He arranged *Oenothera* species into two groups, *Strictae* and *Cernuae*. The former have straight stem tips, and the sepal tips are not subterminal. The latter have nodding to strongly bent stem tips and subterminal sepal tips.

1. GROUP STRICTAE

O. biennis L.

Complexes *albicans* ♀, *rubens* (♀) ♂   ⊙6, ⊙8
Self-pollinating
Leaves broad, stem tips and hair papillae green

O. rubricaulis Klebahn

Complexes *tingens* ♀, *rubens* (♀) ♂   ⊙6, ⊙8
Self-pollinating
Broad leaves, stem tips and papillae red pigmented

The first of these is a typical *biennis II* race. The beta complex of de Vries' *biennis* (*rubens*) is one interchange removed from one of the commoner arrangements found among the beta *biennis II* complexes resident in New York and Pennsylvania (the β *Elma V* arrangement), and its alpha complex (*albicans*) is three interchanges removed from an alpha *biennis II* complex (α *Buck Creek*) found in North Carolina. The alpha complex is phenotypically and segmentally a *strigosa* complex, and the beta complex is of the *biennis* type, as is true of all *biennis II* races. The green stem tips and papillae are characteristic of most *biennis II* races. The European *biennis* has probably been derived directly from North America.

The second species, *rubricaulis*, presents a different picture. It has probably arisen in Europe as the result of hybridization. Since it shares its *rubens* complex with *biennis*, the latter was probably one of its ancestors. The other ancestor is definitely a race of *parviflora I*. *Tingens*, the alpha complex of *rubricaulis*, is identical in segmental arrangement with α *Trois Pistoles I* and α *Trois Pistoles II*, alpha complexes of races growing in the province of Quebec, along the St. Lawrence River. The red stem tips and red papillae of *rubricaulis* have undoubtedly been derived from its *parviflora I* ancestor, since these traits are characteristic of, and often displayed with intensity by, the latter.

O. lamarckiana Ser.

Complexes *velans* ♀ ♂, *gaudens* ♀ ♂   ⊙12, 1 pair
Large flowered, open pollinated, leaves broad, roughly crinkled, stem tips and papillae red, bud cones red-striped

O. coronifera Renner (1956)

Complexes *quaerens* ♀, *paravelans* ♀ ♂   ⊙12, 1 pair
Self-pollinating, flowers of moderate size, leaves very broad, stem tips and papillae red, buds red-striped

Paravelans has the same zygotic lethal as *velans*, and the same segmental arrangement. *Quaerens* is a minimum of three interchanges removed from *gaudens*.

Quaerens is a beta *biennis II* complex, as shown by its phenotypic effect, and its segmental arrangement (Rossmann 1963) which is identical with that found in the beta complexes of a number of *biennis II* races from New York, Ontario, Quebec and New Brunswick (e.g., β *Corning I*). *O. coronifera*, therefore, has arisen by hybridization in Europe, probably between *lamarckiana* as female and an unknown *biennis II* race as male.

If *quaerens* is a beta *biennis II* complex, how is it that it is now an egg complex in *coronifera*? This is probably to be attributed to the "Renner effect". Beta *biennis II* complexes are able in some races to compete with some success in embryo sac development, and *quaerens* may have been capable of such behavior in the unknown race to which it belongs. In strains of de Vries' *biennis* and in *suaveolens*, this type of behavior is exemplified. In competition with *velans*, *quaerens* is partially successful, so that both complexes are capable of transmission through the eggs. In the development of the male gametophyte, however, *quaerens* is apparently unable to compete with *velans*.

O. nuda Renner (1956)

Complexes *calvens* ♀, *glabrans* ♂   ⊙14
Self-pollinating
Leaves of moderate width, glabrous, with red midribs, stem tips, papillae and fruits green *biennis*-like. *Glabrans* has the same lethal as *rubens* and *gaudens*. Plastids are of the *lamarckiana* type

Glabrans is one interchange removed from *rubens* (Jean, Lambert and Linder, 1966, see Appendix I). It is identical segmentally with β *Etna*, etc. *Calvens* has the commonest segmental arrangement found among the alpha *biennis II* complexes. *Nuda*, therefore, is a typical member of the *biennis II* assemblage and is probably a migrant from North America.

O. suaveolens Desf.

Complexes *albicans* ♀, *flavens* ♀♂   ⊙12, 1 pair
Self-pollinating
Broad leaved, plant free of red pigmentation. The *flavens* lethal is occasionally lost by crossing over, resulting in production of seven-paired, alethal *lutescens*

Suaveolens has probably arisen in Europe as a result of hybridization. It shares *albicans* with *biennis* from which it probably received this complex, *biennis* having been the female parent (*biennis* and *suaveolens* have the same class of plastids). *Flavens* is a typical *biennis* complex in its phenotypic effect, but segmentally it is much closer to the alpha *biennis I* complexes than to those of the beta *biennis II* races. It probably traces back to a *biennis I* race in the southeastern states where the alpha complexes are often closely related segmentally to the ancestral arrangement for the subgenus *Oenothera*, from which it differs by only two interchanges. It is not surprising that an alpha *biennis I* complex is able to function, as *flavens* does, in both egg and sperm, since many alpha *biennis I* complexes are able to do this.

O. conferta Renner (Renner and Hirmer 1956)

Complexes *convelans* ♀♂, *aemulans* ♀♂   ⊙12, 1 pair

Self-pollinating

Similar to *lamarckiana*, but with smaller flowers, weaker and lower stems, narrower, less crinkled leaves. Red papillae, red-striped buds, white midribs.

Convelans has the same zygotic lethal and the same segmental arrangement as *velans*. *Aemulans* has an arrangement of segments identical with that of β *Lawrenceville III* from New York State. *O. conferta*, therefore, has probably been synthesized in Europe by a cross of *lamarckiana* with an unidentified *biennis II* race.

The above are *biennis* forms and, like typical *biennis II* forms, they have one complex of the *biennis* type, the other of the *strigosa* (or in the case of *lamarckiana*, *coronifera* and *conferta*, perhaps of the *hookeri* type). That they are closely related is shown by the frequency with which they share complexes, or have nearly identical complexes. Thus, *biennis* and *rubricaulis* have *rubens* in common, and *biennis* and *suaveolens* share the complex *albicans*. Since *biennis* is by far the most widespread species in Europe, it is more likely that *biennis* has served as a parent in the origin of *rubricaulis* and *suaveolens* than that it has been synthesized by the crossing of the latter. *Biennis* is also closely related to *lamarckiana*—the *rubens* complex of the former and the *gaudens* complex of the latter have the same zygotic lethal and the same segmental arrangement, although they are not entirely identical genically, since *gaudens* produces more coarsely crinkled foliage than *rubens*. *Lamarckiana* probably had as one parent a *biennis II* race that was closely related to the *biennis* of de Vries but not identical with it.

Lamarckiana is not only closely related to *biennis*, it is also close to most of the other species listed above. The complex *glabrans* of *nuda* has the same zygotic lethal as *gaudens* and *rubens*, and is only one interchange removed from them. *Velans* has the same lethal as *paravelans* of *coronifera* and *convelans* of *conferta*, and also has the same segmental arrangement. The other complex in each of these species is *gaudens*-like but does not have the same lethal or segmental arrangement. It is possible that *lamarckiana* has served as one parent in the production of *nuda*, *coronifera* and *conferta*, contributing complexes to these forms, which may subsequently have experienced minor genetic alteration.

It is clear, therefore, that these seven species are closely related, and it is probable that some of them are derivatives of the *O. biennis* of de Vries or of closely related forms, possibly of *lamarckiana*.

Renner also recognized two *strictae* growing wild that appear to belong to the *strigosa*, rather than the *biennis* category.

O. hungarica Borbas (= *Bauri Boedijn*)

    Complexes *laxans* ♀, *undans* ♂   ⊙14
    Self-pollinating
    Leaves rather narrow with strigose pubescence and red midribs, buds
      red-striped, silky, stems with fine red papillae

While the segmental arrangement of *laxans* has not been found in North America, it is a typical alpha *strigosa* complex. Like the other alpha *strigosa* complexes, it has not experienced the interchange 1·2 3·4 → 1·4 3·2. It still retains 1·2 but 3·4 has exchanged with 7·10 of the ancestral arrangement to give 3·10 7·4. It has also suffered the following exchanges: 5·6 9·8 → 5·8 9·6; 11·12 13·14 → 11·14 13·12. It is only two interchanges removed from α *Castle Rock*. In having 5·8 9·6 it is like a large number of the alpha *strigosa* complexes found in Colorado and elsewhere.

The *undans* complex is a typical beta *strigosa* complex and has a segmental arrangement identical with that of β *Castle Rock*, an arrangement found in several races from Colorado. It has the 1·4 3·2 typical of the beta *strigosa* and has experienced the exchanges 5·6 7·10 → 5·10 7·6 and 9·8 13·14 → 9·13 8·14.

There can be no question, therefore, that *O. hungarica* is a typical member of the *strigosa* alliance, very close to *O. cockerelli*. While one wonders how a race growing in the Colorado area could have reached Europe, the presence of a complex (*laxans*) that is definitely of the alpha *strigosa* type rules out the likelihood that it was derived from eastern North America, since alpha strigosas (those derived from Population 4) are not found in this area—the *strigosa* complexes of the *parviflora II*, *biennis I* and *biennis II* races are of the beta type (derived from Population 3). The probability is, therefore, that *hungarica* was introduced accidentally into Europe from the Colorado area. It was first described from material collected at Friedrichshagen in Germany.

O. mollis Renner (1956)

    Complexes *simulans* ♀, *planans* ♂   ⊙14
    Self-pollinating
    Leaves rather narrow, soft-haired, buds and young fruits lightly striped
      with red, papillae red, fruits slender

The segmental arrangements of the complexes of *mollis* have not, to my knowledge, been determined. The egg complex is called *simulans* because of its close resemblance to *albicans*, and its pollen complex, *planans*, according to Renner, resembles *truncans*, one of the complexes of the *grandiflora* of de Vries. It is probable, therefore, that *mollis* arose in Europe by a cross between *biennis* or *suaveolens* (or other related strain) as female and the *biennis I* form known as de Vires' *grandiflora*, or a race related to it, as male.

2. GROUP CERNUAE

O. parviflora L. (= *muricata* L. in part)

Complexes *augens* ♀, *subcurvans* ♂   ⊙14
Self-pollinating
Stem tips scarcely nodding, stems red below, green at tip, papillae
    green, leaves rather broad, almost glabrous, red-nerved, purple-
    margined when young, buds stout with subterminal sepal tips

The *"muricata"* of the Rhine valley is included here.

Augens is two interchanges removed from an alpha *parviflora I* complex, (α *St. Stephen* from New Brunswick, Canada): *subcurvans* is two
interchanges removed from *curvans* of de Vries' *muricata*.

O. syrticola Bartlett (= *muricata* L. in part)

Complexes *rigens* ♀, *curvans* (♀)♂   ⊙14
Self-pollinating
Leaves narrow, softly hairy, stems erect below, nodding above, finely
    pubescent with fine red papillae, buds slightly reddened, sepal tips
    slightly subterminal

This is the *muricata* of de Vries, and also of the Danube valley region.
It definitely belongs to *parviflora II*, and is very similar (at least this
is true of de Vries' material) to races found in North America—e.g.
Sunbury.

O. silesiaca Renner (1942b, 1943c) (= *muricata* L. in part)

Complexes *subpingens* ♀, *subcurvans* ♂   ⊙14
Self-pollinating
Leaves narrow, long, dark green, red-nerved, almost glabrous, purple-
    margined when young. Stem often strongly nodding, sepal tips
    spreading
From East Prussia and Silesia

Subcurvans is identical segmentally with the beta complex of *parviflora* L. The alpha complex, *subpingens*, is only one interchange removed

from α *angustissima* from New York State, and is identical with *pingens* of *atrovirens*.*

These three species, which belong to Linnaeus' *muricata*, are in all likelihood derived from *North America*.

O. ammophila Focke

Complexes *rigens* ♀, *percurvans* ♂   ☉12, 1 pair
Self-pollinating
Stems very strongly bent, often from the ground, with large red papillae. Bud cones red, young fruits red-striped, sepal tips subterminal. Leaves narrow, soft-hairy
Widespread throughout northern Europe

var. *germanica* Boedijn. Calyx brownish red, midribs red

The egg complex of *ammophila* is identical with that of *syrticola*, the pollen complex is quite different segmentally from other complexes so far analyzed but belongs definitely to the group of complexes descended from Population 1.

O. rubricuspis Renner (1950)

Complexes *paenepingens* ♀, *praecurvans* ♂   ☉14
Self-pollinating
Stems very strongly nodding, tall, red below, green above with red papillae. Leaves long, narrow, dark green with red midribs and margins, almost glabrous
Sepal tips spreading

The complexes have not been analyzed segmentally.

Since the North American groups *parviflora I* and *II* are both native to the northeastern coastal areas of the continent, and since one species recognized by Renner (*syrticola*) is very close to the American race "*Sunbury*", it is likely that at least some of the European races with bent stem tips and subterminal sepal tips have come directly from North America, probably in ballast. Even though *ammophila*, which is widespread in Europe, has a complex in common with *syrticola* and with *Sunbury* (*rigens*) this may not be the result of hybridization in Europe, since the other *ammophila* complex (*percurvans*) is also of the *parviflora* type, i.e., it produces bent stem tips and subterminal sepal tips. *Ammophila*, therefore, probably represents another introduction from America,

* Renner carried out intensive studies with *atrovirens*, a cruciate race which was derived from de Vries' culture via Gescher of Münster (Renner, 1937b). de Vries' material came from Lake George, New York. Apparently, Renner did not find *atrovirens* in the wild, so did not include it in his list of recognized species.

a race closely related to the ancestor of *syrticola*. The same argument can be applied to all the other members of the group *Cernuae*. In four of the five species of this group recognized by Renner, both complexes are of the *parviflora II* type—in one instance the alpha complex (*paenepingens*) may be of the *biennis* (α *parviflora I* type) type. This suggests that all five may be original American races. It may be asked, however, whether a European origin through hybridization is not possible for some of them, since the egg complexes of four of the *Cernuae*, as is true of the *parviflora II* group in America, are *strigosa*-like, and most European races of *biennis*, as well as the *strigosa* races (*hungarica* and *mollis*) carry a *strigosa*-type egg complex (e.g., *albicans*). However, not one of the alpha complexes of the *Cernuae* is close to the alpha *strictae* segmentally, and all are closely related to one another. This would seem to rule out members of the *strictae* as ancestors and to render an American origin more probable.

It is certainly significant that the affinities of many of the races in Europe are with the oenotheras native to the eastern coast of North America. This is true of all of the *Cernuae*, and also of *biennis* and *rubricaulis* of the *Strictae*. *Lamarckiana* may be an exception inasmuch as one of its complexes (*velans*) may have come from some other region. This complex, in turn, may have given rise to *paravelans* of *coronifera* and *convelans* of *conferta*. *O. hungarica* may also be an exception, probably derived from the Colorado area. The other *biennis*-like *Strictae* in Europe seem to have been derived by hybridization in Europe involving *biennis* or *lamarckiana*.

In addition to the races listed above, Renner found a number of strains growing wild, and apparently well established, that were undoubtedly of recent hybrid origin, since both of their complexes could be traced to other races growing in Europe. To some of these he gave specific names, as follows:

O. drawertii Renner (1950)

    Complexes *laxans* ♀, *flavens* ♂   ⊙4, ⊙4, ⊙6
    Apparently from *hungarica* × *suaveolens*
    Found in France

    Typical *laxans* should give ⊙4, ⊙10 with *flavens*.

O. Holscheri Renner (1950)

    Complexes *rubens* ♀, *undans* ♂   ⊙12, 1 pair
    Apparently from *rubricaulis* × *hungarica*

    Rubens is unable to compete with *undans* in the pollen of this plant.

O. issleri Renner (1950)

Complexes *rubens* ♀, *curvans* ♂   ⊙14
From *syrticola* × *biennis*, or more likely from (*syrticola* × *biennis*)
rigens·rubens × *syrticola*

The cross *syrticola* × *biennis* would yield a diplarrhene hybrid (*curvans·rubens*) with two pollen complexes, which would therefore be transmitted poorly through the female gametophyte. The hybrid *rigens·rubens* however has ⊙6, 4 pairs, so that it could produce gametes with a modified *rubens* complex in which one or more *rigens* chromosomes had been exchanged for *rubens* chromosomes. Such a *rubens* complex might be successful in female gametophyte development.

O. Wienii Renner (1950)

Complexes *tingens* ♀, *undans* ♂
From *rubricaulis* × *hungarica*

The following hybrids found in nature by Renner were not given specific names:

albicans·percurvans from *biennis* × *ammophila*
rubens·subcurvans probably from (*parviflora* × *biennis*) × *parviflora*
rubens·percurvans probably from (*ammophila* × *biennis*) × *ammophila*

A recent taxonomic treatment of the European oenotheras is that of Raven (1968), in Volume 2 of the "Flora Europaea". His list of species conforms for the most part to that of Renner (see Table 19.I).

One of his species, however, (*chicagöensis*) may be open to question, first because it has doubtfully been found in the wild in Europe, and second because it should be called *chicaginensis* rather than *chicagöensis*. This race is derived from material originally collected near Chicago, Illinois, by de Vries, and was called by him *biennis Chicago*. For a time, Renner adopted the term *chicagöensis* in his work, although he did not use this name in any of his publications. The only references to *chicagöensis* in the literature are found in footnotes in the 1930 and 1931 papers by Cleland and Blakeslee (p. 185), where it is mentioned incidentally in connection with acknowledgement of permission by Professor Renner to use the terms *excellens* and *punctulans* which he had assigned to its complexes. The race is not mentioned further in these papers and no description is given.

The first publication in which this race is discussed appeared in 1933 (Renner and Cleland). In that paper and in all subsequent references in

TABLE 19.I

Classification of European members of the subgenus *Oenothera* according to
Renner (1942b, 1950, 1956) and Raven (1968)

Renner	Raven
Strictae	*Biennis* group
biennis	*biennis*
rubricaulis	*rubricaulis*
lamarckiana	*erythrosepala*
nuda	
	chicagöensis
coronifera	*coronifera*
suaveolens	*suaveolens*
conferta	
hungarica	*strigosa*
	renneri
mollis	
Cernuae	*Parviflora* group
parviflora	*parviflora*
silesiaca	*silesiaca*
	atrovirens
ammophila	*ammophila*
syrticola	*syrticola*
rubricuspis	*rubricuspis*

SPECIES OF KNOWN HYBRID ORIGIN
drawertii
holscheri
issleri
Wienii

the literature, the term *chicaginensis* has been used. The nearest approach
to a description of the race is found in de Vries (1909, p. 3; 1913, pp. 32, 52,
Fig. 18), who used, of course, the designation *O. biennis Chicago*. From
this it would seem that there is no basis for the taxonomic acceptance of
"chicagöensis", especially as it is known only as investigative material
in experimental gardens, and has probably not been found in the wild.
It is true that Linder (1957) (who calls it *chicaginensis*) claims that it is
widespread in certain parts of France, and Rostanski (1965) has described it from Silesia, but in the absence of cytogenetic analysis, this
determination is questionable.*

* Through the kindness of Dr. Rostanski of Wroclow, Poland, I grew in 1970 three
strains which he has attributed to *chicaginensis*. They all differ strikingly from the
chicaginensis of Renner, both phenotypically and in chromosome configuration.

Attention may also be called to the fact that Raven accepts *erythro-sepala* as the name for the species to which de Vries' *lamarckiana* belongs. *O. lamarckiana* is mentioned only in the index, as the equivalent of *erythrosepala*.

Mention may also be made of a paper by Scholz (1956) who, in describing oenotheras in the vicinity of Berlin, named material which he considered similar or identical with that described by Renner as *O. mollis*, under the designation of *O. renneri*.

In concluding this brief discussion of the taxonomy of the European oenotheras, a paper by Rostanski (1968) may be cited who claims that *O. biennis, O. suaveolens* and *O. rubricaulis* are not American in origin but are indigenous to the Eurasian continent. He presents as evidence: (1) the wide distribution of *biennis* in Europe as early as the beginning of the seventeenth century; (2) the uniformity and widespread occurrence of *biennis* in Eurasia at the present time; and (3) the absence of the European *biennis* in North America. I do not find the evidence for this hypothesis very convincing. In the first place, the evidence for the presence of *biennis* in Europe in the early 17th century is very scanty or non-existent. While the descriptions of Bauhin (1623), Parkinson (1629) and Morison (1669) indicate that *Oenothera* had already reached Europe, they were not of sufficient accuracy to permit identification as to species. For example, Bauhin's plant had leaves that were too narrow and flowers with hypanthia that were too long for *biennis*. Nor are the illustrations published prior to 1700 of much value in this regard.

As stated in the Introduction (p. vii) the first published illustration of a plant that has been interpreted as an *Oenothera* is that by Prosper Alpinus in his "De Plantis Exoticis" (1627). This plant, however, appeared to have five-petalled flowers: either it was very inaccurately drawn, or it was not an *Oenothera*. The figure published by Parkinson in 1629 and 1640 is so crude that it is scarcely recognizable as an *Oenothera*. The same is true of Morison's figure (1680, reproduced as Fig. 12 in Gates 1915). The earliest illustration that I know of which clearly portrays *Oenothera* is published in Tournefort's Institutiones (1700) (Fig. 19.1). Since this plate depicts only a flower and floral parts, however, it cannot be related with confidence to any modern *Oenothera*. There can be no doubt about the existence of *Oenothera* in Europe prior to the end of the seventeenth century, but one cannot identify with certainty the species that were present at that time.

In the second place, both Renner (1950) and Rostanski (1968) comment on the uniformity of *biennis* throughout its range. de Vries' *biennis*, therefore, may be taken as representative of the species as a whole. This race has been extensively studied, genetically and cytologically, its

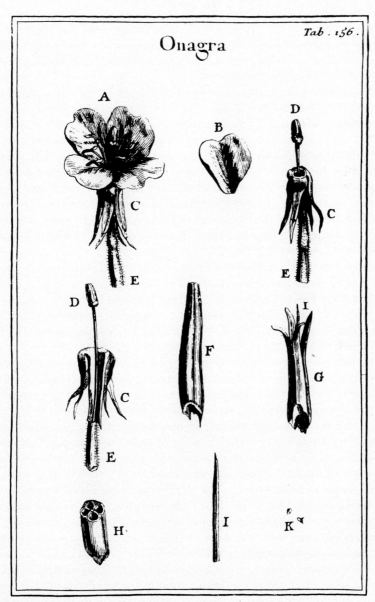

FIG. 19.1. The first accurate portrayal of *Oenothera*, Tournefort's figure 156, published in 1700 (Institutiones Rei Herbariae).

complexes have been fully analyzed segmentally. There can be little doubt that it is a typical member of *biennis II*, i.e., of *biennis* subsp. *caeciarum* Munz (see Chapter 18). It is probable that the ancestor of de Vries' *biennis* still exists among the hundreds of *biennis II* races somewhere in eastern North America. One cannot be sure that the European *biennis* is absent in North America.

In this connection, it should be recalled that the only way a plant can acquire ⊙6, ⊙8 from an original ⊙14 is by the exchange of corresponding segments within a single complex. Since most races of *biennis II* are characterized by the presence of ⊙14, de Vries' *biennis* no doubt arose, probably in North America, by such an interchange. The weight of evidence, in general, seems to favor the view that the *biennis* of de Vries, and probably the whole of *biennis* L., is derived from North America, and is to be regarded as typical of *biennis II* (*O. biennis* subsp. *caeciarum* Munz); and that it arrived in Europe at an early date, probably in ballast. Ships returning in ballast from the colonies in North America became numerous during the seventeenth century.

Finally, our brief excursion into the realm of taxonomy leads me to make a brief comment on the methods of *Oenothera* taxonomy, from the standpoint of the cytogeneticist. Unless taxonomic determinations in this genus are based on the results of cytogenetic research, they are likely to have reduced phylogenetic validity. A few investigators (Gates, 1936) have followed the pattern of giving specific names to every strain grown in the garden or observed in the wild which seemed a bit different from what had been seen before. Such investigators are actually picking out those individual true-breeding lines that have come to their attention from among the hundreds or thousands of such strains found in Nature, and giving them specific names. If one followed this procedure consistently, and made a thorough survey of the *Oenothera* populations in various areas of North America, one would end up with hundreds, if not thousands, of species. Obviously, the individual true-breeding lines so characteristic of *Oenothera* are not worthy of being called species. This has been recognized by Munz (1965) who has given specific rank to large groups of strains which have had a common origin, i.e., according to our interpretation, hybridizations between specific ancestral populations. To be sure, the situation in Europe, where *Oenothera* history covers only a few hundred years, is not nearly as complicated as it is in North America where the genus has been evolving for hundreds of thousands of years. There is more hope, therefore, that careful analysis, based on both phenotypic and cytogenetic studies, will show with considerable clarity the process by which present-day European forms have developed; will distinguish between the forms that have been transported from

North America and those that have originated in Europe; and in the case of the latter, will identify the more important European progenitors—the pioneers in *Oenothera*'s New World. It is particularly important, therefore, that the segmental arrangements of the Renner complexes found in the European population be determined. A system of classification that fails to take into consideration the segmental arrangement of individual complexes is apt to arrive at erroneous conclusions.

In short, one cannot assign material to a species which has already been named, described and analyzed from the standpoint of the segmental arrangement and the genetic constitution of its complexes, until one has analyzed the new material and shown that it conforms cytogenetically as well as phenotypically to this species.

Conversely, even if it is shown that the complexes of the material under study are different segmentally from those of other races, this in itself does not justify description of the material as a new species. Only

TABLE 19.II

Mode of origin and complex-composition of taxonomic entities found in Europe and recognized by Renner as species

Probably derived directly from North America	Probably synthesized in Europe as a result of hybridization
Typical biennis II (α *biennis II* · β *biennis II*) European biennis nuda	α *parviflora I* · β *biennis II* rubricaulis
Typical strigosa (α *strigosa* · β *strigosa*) hungarica	*hookeri-like complex* (*velans*), β *biennis II* (*gaudens*) lamarckiana
Typical parviflora II (α *parviflora II* · β *parviflora II*) parviflora syrticola silesiaca ammophila rubricuspis[a]	*velans* · β *biennis II* coronifera conferta α *biennis II* · α *biennis I* suaveolens[b] α *biennis II* · β *biennis I* mollis

[a] This race may be a *parviflora I* rather than a *parviflora II*. The almost glabrous, dark-green foliage and presence of anthocyanin pigmentation suggest this.

[b] Many alpha *biennis I* complexes can be transmitted through the sperm as well as through the egg. Since *flavens* is much closer in segmental arrangement to the alpha *biennis I* complexes than to the beta *biennis II* complexes, it has probably been derived from *biennis I*.

if it proves to be distinctly different from all known races or species, phenotypically and cytogenetically, suggesting an independent origin, is one justified in giving it specific rank.

Ideally, species of *Oenothera* should be distinguished in the light of what is known of their mode of origin, so that nomenclature will conform to phylogenetic realities. Races with a common evolutionary history should be grouped into a single species; those that have arisen by different routes should be put into different species. Renner and his school have shown the degree and kind of relationship, genetic and cytological, that existed between the various lots of material with which they have worked. As a result of these analyses, plus those performed in our own and other laboratories, the species listed above as recognized by Renner appear to have arisen as shown in Table 19.II. The evidence indicates that these fall into eight groups from the standpoint of the classes of complexes that make them up. Eight or possibly nine distinct and separate origins are indicated. Three of these groups correspond to North American species as defined by Munz (1965), and probably represent introductions from this area. The other groups combine complexes that trace back to different American species: they represent, therefore, different types of hybridization that have occurred, in all probability, in Europe.

Not being a taxonomist, and not having made a firsthand study of the European *Oenothera* flora, I am not herewith proposing a revision of the nomenclature of this group. I recommend to taxonomists, however, the desirability of arranging the described forms into species, each of which includes races which have been shown by cytogenetic analysis to have had a similar origin. If this is done, classification will reflect the historical development of the European *Oenothera* flora and will conform with what is known generally regarding phylogenetic relationships in the genus.

General Summary

The intensive study of *Oenothera* began in 1886 with de Vries' discovery of a race which he assigned to *Oenothera lamarckiana* Ser. From that time until his death in 1935, de Vries devoted most of his research efforts to a study of this genus.

de Vries was dissatisfied with two aspects of Darwin's theory of evolution: (1) the postulated minuteness of evolutionary steps, so that evolution is a process that cannot be observed directly; (2) the hypothesis of pangenesis with its suggestions of the inheritance of acquired characters. He was therefore searching for evidence that evolutionary steps can be large enough to be seen, which evidence he thought was presented by *Oenothera*. At the same time he developed a modified theory of pangenesis which he called "intracellular pangenesis", to explain the mechanism of hereditary transmission. Throughout his career he attempted to explain the breeding behavior of *Oenothera* in terms of this theory, which prevented him from giving proper consideration to the developing science of cytogenetics, and as a result led him to some invalid conclusions.

de Vries concluded that *lamarckiana* is a pure species which is passing through a mutable period in which one or more "pangenes" have become labile and capable of mutating under appropriate environmental conditions. de Vries discovered over the years a large number of aberrants which he interpreted as mutants of this type. One of these proved to be a tetraploid, one a triploid, a few were diploids, but most were trisomics. Aberrancy in the case of heteroploids was not considered to be the result of altered chromosome numbers; rather, extra chromosomes were considered a result of the full mutation of labile pangenes. de Vries believed that his mutations represented evolution at work, giving rise to new elementary species (in the case of the tetraploid from *gigas*) or varieties. Apart from the labile pangenes, *lamarckiana* breeds true, and this de Vries considered to be evidence of the purity of the race. The fact that this race produces two kinds of progeny when outcrossed was at first ascribed to a high frequency of mutation of labile *laeta* pangenes to the inactive *velutina* condition under the influence of the other parent. Later he accepted the idea of a single ancestral mutation, so that *lamarckiana*, while pure in most respects, is heterozygous for the *laeta* pangene.

He did not present a clear-cut explanation for the production of unlike reciprocals in crosses involving heterogamous races. Thus, de Vries ascribed not only the appearance of aberrants but also the peculiar breeding behavior of *Oenothera* to mutation.

His conclusion that the races with which he worked are essentially pure species was opposed by such workers as Heribert-Nilsson and Davis who thought that these races were of hybrid origin, the mutations being the result of the segregation of heterozygous genes.

CHAPTER 3

The true genetical situation in the genus was discovered and analyzed by Renner who found that most races behave as though all genes belong to a single linkage group, for which they are highly heterozygous. They produce, therefore, only two kinds of egg or sperm, genetically speaking. Balanced lethals prevent the development of homozygotes; only heterozygotes can develop as the result of selfing (most races are self-pollinators). The race therefore breeds true and each set of genes is carried on indefinitely, a so-called Renner complex. Renner found, however, that genes are no longer linked into a single group in many hybrids, different hybrids having different numbers and kinds of linkage groups.

CHAPTER 4

The cytological basis for this behavior was found by Cleland. In most races, all chromosomes are linked end to end in diakinesis to form a closed circle. Adjacent chromosomes pass to opposite poles as a rule. Assuming the alternation of paternal and maternal chromosomes, all paternal chromosomes (and genes) go to one pole, the entire maternal genome to the other. In hybrids, however, this may not be true. Fifteen possible arrangements into circles and pairs are possible when 14 chromosomes are present. All 15 arrangements have been found many times in hybrids, each hybrid having one and only one of these arrangements. Renner and Oehlkers, collaborating with Cleland, demonstrated that the number of linkage groups in a hybrid corresponds to the number of chromosome groups (circles or pairs) present. Thus, extensive linkage of genes was proved to be the result of chromosome linkage into closed circles.

CHAPTER 5

The clue to the cause of circle formation in *Oenothera* was furnished by Belling who discovered reciprocal translocation in *Datura* and suggested that this is the basis for the cytological peculiarities in *Oeno-thera*. This suggestion was tested independently by Sturtevant and Emerson, and by Cleland and Blakeslee who found that it was the correct explanation. Segments have been exchanged in wholesale fashion be-tween non-homologous chromosomes. As a result, every complex has its own segmental arrangement. The shuffling of ends has therefore been very extensive. Every pairing end has been found associated with every other end. All 91 possible associations of 14 ends in twos have been found among the complexes analyzed. Cleland and his students have analyzed the end arrangements of 438 complexes in terms of a standard arrange-ment and have found 162 different arrangements.

CHAPTER 6

The methods by which analyses of segmental arrangements are carried out are illustrated.

CHAPTER 7

Retention and accumulation of translocations would not have been possible had it not been for the fact that the ancestors of *Oenothera* had chromosomes that were identical in size and shape, with centromeres centrally placed between equal arms, and surrounded by a region that is largely heterochromatic in nature. Because of this structure, there was a tendency for breaks to occur in the heterochromatic region close to the centromeres, thus resulting in equal exchange. Exchange chromosomes were therefore unmodified in external morphology as compared with non-exchange chromosomes.

In spite of extensive histories of interchange, therefore, all chromo-somes in an *Oenothera* plant tend to be essentially identical in size and morphology. This is especially true of circle chromosomes. From the standpoint of the mechanics of meiosis, differences in size and shape as between non-homologous chromosomes are of no importance, as long as the chromosomes are paired. The situation is different, however, in the case of circle chromosomes. If chromosomes of differing morphology are associated in a circle, the distances from one centromere to the next will differ around the circle. As a result, the forces operating to separate adjacent centromeres will vary in strength and this will make the regular separation of adjacent chromosomes difficult. Plants in which this situation prevails, as in *Rhoeo*, ordinarily show a large proportion of cells in which non-disjunction occurs—adjacent chromosomes going to the same pole. If the distances between centromeres are equal around the circle, however, regular disjunction will be the rule. Circle-bearing forms of *Oenothera* are not handicapped as they are in *Rhoeo*.

Presence of heterochromatin near the centromeres means that in meristematic tissues, the center of the chromosome remains condensed during interphase, thus forming prochromosomes. In prophase, progressive coiling occurs from the center outward so that by mid-prophase the chromosome seems to be composed of a proximal thicker region and a thinner distal portion, the former growing at the expense of the latter throughout prophase.

CHAPTER 8

Early stages of meiosis in *Oenothera* are so difficult to study that they are imperfectly understood. The few attempts that have been made to interpret the events during leptophase and zygophase must be treated with skepticism. There is evidence that synapsis is confined to the ends of the chromosome, that it proceeds only a short distance from the tips. Presumably, the condensation that begins at the centromere and extends outwardly occurs in meiosis as well as mitosis, so that a region is soon reached, as chromosomes synapse, where coiling has occurred, thus making further synapsis impossible. This means that most of the thread system in pachyphase and diplophase is univalent.

Condensation occurs rapidly during the relatively short diplophase and by the time the chromosomes enter diakinesis they have already developed major coils.

In diakinesis, rings or pairs are often found to be interlocked. By metaphase I, these have disappeared, but the number of open rings or bivalents in metaphase is less than would be expected if separation of interlocked rings resulted commonly in breakage of chromatids with subsequent failure of reunion. Reunion of broken ends can occur in such a manner as to produce reciprocal translocations and it is possible that this is one mechanism by which interchanges can occur.

CHAPTER 9

Genetic analysis of complexes in *Oenothera* is complicated by several factors. The presence of large circles and balanced lethals eliminates to a large extent independent assortment so that true linkage cannot be distinguished from linkage that results from chromosome cohesion. Furthermore, where segregation can occur, as in hybrids with more than one chromosome group, ratios are apt to be distorted because of competition between megaspores containing different genomes (the so-called

"Renner effect") or between pollen tubes which grow at different speeds when they contain different genomes. In spite of these handicaps, however, Renner and his students, as well as other workers, have made some progress in the analysis of the genic content of different complexes, and have in a few cases been able to assign certain loci to specific chromosomes or chromosome ends.

CHAPTER 10

Crossing over is rarely observed in wild races because paired ends have become homozygous in the course of time. It is more frequently seen in hybrids between wild strains, where heterozygosity is present with respect to certain loci. If such hybrids are carried on in selfed line, however, crossing over tends to disappear through unconscious selection of individuals homozygous for the dominant.

CHAPTER 11

The only case of position effect so far discovered in higher plants was found by Catcheside in *Oenothera blandina*. Treatment with x-rays brought about a reciprocal translocation between 3·4 and 11·12 resulting in the apparent transfer of the *S* and *P* loci into the neighborhood of heterochromatin. As a consequence, these characters show a mottled appearance in the translocation heterozygotes.

CHAPTER 12

Two characters are known in *Oenothera* which quite commonly display a mottling or inconstancy of behavior. These are the cruciate and the missing petal characters. The former has been intensively studied in a variety of materials by Oehlkers and Renner, both of whom ascribe the behavior of this locus to genic mutation—Oehlkers postulating a high level of mutability at this site, Renner adopting a modified form of Winkler's gene conversion hypothesis. The missing petal character has been described by the present author (1970). It has been found in a number of wild races as well as in certain hybrid combinations.

These two characters can be explained without resort to recurrent alteration in gene structure. It is postulated that they represent mutations which have altered the potency of the respective genes in such a way that the genic environment is able to turn them on or off in given cells and tissues. The *cr* gene is one basic to the development of sepals which has acquired enhanced potency, such that it is able, under proper circumstances, to function in petal primordia, turning off the sequence of gene activity that leads to petal formation and substituting the much more complicated sequence that leads to sepal formation. In many cases, petals are more or less intermediate in appearance between typical petals and sepaloid structures, and these have been shown to be mosaics made up mostly of islands of typical petaloid or sepaloid cells.

The missing petal character is the result of mutation of a locus that is basic to the initiation of petals, by which potency is reduced to the point where local conditions dependent upon the genome as a whole determine whether the gene shall be turned on or not at the site of an individual petal.

Thus, both cases of inconstancy in behavior can be explained in terms of gene regulation rather than of constantly recurring gene mutation.

CHAPTER 13

Self-incompatibility of the *Nicotiana* type is frequently found in other subgenera of *Oenothera* but is not present in the subgenus *Oenothera*, unless *O. organensis* is considered to belong to this group. However, the work of Steiner and his students suggests that this type of self-incompatibility was once present in some of the ancestors of present-day forms—S-factors are still present in the subgenus, but in unbalanced condition, serving in many cases as male gametophytic lethals. The extent to which this is true is as yet undetermined, but it is probable that the pollen lethals in many races of *Oenothera* were originally S-factors and are derived originally from ancestors in which the typical balanced situation found in *Nicotiana* and in other subgenera of *Oenothera* was present.

CHAPTER 14

From the beginning of cytogenetic work on *Oenothera* it has been observed that certain complex-combinations find it difficult or impossible to develop because of gene-plastid incompatibility. Renner found that plastid primordia are contributed to the zygote by both sperm and egg. In many combinations, the plastids derived from one parent are incapable of developing a normal complement of pigments, whereas those contributed by the other parent are quite normal. When two classes of plastid are present in the zygote they tend to segregate in early embryogeny into different cells, thus producing sectorial or periclinal chimeras.

Renner also found that plastids which fail to turn green in the presence of an incompatible genome combination have not been caused to mutate

as a result. It is possible by appropriate techniques to bring them back
into a genic milieu in which they can function normally and when this
occurs, they develop normally once more.

True mutation may occur in plastids, however, and Renner found
numbers of cases where plastids had mutated in such a way that they
could not turn green in any genic environment.

Renner's classical work has been extended and expanded by some of
his students. Stubbe has made an extensive analysis of the situation in
the subgenus *Oenothera*, and finds that there exist five classes of plastid
from the standpoint of their ability or inability to function in particular
genome combinations. There exist also three classes of genome from the
standpoint of their effect on plastid development. These classes are
characteristic of particular species of *Oenothera* as defined by Cleland
and Munz.

Schötz, utilizing certain mutant plastids as standards, has been able
to show that different strains of plastid multiply at different rates. He
and his students have also analyzed in detail the structure and behavior
of different plastids when brought into association with incompatible
genome combinations, using electron microscopy and also following the
process of bleaching and greening that takes place during early development of particular complex-combinations. Schötz and his group have
also analyzed the pigment content of plastids at different stages in the
development of plants in which disharmony is present; and have undertaken the determination of the biochemical nature of plastid mutations.

Stubbe has pointed out that plastids carry multiple genetic factors
which can mutate separately. Some of these factors may affect such
diverse characters as leaf morphology, starch grain shape in the pollen,
and germinability of pollen. Plastids exhibit continuity from one
generation to another, specificity of behavior and genetic activity
independent of the associated genome combination. They are therefore
independent bearers of genetic information.

CHAPTER 15

The characteristic that first attracted de Vries' interest to *Oenothera*
was its propensity for throwing occasional aberrants. As mentioned
above, most of these turned out to be hyperploids, especially trisomics.
A few apparently involved translocations, very few were the result of
gene mutation.

By far the largest group of aberrants is that of the trisomics. These
are of two types—dimorphic mutants which, when selfed, produce two
classes of plants (their own type, and *lamarckiana*) and sesquiplex

mutants which breed true. Both Catcheside and Emerson have studied
in detail the methods by which these two classes of mutants can arise as
the result of irregularities in the zigzag arrangement of chromosomes in
meiosis.

The mode of origin of one class of mutants is not fully understood, that
of the so-called "half-mutants". These mutants have one unmodified
lamarckiana complex and one modified alethal complex which has
suffered at least two interchanges in its origin from *lamarckiana*. Several
of these mutants have arisen frequently in cultures of *lamarckiana*.

CHAPTER 16

Some attempts have been made to induce mutations in *Oenothera*,
notably by Oehlkers and his students. These have resulted in almost no
gene mutations, the effects being mainly of two kinds—reciprocal translocations and non-disjunctions, the latter a result of disturbance in the
regular zigzag arrangement of chromosomes in metaphase I. Various
agents have been used, notably x-rays and chemicals. Some of the latter
have had especially striking effects, but their principal effect has been
to induce translocations. Gene mutations have been conspicuous by
their rarity.

CHAPTER 17

For many years the question of the origin of *Oenothera lamarckiana*
was actively debated, de Vries and certain followers maintaining that it
was a "pure species" derived originally from the wild, Davis and others
arguing for a hybrid origin. Renner was of the opinion that all circle-
bearing oenotheras have arisen by hybridization. The evidence now
indicates that *lamarckiana* arose in Europe as the result of hybridization
between a *biennis* form derived from eastern North America and a race
that was either a *hookeri* from western North America or a form that
carried a complex closely related to those of the hookeris.

CHAPTER 18

De Vries' primary reason for studying *Oenothera* was to arrive at an
understanding of the process of organic evolution. One is impressed, in
reading the books and papers which he published throughout his long
life, with the fact that the study of hereditary mechanisms was of
secondary importance, something that has to be undertaken in order to
understand the way in which new forms arise. His major interest throughout was focused on the so-called mutants.

It has turned out that his mutants did not have the evolutionary importance that he ascribed to them. Nevertheless the subgenus *Oenothera* has remained an object of evolutionary significance, in that it has been possible to trace in some detail the story of its origin and development and to identify some of the important factors responsible for its evolution. We have found evidence of the existence of four populations that have arisen successively in the center of origin which we believe was in Central America or Mexico. Each of these probably arose during a period of Pleistocene glaciation. Each population no doubt began with paired, alethal chromosomes and open-pollinated flowers. Each experienced certain reciprocal translocations many of which survived because they resulted from equal exchanges and so gave rise to small circles of chromosomes in translocation heterozygotes in which regular disjunction of adjacent chromosomes was possible. Some exchanges occurred early in the development and spread of their respective populations and so became characteristic of large sectors of these populations. Others have taken place more recently and so are found only in restricted areas.

These populations, as they spread, met individuals of the preceding or succeeding populations and crossed with them. Thus, Population 1 was hybridized by Populations 2 and 3, Population 2 and 3 intercrossed, as did Populations 3 and 4. Because the histories of interchange had been different in different populations, hybrids between them were occasionally found in which no chromosome derived from one parent had the same association of ends as any in the other parent. As a result such hybrids had a circle of 14 chromosomes. When one parent had an S-factor, this automatically became a pollen lethal. If the complex with the pollen lethal was able to dominate its associated complex in the competition for development of the embryo sac, a balanced lethal situation was established. Plants with circles of 14 and balanced lethals thus derived their complexes from diverse populations. The hybrids were therefore highly heterozygous and no doubt possessed a high level of hybrid vigor. When selfed, such plants would breed true for the presence of the circle and for the maximum of hybrid vigor. Even before self-pollination was established as the result of relative shortening of the style, much pollination was self-pollination, as insects went from flower to flower on the same plant. Thus, individuals with ⊙14 tended to increase in the population. After the advent of self-pollination, true-breeding races with ⊙14 became permanently established.

Thus, we visualize the origin of circles of 14 as being the result of hybridization between diverse populations. Analysis of the segmental arrangements of the chromosomes of hundreds of complexes has enabled us to classify these complexes and to ascribe the origin of each to the particular population from which it has been derived. As a result of this analysis and classification of complexes we are able to divide the subgenus into ten species. Two of these (*elata*, found in Mexico and Central America, and *hookeri*, in California and adjacent areas) are open pollinated, alethal, with paired chromosomes, or (in outlying areas) small circles. Six of the other species, comprising the bulk of the population in the eastern two-thirds of the continent, have a circle of 14, balanced lethals and self pollination. These are *parviflora I* (= Munz's *O. parviflora* subsp. *angustissima*) derived from the cross between population 1 and 2, *parviflora II* (*O. parviflora* subsp. *parviflora*) from population 1 × 3, *biennis I* (*O. biennis* subsp. *centralis*) and *biennis II* (*O. biennis* subsp. *caeciarum*) from population 2 × 3, *biennis III* (*O. biennis* subsp. *austromontana*) from *biennis I* × *biennis II*, *strigosa* (*O. strigosa*) from population 4 × 3. Two species are regarded as relicts, one (*argillicola*) representing Population 1, the other (*grandiflora*) Population 2.

In *Oenothera*, three characters that in most plants are deleterious have been combined in such a way that together they have become advantageous. Translocations are usually disadvantageous because exchanges are ordinarily unequal which means that centromeres in circles in translocation-heterozygotes are unevenly spaced, resulting in a high level of non-disjunction and consequent sterility. In *Oenothera*, however, the translocations that have survived are the result of equal exchanges, so that centromeres are evenly spaced in the circles and little sterility ensues. Translocations are not weeded out, therefore, and are carried on indefinitely in the population.

Lethals are ordinarily deleterious, but balanced lethals in large circles have a compensating advantage in that they maintain a maximum of hybrid vigor at the expense of only 50% zygotic or gametophytic sterility. The sterilizing effect of the lethals, however, is largely overcome by the self-pollinating habit which in other organisms is harmful because it tends to reduce hybrid vigor, but in *Oenothera* cannot have this effect because the lethals prevent homozygosis. Selfing has a compensating beneficial effect in that it ensures heavy pollination and maximum seed set. The sterilizing effect of the lethals is also counteracted by the fact that anthers are introrse, the pollen thus being deposited on the stigma, and also by the fact that the plant produces many flowers—usually two or more per branch per day.

The system of balanced lethals in circles of 14 accompanied by self-pollination is admirably suited to maintaining the population at maximum vigor. It has disadvantages, however. Recessive mutations cannot be brought to light and subjected to the effects of natural selection. They are also ordinarily confined within their race of origin because of the innumerable barriers to gene flow resulting from the self-pollinating habit. It thus becomes difficult for *Oenothera* to evolve with the rapidity that can be achieved by other organisms. To a certain degree the genus has substituted present well-being for the possibility of future evolutionary development.

CHAPTER 19

The nature of the European *Oenothera* flora is discussed in the final chapter. The only person who has studied the European races utilizing both taxonomic and cytogenetic approaches, was Renner, and the classification which he adopted is outlined. Some European races are probably derived directly from North America, others have arisen in Europe as the result of hybridization between introduced races. One cannot always say with complete certainty in specific cases which alternative is the correct one. In any event, *Oenothera* is not indigenous to Europe but has been introduced from the western hemisphere, some introductions having occurred as early as the seventeenth century. One source of North American material was soil carried as ballast in ships returning to Europe and deposited on ballast heaps near some of the important European seaports.

The importance of determining chromosome configurations, and of analyzing segmental arrangements in studies of *Oenothera* taxonomy is stressed. Only when this is done is it possible to determine the origin and relationships of wild races.

Appendix I

Segmental arrangements of complexes of North American and European euoenotheras.

Formulae follow the Cleland system (11 and 12 reversed with respect to the Catcheside system). Trisomics not included.

Complex	State or province	Segmental arrangement							Reference
Hookeri									
[h]Aztec	New Mexico	1·2	3·4	5·6	7·10	9·8	11·12	13·14	Cleland, 1958
[h]Dalton	California	1·2	3·4	5·6	7·10	9·8	11·12	13·14	Cleland, 1935b
[h]Devils Gate	California	1·2	3·4	5·6	7·10	9·8	11·12	13·14	Cleland, 1935b
[h]franciscana E. and S.	California	1·2	3·4	5·6	7·10	9·8	11·12	13·14	Cleland, 1935b
[h]grisea	California	1·2	3·4	5·6	7·10	9·8	11·12	13·14	Cleland, 1958
[h]Hall 21 (= Hall)	California	1·2	3·4	5·6	7·10	9·8	11·12	13·14	Cleland, 1935b
Hall 30 b	California	1·2	3·4	5·6	7·10	9·8	11·12	13·14	Cleland, 1935b
Heusi	California	1·2	3·4	5·6	7·10	9·8	11·12	13·14	Cleland, 1935b
[h]Johansen (= T. and G.)	California	1·2	3·4	5·6	7·10	9·8	11·12	13·14	Cleland, 1935b
[h]Las Vegas	Nevada	1·2	3·4	5·6	7·10	9·8	11·12	13·14	Cleland, 1958
longissima Grand Canyon a	Arizona	1·2	3·4	5·6	7·10	9·8	11·12	13·14	Cleland, 1958
[h]Mataguey	California	1.2	3·4	5·6	7·10	9·8	11·12	13·14[a]	Cleland, 1951
[h]Mateo	California	1·2	3·4	5·6	7·10	9·8	11·12	13·14'	Cleland, 1935b
[h]Mono	California	1·2	3·4	5·6	7·10	9·8	11·12	13·14[a]	Cleland, 1935b
McElmo a	Colorado	1·2	3·4	5·6	7·10	9·8	11·12	13·14	Norby (Cleland, 1958)
Palisade b	Colorado	1·2	3·4	5·6	7·10	9·8	11·12	13·14	Norby (Cleland, 1958
[h]Purgatoire River	Colorado	1·2	3·4	5·6	7·10	9·8	11·12	13·14	Norby (Cleland, 1958)
Ramsey a	Arizona	1·2	3·4	5·6	7·10	9·8	11·12	13·14	Cleland, 1958
Raton a	New Mexico	1·2	3·4	5·6	7·10	9·8	11·12	13·14	Cleland, 1958
[h]Silverton	Colorado	1·2	3·4	5·6	7·10	9·8	11·12	13·14	Cleland, 1958
Taos a	New Mexico	1·2	3·4	5·6	7·10	9·8	11·12	13·14	Cleland, 1958
McElmo b	Colorado	1·2	3·4	5·6	7·10	9·11	8·12	13·14	Norby (Cleland, 1958)
[h]Cocheropa Creek	Colorado	1·4	3·2	5·6	7·10	9·8	11·12	13·14	Norby (Cleland, 1958)
Heusi b	California	1·4	3·2	5·6	7·10	9·8	11·12	13·14	Cleland, 1935b
longissima Grand Canyon b	Arizona	1·4	3·2	5·6	7·10	9·8	11·12	13·14	Cleland, 1958
[h]Moffat	Colorado	1·4	3·2	5·6	7·10	9·8	11·12	13·14	Norby (Cleland, 1958)
Palisade a	Colorado	1·4	3·2	5·6	7·10	9·8	11·12	13·14	Norby (Cleland, 1958)
Raton b	New Mexico	1·4	3·2	5·6	7·10	9·8	11·12	13·14	Cleland, 1958
Salida III a	Colorado	1·4	3·2	5·6	7·10	9·8	11·12	13·14	Norby (Cleland, 1958)
[h]San Miguel	Mexico	1·4	3·2	5·6	7·10	9·8	11·12	13·14	Cleland, 1958
[h]franciscana de V.	California	1·2	3·4	5·6	7·8	9·10	11·12	13·14	Cleland, 1935b
[h]franciscana Shull	California	1·2	3·4	5·6	7·8	9·10	11·12	13·14	Cleland and Hammond, 1950
Hall 30 a	California	1·2	3·4	5·6	7·8	9·10	11·12	13·14	Cleland, 1935b
Hall 34 a	California	1·2	3·4	5·6	7·8	9·10	11·12	13·14	Cleland, 1935b
[h]hookeri de V.	California	1·2	3·4	5·6	7·8	9·10	11·12	13·14	Cleland and Blakeslee, 1931

[a] Plus extras.

Complex	State or province	Segmental arrangement							Reference
hParras	Mexico	1·2	3·4	5·6	7·8	9·10	11·12	13·14	Cleland, 1958
Las Cruces a	New Mexico	1·2	3·4	5·9	7·8	6·10	11·12	13·14	Cleland, 1958
Las Cruces b	New Mexico	1·4	3·2	5·10	7·6	9·8	11·12	13·14	Cleland, 1958
Ramsey b	Arizona	1·2	3·4	5·10	7·6	9·8	11·12	13·14	Cleland, 1958
α Albuquerque B	New Mexico	1·2	3·4	5·6	7·10	9·13	11·12	8·14	Cleland, 1958
β Albuquerque B	New Mexico	1·2	3·11	5·6	7·10	9·4	8·12	13·14	Cleland, 1958
α Jamesii Norman	Oklahoma	1·2	3·4	5·11	7·10	9·8	6·12	13·14	Cleland, 1958
β Jamesii Norman	Oklahoma	1·4	3·11	5·7	6·10	9·8	2·12	13·14	Cleland, 1958
hJensen	Utah	1·4	3·11	5·6	7·10	9·8	2·12	13·14	Cleland, 1958
hLinda Vista	New Mexico	1·4	3·2	5·9	7·10	6·8	11·12	13·14	Cleland, 1958
La Verken a	Utah	1·4	3·2	5·12	7·10	9·8	11·6	13·14	Cleland, 1958
Salida I a	Colorado	1·4	3·2	5·8	7·10	9·6	11·12	13·14	Cleland, 1958
Salida I b	Colorado	1·4	3·2	5·8	7·12	9·6	11·10	13·14	Cleland, 1958
Elata									
hChapultepec	Mexico	1·4	3·2	5·6	7·10	9·14	11·12	13·8	Steiner, 1951
hPuertoaereo	Mexico	1·4	3·2	5·6	7·10	9·14	11·12	13·8	Steiner, 1955
hTexmelucan	Mexico	1·4	3·2	5·6	7·10	9·14	11·12	13·8	Steiner, 1951
hToluca	Mexico	1·4	3·2	5·6	7·10	9·14	11·12	13·8	Steiner, 1951
hZimapan	Mexico	1·4	3·2	5·6	7·10	9·14	11·12	13·8	Steiner, 1951
hAcatzingo	Mexico	1·4	3·2	5·9	7·10	6·8	11·12	13·14	Steiner, 1955
hCholula	Mexico	1·4	3·2	5·9	7·10	6·8	11·12	13·14	Steiner, 1955
hPuebla	Mexico	1·4	3·2	5·9	7·10	6·8	11·12	13·14	Steiner, 1955
hGuatemala	Guatemala	1·2	3·10	5·9	7·4	6·8	11·12	13·14	Steiner, 1951
hChichicastinango	Guatemala	1·2	3·10	5·9	7·4	6·8	11·12	13·14	Steiner, 1951
α Strigosa									
α Minturn	Colorado	1·10	3·4	5·8	7·12	9·6	11·2	13·14	Norby (Cleland, 1958)
α Platteville	Colorado	1·10	3·4	5·8	7·12	9·6	11·2	13·14	Cleland, 1954b
α Littleton I	Colorado	1·10	3·4	5·8	7·2	9·11	6·12	13·14	Cleland, 1954b
α Longmont	Colorado	1·10	3·4	5·8	7·2	9·11	6·12	13·14	Cleland, 1954b
α Yampa Valley	Colorado	1·10	3·4	5·8	7·2	9·11	6·12	13·14	Cleland, 1954b
curtans	Colorado	1·7	3·4	5·8	2·10	9·11	6·12	13·14	Cleland and Hammond, 1950
α Forsberg	Colorado	1·7	3·4	5·8	2·10	9·11	6·12	13·14	Norby (Cleland, 1958)
α Littleton II (tentative)	Colorado	1·7	3·4	5·8	2·10	9·11	6·12	13·14	Norby (Cleland, 1958)
α North Colorado Springs	Colorado	1·7	3·4	5·8	2·10	9·11	6·12	13·14	Norby (Cleland, 1958)
α Palmer Lake	Colorado	1·7	3·4	5·8	2·10	9·11	6·12	13·14	Cleland, 1954b
α Sky View	Colorado	1·7	3·4	5·8	2·10	9·11	6·12	13·14	Norby (Cleland, 1958)
α Turkey Creek Pass	Colorado	1·7	3·4	5·12	2·10	9·11	6·8	13·14	Norby (Cleland, 1958)
α Tinytown	Colorado	1·7	3·4	5·9	2·10	6·12	11·8	13·14	Norby (Cleland, 1958)
α West Estes Park	Colorado	1·7	3·4	5·14	2·10	9·12	11·8	13·6	Norby (Cleland, 1958)
α Sutherland	Nebraska	1·7	3·4	5·9	6·8	10·12	11·2	13·14	Cleland, 1954b
α Loveland	Colorado	1·10	3·4	5·12	7·2	9·11	6·8	13·14	Cleland, 1954b
α Ashland C	Wisconsin	1·10	3·4	5·12	7·2	9·6	11·8	13·14	Cleland, 1954b
α Fargo	North Dakota	1·10	3·13	5·12	7·2	9·6	11·8	4·14	Cleland, 1954b
α Iowa VI	Iowa	1·2	3·13	5·12	7·10	9·6	11·8	4·14	Cleland and Hammond, 1950
α Mason City	Iowa	1·2	3·13	5·12	7·10	9·6	11·8	4·14	Cleland, 1954b
α Gothenburg	Nebraska	1·10	3·12	5·8	7·2	9·6	11·4	13·14	Cleland (cited by Steiner, 1952)
α Heber	Utah	1·10	3·12	5·9	7·2	6·8	11·4	13·14	Cleland, 1954b
α Iowa II	Iowa	1·2	3·14	5·8	7·11	9·6	10·12	13·4	Cleland and Hammond, 1950
α Iowa XII	Iowa	1·2	3·14	5·8	7·11	9·6	10·12	13·4	Cleland and Hammond, 1950
α Spooner	Wisconsin	1·2	3·14	5·8	7·11	9·6	10·12	13·4	Cleland, 1954b
α Nebraska	Nebraska	1·2	3·7	5·8	11·14	9·6	10·12	13·4	Cleland, 1954b

Complex	State or province	Segmental arrangement						Reference
α Haskett	Manitoba	1·2	3·11 5·8	7·12	9·6	4·10 13·14		Cleland, 1954b
α Leonard	North Dakota	1·2	3·11 5·8	7·12	9·6	4·10 13·14		Cleland, 1954b
α Castle Rock	Colorado	1·2	3·10 5·8	7·4	9·12 11·6	13·14		Cleland, 1954b
α Birch Tree III	Missouri	1·5	3·13 2·6	7·12	9·14 11·10	4·8		Cleland, 1954b
α Clarks	Nebraska	1·13	3·4 5·6	7·8	9·11 10·12	2·14		Cleland, 1954b
α Granger	Washington	1·12	3·13 5·9	7·10	2·4 11·14	6·8		Cleland, 1954b
α Yakima	Washington	1·8	3·14 5·11	7·10	9·12 2·6	13·4		Cleland, 1954b
α St. Anthony	Idaho	1·2	3·9 5·6	7·13	8·12 11·10	4·14		Cleland, 1954b
α Brookston	Colorado	1·2	3·10 5·7	4·6	9·8 11·13	12·14		Cleland, unpublished
α Monett	Arkansas	1·2	3·10 5·12	7·14	9·8 11·6	13·4		Cleland, 1954b
α Cœur d'Alene	Idaho	1·9	3·5 2·8	7·10	4·6 11·14	13·12		Cleland, unpublished
α Greenland I	Arkansas	1·4	3·12 5·2	7·10	9·14 11·8	13·6		Hoff, 1966
α Fruitland	Idaho	1·2	3·5 6·12	7·10	9·4 11·8	13·14		Cleland, 1954b
β Strigosa								
β Brookston	Colorado	1·4	3·2 5·14	7·8	9·10 11·12	13.6		Cleland 1954b
β Fargo	North Dakota	1·4	3·2 5·14	7·8	9·10 11·12	13·6		Cleland, 1954b
β Farmer	South Dakota	1·4	3·2 5·14	7·8	9·10 11·12	13·6		Cleland, unpublished
β Fruitland	Idaho	1·4	3·2 5·14	7·8	9·10 11·12	13·6		Cleland, 1954b
β Granger	Washington	1·4	3·2 5·14	7·8	9·10 11·12	13·6		Cleland, 1954b
β Haskett	Manitoba	1·4	3·2 5·14	7·8	9·10 11·12	13·6		Cleland, 1954b
β Leonard	North Dakota	1·4	3·2 5·14	7·8	9·10 11·12	13·6		Cleland, 1954b
β Minturn	Colorado	1·4	3·2 5·14	7·8	9·10 11·12	13·6		Norby (Cleland, 1958)
β Red Cliff	Colorado	1·4	3·2 5·14	7·8	9·10 11·12	13·6		Norby (Cleland, 1958)
β St. Anthony	Idaho	1·4	3·2 5·14	7·8	9·10 11·12	13·6		Cleland, 1954b
β Sky View	Colorado	1·4	3·2 5·14	7·8	9·10 11·12	13·6		Norby (Cleland, 1958)
stringens	Wyoming	1·4	3·2 5·14	7·8	9·10 11·12	13·6		Cleland and Hammond, 1950
β Tinytown	Colorado	1·4	3·2 5·14	7·8	9·10 11·12	13·6		Norby (Cleland, 1958)
β Turkey Creek Pass	Colorado	1·4	3·2 5·14	7·8	9·10 11·12	13·6		Norby (Cleland, 1958)
β West Colorado	Colorado	1·4	3·2 5·14	7·8	9·10 11·12	13·6		Norby (Cleland 1958)
β Yakima	Washington	1·4	3·2 5·14	7·8	9·10 11·12	13·6		Cleland, 1954b
β Nebraska	Nebraska	1·4	3·2 5·10	7·6	9·8 11·12	13·14		Cleland 1954b
β Clarks	Nebraska	1·4	3·2 5·10	7·6	9·8 11·13	12·14		Cleland, 1954b
β Iowa II	Iowa	1·4	3·2 5·10	7·6	9·8 11·13	12·14		Cleland and Hammond, 1950
β Mason City	Iowa	1·4	3·2 5·10	7·6	9·8 11·13	12·14		Cleland, 1954b
β Spooner	Wisconsin	1·4	3·2 5·10	7·6	9·8 11·13	12·14		Cleland, 1954b
β Castle Rock	Colorado	1·4	3·2 5·10	7·6	9·13 11·12	8·14		Cleland, 1954b
North Colorado Springs	Colorado	1·4	3·2 5·10	7·6	9·13 11·12	8·14		Norby (Cleland, 1958)
β Palmer Lake	Colorado	1·4	3·2 5·10	7·6	9·13 11·12	8·14		Cleland, 1954b
β West Estes Park	Colorado	1·4	3·2 5·10	7·6	9·13 11·12	8·14		Norby (Cleland, 1958)
elongans	Colorado	1·4	3·2 5·10	7·6	9·14 11·12	13·8		Cleland and Hammond, 1950
β Gothenburg	Nebraska	1·4	3·2 5·10	7·6	9·14 11·12	13·8		Cleland, 1954b
β Sutherland	Nebraska	1·4	3·2 5·10	7·6	9·14 11·12	13·8		Cleland, 1954b
β Iowa VI	Iowa	1·4	3·2 5·7	6·10	9·14 11·12	13·8		Cleland and Hammond, 1950
β Iowa XII	Iowa	1·4	3·2 5·7	6·10	9·14 11·13	8·12		Cleland and Hammond, 1950
β Ashland C	Wisconsin	1·4	3·2 5·7	6·10	9·14 11·13	8·12		Cleland, 1954b
β Littleton I	Colorado	1·13	3·2 5·10	7·6	9·8 11·12	4·14		Cleland, 1954b
β Longmont	Colorado	1·13	3·2 5·10	7·6	9·8 11·12	4·14		Cleland, 1954b
β Platteville	Colorado	1·13	3·2 5·10	7·6	9·8 11·12	4·14		Cleland, 1954b

Complex	State or province	Segmental arrangement						Reference
β Yampa Valley	Colorado	1·13	3·2	5·10	7·6	9·8	11·12 4·14	Norby (Cleland, 1958)
β Forsberg	Colorado	1·4	3·2	5·10	7·14	9·8	11·12 13·6	Norby (Cleland, 1958)
β Littleton II	Colorado	1·4	3·2	5·10	7·14	9·$	11·12 13·6	Norby (Cleland, 1958)
β Loveland	Colorado	1·4	3·2	5·13	7·8	9·10	11·12 6·14	Cleland, 1954b
β Heber	Utah	1·3	2·4	5·14	7·8	9·10	11·12 13·6	Cleland, 1954b
β Monett	Missouri	1·4	3·2	5·11	7·10	9·6	8·14 13·12	Cleland, 1954b
β Birch Tree III	Missouri	1·4	3·2	5·11	7·10	9·6	8·14 13·12	Cleland, 1954b
β Cœur d'Alene	Idaho	1·4	3·2	5·9	7·12	10·14 11·8	13·6	Cleland, unpublished
α Biennis I								
α Beaufort	North Carolina	1·4	3·2	5·6	7·10	9·8	11·12 13·14	Stinson and Steiner, 1955
acuens	Alabama	1·4	3·2	5·6	7·10	9·8	11·12 13·14	Cleland, 1935a
excellens	Illinois	1·2	3·4	5·6	7·10	9·8	11·12 13·14	Cleland and Blakeslee, 1931
α Greenland II	Arkansas	1·2	3·4	5·6	7·10	9·8	11·12 13·14	Hoff, 1966
α Athens B	Georgia	1·2	3·4	5·14	7·10	9·8	11·12 13·6	Steiner, 1952
α Athens D	Georgia	1·2	3·4	5·14	7·10	9·8	11·12 13·6	Steiner, 1952
α Baltimore	Maryland	1·2	3·4	5·14	7·10	9·8	11·12 13·6	Cleland, 1958
α Belleville	Illinois	1·2	3·4	5·14	7·10	9·8	11·12 13·6	Cleland, 1958
α Best	Arkansas	1·2	3·4	5·14	7·10	9·8	11·12 13·6	Hoff, 1966
α Birch Tree I	Missouri	1·2	3·4	5·14	7·10	9·8	11·12 13·6	Cleland, 1958
α Birch Tree II	Missouri	1·2	3·4	5·14	7·10	9·8	11·12 13·6	Cleland, 1958
α Bloomington II	Indiana	1·2	3·4	5·14	7·10	9·8	11·12 13·6	Stinson (Cleland, 1958)
α Cambridge I (tentative)	Ohio	1·2	3·4	5·14	7·10	9·8	11·12 13·6	Cleland, 1958
α Camp Peary E	Virginia	1·2	3·4	5·14	7·10	9·8	11·12 13·6	Stinson and Steiner, 1955
α Camp Peary L	Virginia	1·2	3·4	5·14	7·10	9·8	11·12 13·6	Stinson and Steiner, 1955
α Cardiff Delta	Pennsylvania	1·2	3·4	5·14	7·10	9·8	11·12 13·6	Cleland, unpublished
α Chevy Chase	Maryland	1·2	3·4	5·14	7·10	9·8	11·12 13·6	Cleland and Hammond, 1950
α Decatur	Arkansas	1·2	3·4	5·14	7·10	9·8	11·12 13·6	Hoff, 1966
α Delaware	Ohio	1·2	3·4	5·14	7·10	9·8	11·12 13·6	Cleland, 1958
α Dutton	Arkansas	1·2	3·4	5·14	7·10	9·8	11·12 13·6	Hoff, 1966
α Fayetteville	Arkansas	1·2	3·4	5·14	7·10	9·8	11·12 13·6	Hoff, 1966
α Friendship	Indiana	1·2	3·4	5·14	7·10	9·8	11·12 13·6	Cleland, 1958
α Georgetown	Arkansas	1·2	3·4	5·14	7·10	9·8	11·12 13·6	Hoff, 1966
α Gunpowder	Maryland	1·2	3·4	5·14	7·10	9·8	11·12 13·6	Cleland, unpublished
α Hendricksville	Indiana	1·2	3·4	5·14	7·10	9·8	11·12 13·6	Cleland, unpublished
α Hilltop (tentative)	?	1·2	3·4	5·14	7·10	9·8	11·12 13·6	Hoff, unpublished
α Hindsville	Arkansas	1·2	3·4	5·14	7·10	9·8	11·12 13·6	Hoff, 1966
α Hopkinsville	Kentucky	1·2	3·4	5·14	7·10	9·8	11·12 13·6	Cleland and Hammond, 1950
α Hot Springs	Arkansas	1·2	3·4	5·14	7·10	9·8	11·12 13·6	Cleland, 1958
α Indianapolis	Indiana	1·2	3·4	5·14	7·10	9·8	11·12 13·6	Hoff, unpublished
α Japton	Arkansas	1·2	3·4	5·14	7·10	9·8	11·12 13·6	Hoff, 1966
α Lake	Virginia	1·2	3·4	5·14	7·10	9·8	11·12 13·6	Cleland, 1958
α Le Moyne	Pennsylvania	1·2	3·4	5·14	7·10	9·8	11·12 13·6	Cleland, unpublished
α Lexington	Kentucky	1·2	3·4	5·14	7·10	9·8	11·12 13·6	Cleland and Hammond, 1950
α Loch Raven	Maryland	1·2	3·4	5·14	7·10	9·8	11·12 13·6	Hammond (Cleland, 1958)
α Londonderry	Ohio	1·2	3·4	5·14	7·10	9·8	11·12 13·6	Cleland, 1958
α Magnolia	Kentucky	1·2	3·4	5·14	7·10	9·8	11·12 13·6	Cleland, 1958

Complex	State or province	Segmental arrangement							Reference
α Mifflintown	Pennsylvania	1·2	3·4	5·14	7·10	9·8	11·12	13·6	Cleland, unpublished
α New Kent	Virginia	1·2	3·4	5·14	7·10	9·8	11·12	13·6	Cleland, unpublished
α Newport News B	Virginia	1·2	3·4	5·14	7·10	9·8	11·12	13·6	Stinson and Steiner, 1955
α Oxford	North Carolina	1·2	3·4	5·14	7·10	9·8	11·12	13·6	Stinson and Steiner, 1955
α Paducah	Kentucky	1·2	3·4	5·14	7·10	9·8	11·12	13·6	Cleland and Hammond, 1950
α Petersburg	Virginia	1·2	3·4	5·14	7·10	9·8	11·12	13·6	Cleland, 1958
α Poplar Bluff	Missouri	1·2	3·4	5·14	7·10	9·8	11·12	13·6	Cleland, 1958
α Princeton	West Virginia	1·2	3·4	5·14	7·10	9·8	11·12	13·6	Cleland, 1958
α Rich Mountain	Arkansas	1·2	3·4	5·14	7·10	9·8	11·12	13·6	Cleland, 1958
α Roanoke B	Virginia	1·2	3·4	5·14	7·10	9·8	11·12	13·6	Stinson and Steiner, 1955
α Tensaw	Alabama	1·2	3·4	5·14	7·10	9·8	11·12	13·6	Steiner, 1952
α Tuscaloosa A	Alabama	1·2	3·4	5·14	7·10	9·8	11·12	13·6	Steiner, 1952
α Tuscaloosa B	Alabama	1·2	3·4	5·14	7·10	9·8	11·12	13·6	Steiner, 1952
α Tuscaloosa C	Alabama	1·2	3·4	5·14	7·10	9·8	11·12	13·6	Steiner, 1952
α Walkerton	Indiana	1·2	3·4	5·14	7·10	9·8	11·12	13·6	Cleland, 1958
α Warrenton	Virginia	1·2	3·4	5·14	7·10	9·8	11·12	13·6	Cleland, 1958
α Warwick A	Virginia	1·2	3·4	5·14	7·10	9·8	11·12	13·6	Stinson and Steiner, 1955
α Wesley	Arkansas	1·2	3·4	5·14	7·10	9·8	11·12	13·6	Hoff, 1966
α Williamsburg	Virginia	1·2	3·4	5·14	7·10	9·8	11·12	13·6	Cleland, 1958
α Williamston	North Carolina	1·2	3·4	5·14	7·10	9·8	11·12	13·6	Stinson and Steiner, 1955
α Winslow	Arkansas	1·2	3·4	5·14	7·10	9·8	11·12	13·6	Cleland, 1958
α Evansville I (6 plants)	Indiana	1·2	3·4	5·14	7·10	9·8	11·12	13·6	Winternheimer (Cleland, 1958)
α Evansville III (2 plants)	Indiana	1·2	3·4	5·14	7·10	9·8	11·12	13·6	Winternheimer (Cleland, 1958)
α Citronella	Alabama	1·2	3·4	5·14	7·9	8·10	11·12	13·6	Cleland, cited by Steiner, 1952
α Dyer	Indiana	1·13	3·4	5·14	7·10	9·8	11·12	2·6	Cleland, 1958
α La Crosse	Virginia	1·2	3·7	5·14	4·10	9·8	11·12	13·6	Cleland, 1958
α Tyronza	Arkansas	1·2	3·4	5·14	7·10	9·11	8·12	13·6	Cleland, 1958
α Bestwater I	Arkansas	1·4	3·8	5·14	7·10	9·2	11·12	13·6	Cleland, 1958
α Bestwater II	Arkansas	1·4	3·8	5·14	7·10	9·2	11·12	13·6	Cleland, 1958
α Cambridge II	Ohio	1·2	3·4	5·6	7·10	9·13	11·12	8·14	Cleland, 1958
α New Concord I	Ohio	1·2	3·4	5·6	7·10	9·13	11·12	8·14	Cleland, 1958
α New Concord II	Ohio	1·2	3·4	5·6	7·10	9·13	11·12	8·14	Cleland, unpublished
α Warsaw	Indiana	1·2	3·4	5·6	7·10	9·13	11·12	8·14	Cleland, unpublished
α Harrisonburg	Virginia	1·4	3·2	5·10	7·14	9·8	11·12	13·6	Cleland, 1958
α Reynoldsii	Tennessee	1·4	3·2	5·10	7·14	9·8	11·12	13·6	Cleland, 1958
α Hebron I	Ohio	1·4	3·2	5·14	7·6	9·8	11·12	13·10	Cleland, 1958
α Hebron III	Ohio	1·4	3·2	5·14	7·6	9·8	11·12	13·10	Cleland, unpublished
α Richmond	Virginia	1·9	3·10	5·14	7·4	2·8	11·12	13·6	Cleland, 1958
α Athens A	Georgia	1·4	3·10	5·13	7·2	9·8	11·12	6·14	Steiner, 1952
α Athens C	Georgia	1·4	3·10	5·13	7·2	9·8	11·12	6·14	Steiner, 1952
α Iowa I	Iowa	1·2	3·8	5·4	7·10	9·13	11·12	6·14	Cleland and Hammond, 1950
β Biennis I									
β Baltimore	Maryland	1·14	3·2	5·9	7·8	6·12	11·10	13·4	Cleland, 1958
β Belleville	Illinois	1·14	3·2	5·9	7·8	6·12	11·10	13·4	Cleland, 1958
β Bestwater II	Arkansas	1·14	3·2	5·9	7·8	6·12	11·10	13·4	Cleland, 1958
β Cardiff Delta	Pennsylvania	1·14	3·2	5·9	7·8	6·12	11·10	13·4	Cleland, unpublished
β Cambridge I	Ohio	1·14	3·2	5·9	7·8	6·12	11·10	13·4	Cleland, 1958
β Cambridge II	Ohio	1·14	3·2	5·9	7·8	6·12	11·10	13·4	Cleland, unpublished

Complex	State or province	Segmental arrangement						Reference
β Chevy Chase	Maryland	1·14 3·2	5·9	7·8	6·12 11·10 13·4			Cleland and Hammond, 1950
β Evansville I (1 plant)	Indiana	1·14 3·2	5·9	7·8	6·12 11·10 13·4			Winternheimer (Cleland, 1958)
β Evansville II (2 plants)	Indiana	1·14 3·2	5·9	7·8	6·12 11·10 13·4			Winternheimer (Cleland, 1958)
β Gunpowder	Maryland	1·14 3·2	5·9	7·8	6·12 11·10 13·4			Cleland, unpublished
β Hopkinsville	Kentucky	1·14 3·2	5·9	7·8	6·12 11·10 13·4			Cleland and Hammond, 1950
β Lexington	Kentucky	1·14 3·2	5·9	7·8	6·12 11·10 13·4			Cleland and Hammond, 1950
β New Concord I	Ohio	1·14 3·2	5·9	7·8	6·12 11·10 13·4			Cleland, 1958
β Paducah	Kentucky	1·14 3·2	5·9	7·8	6·12 11·10 13·4			Cleland and Hammond, 1950
β Roanoke B	Virginia	1·14 3·2	5·9	7·8	6·12 11·10 13·4			Stinson and Steiner, 1955
β Warwick A	Virginia	1·14 3·2	5·9	7·8	6·12 11·10 13·4			Stinson and Steiner, 1955
β Bloomington II	Indiana	1·4	3·9	5·2	7·8	6·12 11·10 13·14		Stinson (Cleland, 1958)
β Dyer	Indiana	1·4	3·9	5·2	7·8	6·12 11·10 13·14		Cleland, 1958
β Evansville II (2 plants)	Indiana	1·4	3·9	5·2	7·8	6·12 11·10 13·14		Winternheimer (Cleland, 1958)
β Evansville III (1 plant)	Indiana	1·4	3·9	5·2	7·8	6·12 11·10 13·14		Winternheimer (Cleland, 1958)
β Friendship	Indiana	1·4	3·9	5·2	7·8	6·12 11·10 13·14		Cleland, 1958
β Le Moyne	Pennsylvania	1·4	3·9	5·2	7·8	6·12 11·10 13·14		Cleland, unpublished
β Londonderry	Ohio	1·4	3·9	5·2	7·8	6·12 11·10 13·14		Cleland, unpublished
β Mifflintown	Pennsylvania	1·4	3·9	5·2	7·8	6·12 11·10 13·14		Cleland, unpublished
β New Kent	Virginia	1·4	3·9	5·2	7·8	6·12 11·10 13·14		Cleland, unpublished
β Newman	Illinois	1·4	3·9	5·2	7·8	6·12 11·10 13·14		Cleland, 1958
β Newport News B	Virginia	1·4	3·9	5·2	7·8	6·12 11·10 13·14		Stinson and Steiner, 1955
β Princeton	West Virginia	1·4	3·9	5·2	7·8	6·12 11·10 13·14		Cleland, 1958
punctulans	Illinois	1·4	3·9	5·2	7·8	6·12 11·10 13·14		Cleland and Hammond, 1950
β Reynoldsii	Tennessee	1·4	3·9	5·2	7·8	6·12 11·10 13·14		Cleland, 1958
β Tyronza	Arkansas	1·4	3·9	5·2	7·8	6·12 11·10 13·14		Cleland, 1958
β Warsaw	Indiana	1·4	3·9	5·2	7·8	6·12 11·10 13·14		Cleland, unpublished
β Williamsburg	Virginia	1·4	3·9	5·2	7·8	6·12 11·10 13·14		Cleland, 1958
β Winslow	Arkansas	1·4	3·9	5·2	7·8	6·12 11·10 13·14		Cleland, 1958
β Hilltop	?	1·4	3·11	5·2	7·8	9·12 6·10 13·14		Hoff, unpublished
β Indianapolis	Indiana	1·4	3·11	5·8	7·6	9·10 12·14 13·2		Cleland, unpublished
β Evansville I (1 plant)	Indiana	1·4	3·6	5·9	7·8	2·12 11·10 13·14		Winternheimer (Cleland, 1958)
β Evansville II	Indiana	1·4	3·6	5·9	7·8	2·12 11·10 13·14		Winternheimer (Cleland, 1958)
β Delaware	Ohio	1·4	3·8	5·7	2·14	9·6 11.10 13·12		Cleland, unpublished
β Birch Tree II	Missouri	1·13 3·2	5·7	4·12	9·6 11·10 8·14			Cleland, cited by Steiner, 1952
β Tuscaloosa A	Alabama	1·13 3·2	5·7	4·8	9·12 11·10 6·14			Steiner, 1952
β Athens C	Georgia	1·13 3·6	5·7	4·8	9·14 11·10 2·12			Steiner, 1952
β La Crosse	Virginia	1·13 3·6	5·2	7·4	9·14 11·10 8·12			Cleland, cited by Steiner, 1952
β Magnolia	Kentucky	1·13 3·9	5·2	7·14	4·6 11·10 8·12			Cleland, cited by Steiner, 1952
β Athens A	Georgia	1·13 3·7	5·2	4·6	9·14 11·10 8·12			Steiner, 1952
β Athens B	Georgia	1·13 3·7	5·2	4·6	9·14 11·10 8·12			Steiner, 1952

Complex	State or province	Segmental arrangement							Reference
β Athens D	Georgia	1·13	3·7	5·2	4·6	9·14	11·10	8·12	Steiner, 1952
β Warrenton	Virginia	1·13	3·7	5·2	4·6	9·14	11·10	8·12	Cleland, cited by Steiner, 1952
truncans	Alabama	1·13	3·7	5·2	4·6	9·14	11·10	8·12	Cleland and Hammond, 1950
β Rich Mountain	Arkansas	1·4	3·7	5·11	6·10	9·2	8·12	13·14	Cleland, cited by Steiner, 1952
β Tensaw	Alabama	1·4	3·7	5·11	6·10	9·2	8·12	13·14	Steiner, 1952
β Tuscaloosa B	Alabama	1·4	3·7	5·11	6·10	9·2	8·12	13·14	Steiner, 1952
β Tuscaloosa C	Alabama	1·4	3·7	5·11	6·10	9·2	8·12	13·14	Steiner, 1952
β Bestwater I	Arkansas	1·2	3·7	5·11	4·10	9·6	8·14	13·12	Cleland, 1958
β Birch Tree I	Missouri	1·4	3·13	5·7	8·12	9·10	11·6	2·14	Cleland. cited by Steiner, 1952
β Camp Peary E	Virginia	1·13	3·8	5·12	7·6	9·14	11·10	2·4	Stinson and Steiner, 1955
β Camp Peary L	Virginia	1·10	3·9	5·13	7·11	2·6	8·12	4·14	Stinson and Steiner, 1955
β Citronelle	Alabama	1·4	3·13	5·9	7·2	8·12	11·6	10·14	Steiner, 1952
β Greene	Ohio	1·4	3·2	5·14	7·11	9·10	8·12	13·6	Cleland, unpublished
β Harrisonburg	Virginia	1·3	4·8	5·7	6·12	9·14	11·10	13·2	Cleland, 1958
β Iowa I	Iowa	1·11	3·2	5·7	6·10	9·14	4·12	13·8	Cleland and Hammond, 1950
β Petersburg	Virginia	1·9	3·2	5·13	7·6	10·12	11·4	8·14	Cleland, unpublished
β Poplar Bluff	Missouri	1·12	3·2	5·9	7·8	4·10	11·6	13·14	Cleland, 1958
β Richmond	Virginia	1·5	3·9	4·14	7·2	8·12	11·6	13·10	Cleland, 1958
β Shullsberg	Wisconsin	1·11	3·8	5·10	7·6	9·14	2·4	13·12	Cleland, unpublished
β Walkerton	Indiana	1·4	3·6	5·11	7·14	9·12	2·8	13·10	Cleland, 1958
β Williamston	North Carolina	1·7	3·2	5·4	6·10	9·12	11·14	13·8	Stinson and Steiner, 1955
α Biennis II									
α Brighton	Ontario	1·6	3·2	5·7	4·10	9·8	11·12	13·14	Geckler, 1950
α Corning I	New York	1·6	3·2	5·7	4·10	9·8	11·12	13·14	Cleland, 1958
α Corning II	New York	1·6	3·2	5·7	4·10	9·8	11·12	13·14	Cleland, unpublished
α Cornwall I	Ontario	1·6	3·2	5·7	4·10	9·8	11·12	13·14	Cleland, unpublished
α Cornwall II	Ontario	1·6	3·2	5·7	4·10	9·8	11·12	13·14	Cleland, unpublished
α Elma I	New York	1·6	3·2	5·7	4·10	9·8	11·12	13·14	Cleland, unpublished
α Elma II	New York	1·6	3·2	5·7	4·10	9·8	11·12	13·14	Sloatman, 1953
α Elma V	New York	1·6	3·2	5·7	4·10	9·8	11·12	13·14	Cleland, 1958
α Essex	New York	1·6	3·2	5·7	4·10	9·8	11·12	13·14	Cleland, unpublished
α Hollis	New Hampshire	1·6	3·2	5·7	4·10	9·8	11·12	13·14	Cleland, 1958
α Ithaca	New York	1·6	3·2	5·7	4·10	9·8	11·12	13·14	Preer, 1950
α La Salle I	New York	1·6	3·2	5·7	4·10	9·8	11·12	13·14	Sloatman, 1953
α La Salle II	New York	1·6	3·2	5·7	4·10	9·8	11·12	13·14	Cleland, unpublished
α La Salle III	New York	1·6	3·2	5·7	4·10	9·8	11·12	13·14	Cleland, 1958
α Lawrenceville III	Pennsylvania	1·6	3·2	5·7	4·10	9·8	11·12	13·14	Cleland, 1958
α Marienville I	Pennsylvania	1·6	3·2	5·7	4·10	9·8	11·12	13·14	Cleland, 1958
α Marienville II	Pennsylvania	1·6	3·2	5·7	4·10	9·8	11·12	13·14	Cleland, 1958
α Moncton	New Brunswick	1·6	3·2	5·7	4·10	9·8	11·12	13·14	Preer, 1950
α Montreal	Quebec	1·6	3·2	5·7	4·10	9·8	11·12	13·14	Cleland, unpublished
α Pamelia II	New York	1·6	3·2	5·7	4·10	9·8	11·12	13·14	Cleland, unpublished
α Rondeau Park B	Ontario	1·6	3·2	5·7	4·10	9·8	11·12	13·14	Preer, 1950
α St. Anne	Quebec	1·6	3·2	5·7	4·10	9·8	11·12	13·14	Geckler, 1950
α Tonawanda I	New York	1·6	3·2	5·7	4·10	9·8	11·12	13·14	Sloatman, 1953
α Tonawanda II	New York	1·6	3·2	5·7	4·10	9·8	11·12	13·14	Sloatman, 1953

Complex	State or province	Segmental arrangement							Reference
α Victorini	Quebec	1·6	3·2	5·7	4·10	9·8	11·12	13·14	Preer, 1950
α Waterbury	Vermont	1·6	3·2	5·7	4·10	9·8	11·12	13·14	Cleland and Hammond, 1950
jugens	New Jersey	1·6	3·2	5·14	4·10	9·8	11·12	13·7	Cleland and Hammond, 1950
α Morgantown	West Virginia	1·4	3·2	5·6	7·10	9·8	11·12	13·14	Preer, 1950
α St. Croix	Wisconsin	1·4	3·2	5·6	7·10	9·8	11·12	13·14	Geckler, 1950
α Williamsville	New York	1·4	3·2	5·6	7·10	9·8	11·12	13·14	Sloatman, 1953
α Stanley Bridge I	Prince Edward Island	1·3	2·4	5·7	6·10	9·8	11·12	13·14	Alford, unpublished
α Indian River	Michigan	1·4	3·2	5·7	6·10	9·8	11·12	13·14	Preer, 1950
α Buck Creek	North Carolina	1·4	3·10	5·7	2·6	9·8	11·12	13·14	Preer, 1950
α Wakefield	Michigan	1·7	3·10	5·4	2·6	9·8	11·12	13·14	Cleland, unpublished
α Bloomington III	Indiana	1·4	3·6	5·7	2·14	9·8	11·10	13·12	Stinson (Cleland, 1958)
α Lenoraie	Quebec	1·4	3·12	5·2	7·11	9·8	6·10	13·14	Preer, 1950
α Micaville	North Carolina	1·5	3·9	2·6	7·10	4·8	11·12	13·14	Preer, 1950
α Omaha	Nebraska	1·6	3·2	5·12	7·10	9·4	11·8	13·14	Cleland, 1958
α Pamelia I	New York	1·4	3·2	5·9	7·8	6·10	11·12	13·14	Cleland, unpublished
β Biennis II									
β Brighton	Ontario	1·2	3·8	5·6	7·10	9·12	11·13	4·14	Geckler, 1950
β Hollis	New Hampshire	1·2	3·8	5·6	7·10	9·12	11·13	4·14	Cleland, 1958
β Ithaca	New York	1·2	3·8	5·6	7·10	9·12	11·13	4·14	Preer, 1950
maculans	New Jersey	1·2	3·8	5·6	7·10	9·12	11·13	4·14	Cleland and Hammond, 1950
β Rondeau Park B	Ontario	1·2	3·8	5·6	7·10	9·12	11·13	4·14	Preer, 1950
β Tonawanda II	New York	1·2	3·8	5·6	7·10	9·12	11·13	4·14	Sloatman, 1953
β Lawrenceville III	Pennsylvania	1·2	3·11	5·6	7·10	9·4	8·14	13·12	Cleland, 1958
β Horseheads I	New York	1·2	3·11	5·6	7·10	9·12	4·14	13·8	Cleland, unpublished
β Elma V	New York	1·2	3·12	5·6	7·10	9·4	11·13	8·14	Cleland, 1958
β Etna	New York	1·2	3·12	5·6	7·10	9·4	11·13	8·14	Preer, 1950
β Marienville I	Pennsylvania	1·2	3·12	5·6	7·10	9·4	11·13	8·14	Cleland, 1958
β Marienville II	Pennsylvania	1·2	3·12	5·6	7·10	9·4	11·13	8·14	Cleland, 1958
β Pamelia II (tentative)	New York	1·2	3·12	5·6	7·10	9·4	11·13	8·14	Cleland, unpublished
β Pamelia II	New York	1·2	3·12	5·6	7·10	9·4	11·13	8·14	Cleland, unpublished
β Corning I	New York	1·2	3·12	5·6	7·10	9·4	11·13	8·14	Cleland, unpublished
β Corning II	New York	1·2	3·12	5·6	7·10	9·4	11·13	8·14	Cleland, 1958
β Cornwall I	Ontario	1·2	3·12	5·6	7·10	9·4	11·13	8·14	Cleland, unpublished
β Cornwall II	Ontario	1·2	3·12	5·6	7·10	9·4	11·13	8·14	Cleland, unpublished
β Elma II	New York	1·2	3·12	5·6	7·10	9·4	11·13	8·14	Sloatman, 1953
β Moncton	New Brunswick	1·2	3·12	5·6	7·10	9·4	11·13	8·14	Preer, 1950
β Victorini	Quebec	1·2	3·12	5·6	7·10	9·4	11·13	8·14	Preer, 1950
β Lanoraie	Quebec	1·2	3·12	5·6	7·4	9·4	10·14	8·14	Preer, 1950
β Buck Creek	North Carolina	1·2	3·5	6·8	7·12	9·13	10·14	11·4	Preer, 1950
β Hebron II	Ohio	1·2	3·5	6·8	7·12	9·13	10·14	11·4	Cleland, unpublished
β La Salle I	New York	1·2	3·5	6·8	7·12	9·13	10·14	11·4	Cleland, unpublished
β La Salle II	New York	1·2	3·5	6·8	7·12	9·13	10·14	11·4	Cleland, unpublished
β La Salle III	New York	1·2	3·5	6·8	7·12	9·13	10·14	11·4	Cleland, unpublished
β Micaville	North Carolina	1·2	3·5	6·8	7·12	9·13	10·14	11·4	Preer, 1950
β Wakefield	Michigan	1·2	3·5	6·8	7·12	9·13	10·14	11·4	Cleland, unpublished
β Bloomington III	Indiana	1·9	3·8	5·6	7·10	4·14	11·13	2·12	Stinson (Cleland, 1958)

Complex	State or province	Segmental arrangement							Reference
α Indian River	Michigan	1·7	3·12	5·9	2·10	4·14	11·8	13·6	Preer, 1950
β Newcastle	New Brunswick	1·2	3·7	5·6	4·14	9·11	10·12	13·8	Alford, unpublished
β Omaha	Nebraska	1·2	3·5	6·8	7·12	9·14	11·4	13·10	Cleland, 1958
β St. Anne	Quebec	1·13	3·4	5·8	7·11	9·14	2·6	10·12	Geekler, 1950
β Waterbury	Vermont	1·10	3·12	5·9	7·2	4·14	11·6	13·8	Alford, unpublished
α Biennis III									
α White Top	Virginia	1·4	3·2	5·6	7·10	9·8	11·12	13·14	Cleland, 1958
α Coudersport I	Pennsylvania	1·13	3·4	5·10	7·14	9·8	11·12	2·6	Cleland, unpublished
α Coudersport II	Pennsylvania	1·13	3·4	5·10	7·14	9·8	11·12	2·6	Cleland, unpublished
α Coudersport III	Pennsylvania	1·13	3·7	5·14	4·10	9·8	11·12	2·6	Cleland, unpublished
α Marienville III	Pennsylvania	1·13	3·7	5·14	4·10	9·8	11·12	2·6	Cleland, 1958
α Smethport	Pennsylvania	1·13	3·7	5·14	4·10	9·8	11·12	2·6	Cleland, 1958
α Newfound Gap	Tennessee	1·9	3·7	5·14	4·10	2·8	11·12	13·6	Cleland, 1958
α Pineola	North Carolina	1·9	3·7	5·14	4·10	2·8	11·12	13·6	Cleland, 1958
α Blowing Rock	North Carolina	1·13	3·4	5·14	7·8	9·10	11·12	2·6	Preer, 1950
α Mitchell	North Carolina	1·13	3·10	5·14	7·4	9·8	11·12	2·6	Preer, 1950
α Linville	North Carolina	1·11	3·4	5·9	7·14	6·8	2·12	13·10	Preer, 1950
α Mountain Lake	Virginia	1·4	3·2	5·14	7·8	9·10	11·12	13·6	Preer, 1950
β Biennis III									
β Blowing Rock	North Carolina	1·2	3·8	5·6	7·10	9·12	11·13	4·14	Preer, 1950
β Mitchell	North Carolina	1·2	3·8	5·6	7·10	9·12	11·13	4·14	Preer, 1950
β Mountain Lake	North Carolina	1·2	3·8	5·6	7·10	9·12	11·13	4·14	Preer, 1950
β Linville	North Carolina	1·2	3·12	5·6	7·10	9·4	11·13	8·14	Preer, 1950
β Pineola	North Carolina	1·2	3·12	5·6	7·10	9·4	11·13	8·14	Cleland, 1958
β White Top	Virginia	1·2	3·12	5·6	7·9	4·10	11·13	8·14	Cleland, 1958
β Coudersport II	Pennsylvania	1·2	3·12	5·6	7·10	9·11	4·14	13·8	Cleland, unpublished
β Coudersport III	Pennsylvania	1·2	3·12	5·6	7·10	9·11	4·14	13·8	Cleland, 1958
β Horseheads II	New York	1·2	3·12	5·6	7·10	9·11	4·14	13·8	Cleland, unpublished
β Marienville III (tentative)	Pennsylvania	1·2	3·12	5·6	7·10	9·11	4·14	13·8	Cleland, 1958
β Smethport	Pennsylvania	1·2	3·12	5·6	7·10	9·11	4·14	13·8	Cleland, 1958
Grandiflora complexes									
hAlabama	Alabama	1·2	3·4	5·6	7·10	9·8	11·12	13·14	Steiner, 1952
hDixie Landing	Alabama	1·2	3·4	5·6	7·10	9·8	11·12	13·14	Steiner, 1952
Martins Branch I a	Alabama	1·2	3·4	5·6	7·10	9·8	11·12	13·14	Cleland, unpublished
Seabury Creek I a	Alabama	1·2	3·4	5·6	7·10	9·8	11·12	13·14	Cleland, unpublished
Stockton I a	Alabama	1·2	3·4	5·6	7·10	9·8	11·12	13·14	Cleland, unpublished
α Biloxi	Mississippi	1·2	3·4	5·10	7·6	9·8	11·12	13·14	Steiner, 1952
α Parviflora I									
α angustissima	New York	1·13	3·4	5·8	7·14	9·2	11·12	6·10	Cleland, 1958
α Clifton Forge	Virginia	1·13	3·7	5·14	4·10	9·8	11·12	2·6	Stinson (Cleland, 1958)
α Gala I	Virginia	1·9	3·7	5·14	4·10	2·8	11·12	13·6	Cleland, 1958
α Roanoke A	Virginia	1·9	3·7	5·14	4·10	2·8	11·12	13·6	Stinson (Cleland, 1958)
α Port Hope	Ontario	1·6	3·2	5·7	4·10	9·8	11·12	13·14	Geekler, 1950
α St. George	New Brunswick	1·9	3·10	5·12	7·4	2·8	11·6	13·14	Geekler, 1950
α St. Lawrence	Quebec	1·13	3·4	5·9	7·14	6·8	11·12	2·10	Geekler, 1950
α St. Stephen	New Brunswick	1·2	3·12	5·6	7·10	9·11	4·14	13·8	Cleland, unpublished
α Stanley Bridge II	Prince Edward Island	1·5	3·4	2·8	7·12	9·10	11·6	13·14	Cleland, unpublished
α Trois Pistoles I	Quebec	1·7	3·4	5·12	2·8	9·10	11·6	13·14	Cleland, unpublished

Complex	State or province	Segmental arrangement							Reference
α Trois Pistoles II	Quebec	1·7	3·4	5·12	2·8	9·10	11·6	13·14	Cleland, unpublished
β Parviflora I									
β Clifton Forge	Virginia	1·4	3·2	5·12	7·10	9·6	11·8	13·14	Stinson (Cleland, 1958)
β Iron Mountain	Michigan	1·6	3·8	5·7	2·12	9·4	11·13	10·14	Geckler, 1950
α Parviflora II									
α cruciata II	Vermont?	1·2	3·4	5·6	7·11	9·10	8·14	13·12	Cleland, 1958
α fascians (tentative)	Massachusetts	1·2	3·4	5·6	7·11	9·10	8·14	13·12	Cleland, 1937
α Kennebunk	Maine	1·2	3·4	5·6	7·11	9·10	8·14	13·12	Cleland, unpublished
rigens	Europe	1·2	3·4	5·6	7·11	9·10	8·14	13·12	Cleland, 1937
α Saybrook	Connecticut	1·2	3·4	5·6	7·11	9·10	8·14	13·12	Cleland, unpublished
α Sunbury	Pennsylvania	1·2	3·4	5·6	7·11	9·10	8·14	13·12	Geckler, 1950
α York Beach	Maine	1·2	3·4	5·6	7·11	9·10	8·14	13·12	Geckler, 1950
α Manistique	Michigan	1·12	3·2	5·6	7·11	9·10	8·14	13·4	Geckler, 1950
accelerans	Massachusetts	1·11	3·13	5·6	7·2	9·12	8·14	4·10	Cleland and Hammond, 1950
α Tidestromii	Maryland	1·12	3·13	5·6	7·2	9·11	8·14	4·10	Cleland, 1958
pubens	Massachusetts	1·12	3·13	5·6	7·2	9·11	8·14	4·10	Cleland, 1958
α Ashland A	Wisconsin	1·4	3·9	5·10	7·6	8·12	11·2	13·14	Geckler, 1950
α Rock Falls III	Ontario	1·2	3·11	5·8	7·12	9·6	4·10	13·14	Cleland, unpublished
α Twin Lakes II	Ontario	1·2	3·11	5·8	7·12	9·6	4·10	13·14	Cleland, unpublished
β Parviflora II									
β Ashland A	Wisconsin	1·3	2·14	5·4	7·9	6·8	11·10	13·12	Geckler, 1950
curvans	Europe	1·14	3·2	5·13	7·12	9·8	11·10	4·6	Cleland, and Hammond, 1950[a]
β Manistique	Michigan	1·13	3·7	5·2	6·10	9·8	11·14	4·12	Geckler, 1950
β Rondeau Park A	Ontario	1·6	3·12	5·7	4·8	9·13	11·10	2·14	Geckler, 1950
Argillicola complexes									
hargillicola A (Huntingdon A)	Pennsylvania	1·4	3·6	5·2	7·10	9·8	11·12	13·14	Cleland and Hammond, 1950
hargillicola B (Huntingdon B)	Pennsylvania	1·8	3·4	5·6	7·10	9·2	11·12	13·14	Stinson, 1953
hDouthat 3	Pennsylvania	1·2	3·4	5·6	7·10	9·8	11·12	13·14	Stinson, 1953
hDouthat 1	Pennsylvania	1·4	3·2	5·6	7·10	9·8	11·12	13·14	Stinson, 1953
hDouthat 2	Pennsylvania	1·4	3·2	5·6	7·10	9·8	11·12	13·14	Stinson, 1953
hDouthat 4a	Pennsylvania	1·13	3·4	5·6	7·10	9·8	11·12	2·14	Stinson, 1953
hDouthat 4b	Pennsylvania	1·4	3·12	5·6	7·10	9·8	11·2	13·14	Stinson, 1953
Other complexes from Europe									
aemulans (conferta)		1·2	3·11	5·6	7·10	9·4	8·14	13·12	Renner and Hirmer, 1956
albicans (biennis, suaveolens)		1·4	3·6	5·7	2·14	9·8	11·10	13·12	Cleland and Hammond, 1950[a]
angustans (argillicola)		1·4	3·2	5·6	7·8	9·10	11·12	13·14	Mickan, 1936
augens (parviflora)		1·2	3·13	5·6	7·10	9·14	11·4	8·12	Renner, 1956
hblandina		1·2	3·4	5·6	7·10	9·14	11·12	13·8	Catcheside, 1940
blandina-A		1·2	3·12	5·6	7·10	9·14	11·4	13·8	Catcheside, 1940
α Cantabrigiana (England)		1·2	3·12	5·6	7·11	9·4	8·14	13·10	Cleland, 1958
gaudens (lamarckiana)		1·2	3·12	5·6	7·11	9·4	8·14	13·10	Cleland and Hammond, 1950[b]
β Poznan I (Poland)		1·2	3·12	5·6	7·11	9·4	8·14	13·10	Cleland, unpublished

[a] Catcheside determined *albicans* in 1940 as having 11.13 10.12 and *curvans* as having 7.11 10.12.

[b] Catcheside determined gaudens in 1940 as having 7.12 3.11.

Complex	State or province	Segmental arrangement							Reference
β Poznan II (Poland)		1·2	3·12	5·6	7·11	9·4	8·14	13·10	Cleland, unpublished
β Poznan III (Poland)		1·2	3·12	5·6	7·11	9·4	8·14	13·10	Cleland, unpublished
rubens (biennis of de Vries)		1·2	3·12	5·6	7·11	9·4	8·14	13·10	Catcheside, 1940
glabrans (nuda)		1·2	3·12	5·6	7·10	9·4	8·14	11·13	Jean, Lambert, and Linder, 1966[a]
β Cantabrigiana		1·4	3·2	5·10	7·14	9·8	11·12	13·6	Cleland, 1958
[h]decipiens		1·2	3·4	5·6	7·14	9·10	11·12	13·8	Catcheside, 1939
[h]deserens		1·2	3·4	5·6	7·14	9·10	11·12	13·8	Renner, 1943a
dilatans (argillicola)		1·4	3·2	5·6	7·10	9·8	11·12	13·14	Mickan, 1936
flavens (suaveolens)		1·4	3·2	5·6	7·8	9·10	11·12	13·14	Cleland and Blakeslee, 1931
flavens Grado		1·4	3·2	5·6	7·10	9·8	11·12	13·14	Stubbe, 1953
flavens Fünfkirchen		1·4	3·2	5·6	7·10	9·8	11·12	13·14	Stubbe, 1953
flavens Friedrichshagen		1·4	3·2	5·6	7·10	9·11	8·12	13·14	Stubbe, 1953
flectens (atrovirens)		1·4	3·2	5·7	6·10	9·8	11·14	13·12	Catcheside, 1940
M-gaudens		1·12	3·2	5·6	7·11	9·4	8·14	13·10	Renner, 1942b
latifrons		1·2	3·4	5·6	7·10	9·13	11·12	8·14	Catcheside, 1940
laxans (hungarica)		1·2	3·10	5·8	7·4	9·6	11·14	13·12	Baerecke, 1944
paravelans (coronifera)		1·2	3·4	5·8	7·6	9·10	11·12	13·14	Rossman, 1963a
percurvans (ammophila)		1·14	3·5	6·8	7·10	9·2	11·4	13·12	Baerecke, 1944
pingens (atrovirens)		1·13	3·4	5·8	7·14	9·6	11·12	2·10	Baerecke, 1944
[h]purpurata		1·2	3·4	5·6	7·10	9·8	11·12	13·14	Catcheside, 1940
quaerans (coronifera)		1·2	3·12	5·6	7·10	9·11	4·14	13·8	Rossmann, 1963
rubricalyx-α		1·2	3·4	5·6	7·14	9·10	11·12	13·8	Sikka, 1939 Catcheside, 1940
subcurvans (silesiaca, parviflora)		1·14	3·10	5·4	7·12	9·2	11·8	13·6	Baerecke, 1944
subpingens (silesiaca)		1·13	3·4	5·8	7·14	9·6	11·12	2·10	Baerecke, 1944
tingens (rubricaulis)		1·7	3·4	5·12	2·8	9·10	11·6	13·14	Baerecke, 1944
undans (hungarica)		1·4	3·2	5·10	7·6	9·13	11·12	8·14	Baerecke, 1944
velans (lamarckiana)		1·2	3·4	5·8	7·6	9·10	11·12	13·14	Cleland and Blakeslee, 1931; Emerson and Sturtevant, 1931
convelans (conferta)		1·2	3·4	5·8	7·6	9·10	11·12	13·14	Renner and Hirmer, 1956
calvans (nuda)		1·6	3·2	5·7	4·10	9·8	11·12	13·14	Jean, Lambert and Linder, 1966
Unclassified									
β Biloxi	Mississippi	1·10	3·5	8·12	7·14	9·6	11·2	13·4	Steiner, 1952
α Camas	Washington	1·9	3·5	4·6	7·10	2·8	11·14	13·12	Cleland, 1958
β Camas	Washington	1·4	3·2	5·10	7·14	9·8	11·12	13·6	Cleland and Hammond, 1950
α Elgin	Illinois	1·14	3·4	5·8	7·11	9·6	10·12	13·2	Cleland, 1958
β Elgin	Illinois	1·2	3·12	5·6	7·10	9·4	11·13	8·14	Cleland, unpublished
β Kooskie	Idaho	1·4	3·2	5·14	7·8	9·10	11·12	13·6	Norby (Cleland, 1958)
La Follette b	Tennessee	1·14	3·2	5·9	7·8	6·12	11·10	13·4	Cleland, unpublished
La Follette a	Tennessee	1·14	3·2	5·9	7·11	6·8	10·12	13·4	Cleland, 1958
N		1·2	3·4	5·6	7·14	9·10	11·12	13·8	Emerson and Sturtevant, 1931
neo-acuens		1·13	3·2	5·6	7·10	9·8	11·12	4·14	Cleland, 1950
β Portland	Oregon	1·7	3·4	5·14	2·10	9·11	8·12	13·6	Norby (Cleland, 1958)
α Wolf Creek	Colorado	1·4	3·2	5·6	7·10	9·8	11·12	13·14	Norby (Cleland, 1958)
α Charlottesville	Virginia	1·13	3·7	5·2	4·6	9·14	11·10	8·12	Cleland, 1958
β Charlottesville	Virginia	1·4	3·6	5·12	7·10	9·8	11·2	13·14	Cleland, 1958

[a] These authors follow Catcheside's system (3.11 12.13 instead of 3.12 11.13).

Appendix II

NOTES ON MICROTECHNIQUE USED IN
PREPARATION OF MICROSPOROCYTES OF *Oenothera*

In the earlier years of our investigations we used sectioned material in the study of pollen mother cells. Sectioning methods are still indicated where it is important to obtain metaphase I figures in which the zigzag arrangement of circle chromosomes has not been disturbed. Since 1939, however, a smear method has been used exclusively.

The Section Method

Remove sepals and cut away most of receptacle, leaving just enough to hold anthers together. Dip briefly into 30% alcohol to cut the wax covering, then fix in Bouin (75 parts sat. aq. sol. picric acid, 25 parts formaldehyde, 5 parts glacial acetic acid) to which is added 1% chromic acid and 1% lactose. Fix 3–5 h, rinse, run up to 70% ethyl alcohol at 1-h intervals, wash in changes of 70% until all color is gone. (It is important in plant material to get rid of all traces of Bouin, otherwise, chromosomes are softened.)

Run up slowly to absolute alcohol. Run cedar oil into tilted vial. Alcohol with buds will float on top. Allow buds to sink gradually into cedar oil. Replace with fresh cedar oil. Transfer to 1 : 1 cedar oil and xylol for an hour, then into pure xylol (change xylol once). Embed in paraffin without delay (too long exposure to xylol renders material hard and brittle). The transfer to paraffin is accomplished by pipetting warm (not hot) paraffin carefully onto surface of xylol. Place on a warm (not hot) surface without cork. Add slightly more paraffin than the amount of xylol present. After paraffin has dissolved, pipette off a portion of the mixture and add more warm paraffin. When proportion of paraffin becomes great enough, it begins to crystallize out throughout the solution (it is then approximately 80% paraffin, 20% xylol.) Place in paraffin oven until melted, pour off and replace with pure paraffin, change paraffin once. Leave in oven a minimum time before casting block, which can be done in paper boats floated on ice water. Section at 14 μ and stain with c.p. Heidenhain's haematoxylin, using no counterstain. Mount in gum Damar (which does not oxidize and cause discoloration; slides 50 years old are still well stained).

The Smear Method

Smearing of *Oenothera* anthers is difficult because the walls are thick, loculi are long and slender, usually not more than two sporocytes abreast. Cells may be tapped out in acetocarmine or aceto-orcein but the chromosomes ordinarily stain very poorly in fresh cells (on rare occasions, a cell or island of sporocytes will lack starch grains, in which case the chromosomes stain beautifully. The difference in behavior is not because starch grains ordinarily hide the chromosomes from view—chromosomes are visible in starch-containing cells but do not stain intensely. The different behavior in starch-filled and starch-free cells is apparently the result of a metabolic difference.) If anthers are fixed first in FAA or Carnoy, and cells are later tapped out in acetocarmine, the chromosomes stain well, but the fixation is inferior.

Our method results in smears that are permanent and hold their stain indefinitely. Anthers are laid out in parallel on a slide with the ends aligned in a straight line. A flat-surfaced knife is pressed in a rolling fashion toward the aligned ends in such a position that the edge of the knife comes to coincide with the ends of the anthers. This results in the extrusion of a cluster of sporocytes from the ends of the anthers. The slide is immediately placed face down in a flat dish containing fixative, one end resting on a glass rod. The fixative is a modified Navashin with the following composition:

Solution A:	chromic acid crystals	10 gm
	glacial acetic	100 cc
	water	640 cc
Solution B:	lactose	8 gm
	formalin	200 cc
	water	550 cc

Mix equal portions of A and B just prior to use. Fix overnight.
  Rinse in water:
    1% chromic acid, 10 min
  Rinse in two changes of water:
    30%, 50% ethyl alcohol with I KI, $2\frac{1}{2}$ min in each
    80% alcohol with I KI, 10 min
    50% and 30% alcohol with I KI, $2\frac{1}{2}$ min each
    crystal violet, c.p., 1% boiled and filtered, 20 min
  Rinse in two changes of water:
    30% with I KI, 11 sec
    50% with I KI, 14 sec
    80% with I KI, 20 sec
    90% alcohol, 20 sec

100% alcohol saturated with picric acid, 40 sec
first clove oil, 1 min
second clove oil, 2 min
100% alcohol saturated with picric, 30 sec
first clove oil, 30 sec
second clove oil, 1 min
Five changes of xylol, 5 min each. Mount in gum Damar.

Comments

30% and 50% alcohol with I KI are made up by dilution of 80% alcohol containing 1% I and 1% KI. Iodine is used as a mordant prior to staining and to prevent too rapid loss of stain after staining.

Crystal violet must be c.p., certified. Impure crystal violet (gentian violet) stains cytoplasm.

Picric acid slows rate of fading of crystal violet. Slides made in 1939 are still good.

Timing of steps following staining has been arrived at by trial and error and gives consistently good results. Gentle agitation of slides while in each solution ensures even destaining.

Slides made by this procedure are permanent. When a green filter is used, chromosomes appear black.

Cells smeared in this manner often show disturbances in the regular zigzag arrangement in metaphase I, and probably suffer occasional breakage of intact circles, both resulting from the presence of the knife. On the whole, however, fixation of chromosomes is excellent. Cells are unshrunken. Stages earlier than diplophase, however, can rarely be smeared successfully.

References

Alpinus, Prosper (1627). "De plantis exoticis 1." Guerilium, Venice.

Baerecke, M. (1944). Zur Genetik und Cytologie von *Oenothera ammophila* Focke, *Bauri* Boedijn, *Beckeri* Renner, *parviflora* L., *rubricaulis* Klebahn, *silesiaca* Renner. *Flora* **138**, 57–92.

Bartlett, H. H. (1914). An account of the cruciate-flowered oenotheras of the subgenus *Onagra*. *Am. J. Bot.* **1**, 226–243.

Bartlett, H. H. (1915a). Additional evidence of mutation in *Oenothera*. *Bot. Gaz.* **59**, 81–123.

Bartlett, H. H. (1915b). Mass mutation in *Oenothera pratincola*. *Bot. Gaz.* **60**, 425–456.

Bartlett, H. H. (1915c). The mutations of *Oenothera stenomeres*. *Am. J. Bot.* **2**, 100–109.

Bartlett, H. H. (1915d). The experimental study of genetic relationships. *Am. J. Bot.* **2**, 132–155.

Bartlett, H. H. (1915e). Mutation en masse. *Am. Nat.* **49**, 129–139.

Bartlett, H. H. (1916). The status of the mutation theory, with especial reference to *Oenothera*. *Am. Nat.* **50**, 513–529.

Bauhin, C. (1623). "Pinax theatri botanici." Basel.

Baur, E. (1909). Das Wesen und die Erblichkeitsverhältnisse der "Varietates albomarginatae Hort." von *Pelargonium zonale*. *Zeitschr. ind. Abst.- u. Vererb.* **1**, 330–351.

Belling, J. (1927). The attachment of chromosomes at the reduction division in flowering plants. *J. Genet.* **18**, 177–205.

Belling, J. and Blakeslee, A. F. (1926). On the attachment of nonhomologous chromosomes at the reduction division in certain 25-chromosome Daturas. *Proc. Nat. Acad. Sci. Wash.* **12**, 7–11.

Bhaduri, D. N. (1940). Cytological studies in *Oenothera* with special reference to the relation of chromosomes and nucleoli. *Proc. Roy. Soc. London* **B128**, 353–378.

Blakeslee, A. F. and Cleland, R. E. (1930). Circle formation in *Datura* and *Oenothera*. *Proc. Nat. Acad. Sci. Wash.* **16**, 177–183.

Boedijn, K. (1925). Der Zusammenhang zwischen den Chromosomen und Mutationen bei *Oenothera lamarckiana*. *Rec. trav. botan. Néerl.* **22**, 173–261.

Boedijn, K. (1928). Chromosomen und Pollen der Oenotheren. *Rec. trav. botan. Néerl.* **25A**, 25–35.

Borbás, V. (1903). Az Oenothera hazánkban (in Hungarian). *Magyar Bot. Lapok* **2**, 243–248.

Bridges, C. B. (1919). Duplication. *Anat. Rec.* **15**, 357–358.

Brittingham, W. H. (1931). Variation in the evening primrose induced by radium. *Science*, **74**, 463–464.

Brown, W. V. and Stack, S. M. (1968). Somatic pairing as a regular preliminary to meiosis. *Bull. Torrey Bot. Club* **95**, 369–378.

Catcheside, D. G. (1931) Critical evidence of parasynapsis in *Oenothera*. *Proc. Roy. Soc. London* **B109**, 165–184.

Catcheside, D. G. (1932). The chromosomes of a new haploid *Oenothera*. *Cytologia* **4**, 68–113.

Catcheside, D. G. (1933). Chromosome configurations in trisomic oenotheras. *Genetica* **15**, 177–201.

Catcheside, D. G. (1935). X-ray treatment of *Oenothera* chromosomes. *Genetica* **17**, 313–341.

Catcheside, D. G. (1936). Origin, nature and breeding behavior of *Oenothera lamarckiana* trisomics. *J. Genet.* **33**, 1–23.

Catcheside, D. G. (1937a). Recessive mutations induced in *Oenothera blandina* by X-rays. *Genetica* **18**, 134–142.

Catcheside, D. G. (1937b). The extra chromosomes of *Oenothera lamarckiana lata*. *Genetica* **22**, 564–576.

Catcheside, D. G. (1939). A position effect in *Oenothera*. *J. Genet.* **38**, 345–352.

Catcheside, D. G. (1940). Structural analysis of *Oenothera* complexes. *Proc. Roy. Soc. London* **B128**, 509–535.

Catcheside, D. G. (1947a). The P-locus position effect in *Oenothera*. *J. Genet.* **48**, 31–42.

Catcheside, D. G. (1947b). A duplication and deficiency in *Oenothera*. *J. Genet.* **48**, 99–110.

Catcheside, D. G. (1954). The genetics of *brevistylis* in *Oenothera*. *Heredity* **8**, 125–137.

Clapham, A. R., Tutin, T. G. and Warburg, E. F. (1952). "Flora of the British Isles." Cambridge Univ. Press.

Chrometzka, P. (1955). Zur Kenntnis der Morphologie und des Wuchsstoff-Verhaltens der *Oenothera hybrida* mut. *helix*. *Zeitschr. ind. Abst.- u. Vererb.* **87**, 267–297.

Chrometzka, P. (1956). Zur Kenntnis der Lage des Gens *Hel* der *Oenothera* hybrida mut. *helix*. *Planta* **46**, 643–648.

Cleland, R. E. (1922). The reduction divisions in the pollen mother cells of *Oenothera franciscana*. *Amer. J. Bot.* **9**, 391–413.

Cleland, R. E. (1923). Chromosome arrangements during meiosis in certain oenotheras. *Amer. Nat.* **57**, 562–566.

Cleland, R. E. (1924). Meiosis in pollen mother cells of *Oenothera franciscana sulfurea*. *Bot. Gaz.* **77**, 149–170.

Cleland, R. E. (1925). Chromosome behavior during meiosis in the pollen mother cells of certain oenotheras. *Amer. Nat.* **59**, 475–479.

Cleland, R. E. (1926a). Meiosis in the pollen mother cells of *Oenothera biennis* and *Oenothera biennis sulfurea*. *Genetics* **11**, 127–162.

Cleland, R. E. (1926b). Cytological study of meiosis in anthers of *Oenothera muricata*. *Bot. Gaz.* **82**, 55–70.

Cleland, R. E. (1928). The genetics of *Oenothera* in relation to chromosome behavior, with special reference to certain hybrids. *Zeitschr. ind. Abst.- u. Vererb.* Suppl. Bd. 554–567.

Cleland, R. E. (1929a). Meiosis in the pollen mother cells of the oenotheras, and its probable bearing upon certain genetical problems. *Proc. Intern. Congr. Plant Sci.* **1**, 317–331.

Cleland, R. E. (1929b). Chromosome behavior in the pollen mother cells of several strains of *Oenothera lamarckiana*. *Zeitschr. ind. Abst.- u. Vererb.* **51**, 126–145.

Cleland, R. E. (1929c). Die Zytologie der *Oenothera*-Gruppe *biennis* in ihrem Verhältnis zur Vererbunglehre. *Tübinger Naturwissenschaftliche Abhandlungen* **12**, 50–55.

Cleland, R. E. (1931a). Cytological evidence of genetical relationships in *Oenothera*. *Am. J. Bot.* **18**, 629–640.

Cleland, R. E. (1931b). The probable origin of *Oenothera rubricalyx* "Afterglow" on the basis of the segmental interchange theory. *Proc. Nat. Acad. Sci. Wash.* **17**, 437–440.

Cleland, R. E. (1932). Further data bearing upon circle-formation in *Oenothera*, its cause and its genetical effect. *Genetics* **17**, 572–602.

Cleland, R. E. (1933). Predictions as to chromosome configuration, as evidence for segmental interchange in *Oenothera*. *Am. Nat.* **67**, 407–418.

Cleland, R. E. (1935a). Chromosome configuration in *Oenothera* (*grandiflora* × *lamarckiana*). *Am. Nat.* **69**, 466–468.

Cleland, R. E. (1935b). Cyto-taxonomic studies on certain oenotheras from California. *Proc. Am. Philos. Soc.* **75**, 339–429.

Cleland, R. E. (1936). Some aspects of the cytogenetics of *Oenothera*. *Botan. Rev.* **2**, 316–348.

Cleland, R. E. (1937). Species relationships in *Onagra*. *Proc. Am. Philos. Soc.* **77**, 477–542.

Cleland, R. E. (1940). Analysis of wild American races of *Oenothera* (*Onagra*). *Genetics* **25**, 636–644.

Cleland, R. E. (1942). The origin of ʰ*decipiens* from the complexes of *Oenothera lamarckiana* and its bearing upon the phylogenetic significance of similarities in segmental arrangements. *Genetics* **27**, 55–83.

Cleland, R. E. (1944). The problem of species in *Oenothera*. *Am. Nat.* **78**, 5–28.

Cleland, R. E. (1949a). Phylogenetic relationships in *Oenothera*. *Proc. 8th Intern. Congr. Genetics*, 173–188.

Cleland, R. E. (1949b). A botanical noncomformist. *Sci. Monthly* **68**, 35–41.

Cleland, R. E. (1950a). Studies in *Oenothera* cytogenetics and phylogeny. Introduction and general summary. *Indiana Univ. Publ. Sci. Ser.* **16**, 5–9.

Cleland, R. E. (1950b). The origin of *neo-acuens*. *Indiana Univ. Publ. Sci. Ser.* **16**, 73–81.

Cleland, R. E. (1951). Extra diminutive chromosomes in *Oenothera*. *Evolution* **5**, 165–176.

Cleland, R. E. (1954a). Evolution of the euoenotheras: the strigosas. *Proc. 9th Intern. Congr. Genetics*, 1139–1141.

Cleland, R. E. (1954b). Evolution of the North American euoenotheras: the strigosas. *Proc. Am. Philos. Soc.* **98**, 189–203.

Cleland, R. E. (1957). Chromosome structure in *Oenothera* and its effect on the evolution of the genus. *Proc. Intern. Genet. Symp.* 1956 (*Cytologia*, Suppl. Vol.), 5–19.

Cleland, R. E. (1958). The evolution of the North American oenotheras of the "*biennis*" group. *Planta* **51**, 378–398.

Cleland, R. E. (1960a). A case history of evolution. *Proc. Indiana Acad. Sci.* **69**, 51–64.

Cleland, R. E. (1960b). The S-factor situation in a small sample of an *Oenothera* (*Raimannia*) *heterophylla* population. *Zeitschr. Vererb.* **91**, 303–311.

Cleland, R. E. (1962). Plastid behavior in North American euoenotheras. *Planta* **57**, 699–712.

Cleland, R. E. (1964). The evolutionary history of the North American evening primroses of the "*biennis* group". *Proc. Am. Philos. Soc.* **108**, 88–98.

Cleland, R. E. (1967a). Further evidence bearing upon the origin of extra diminutive chromosomes in *Oenothera hookeri*. *Evolution* **21**, 341–344.

Cleland, R. E. (1967b). The origin of closed circles of five chromosomes in *Oenothera*. *Am. J. Bot.* **54**, 993–997.

Cleland, R. E. (1970). The missing petal character in *Oenothera* and its relation to the cruciate character. *Am. J. Bot.* **57**, 850–855.

Cleland, R. E. and Blakeslee, A. F. (1930). Interaction between complexes as evidence for segmental interchange in *Oenothera*. *Proc. Nat. Acad. Sci. Wash.* **16**, 183–189.

Cleland, R. E. and Blakeslee, A. F. (1931). Segmental interchange, the basis of chromosomal attachments in *Oenothera*. *Cytologia* **2**, 175–233.

Cleland, R. E. and Brittingham, W. H. (1934). A contribution to an understanding of crossing over within chromosome rings in *Oenothera*. *Genetics* **19**, 62–72.

Cleland, R. E. and Hammond, B. L. (1950). Analysis of segmental arrangements in certain races of *Oenothera*. *Indiana Univ. Publ. Sci. Ser.* **16**, 10–72.

Cleland, R. E. and Hyde, B. B. (1963). Evidence of relationship between extra diminutive chromosomes in geographically remote races of *Oenothera*. *Am. J. Bot.* **50**, 179–185.

Cleland, R. E. and Newcomb, M. (1946). The growth of *Oenothera* plants from embryos cultured *in vitro*. *Proc. Indiana Acad. Sci.* **55**, 35.

Cleland, R. E. and Oehlkers, F. (1929). New evidence bearing upon the problem of the cytological basis for genetical peculiarities in the oenotheras. *Am. Nat.* **63**, 497–510.

Cleland, R. E. and Oehlkers, F. (1930). Erblichkeit und Zytologie verschiedener Oenotheren und ihrer Kreuzungen. *Jahrb. wiss. Bot.* **73**, 1–124.

Cleland, R. E., Preer, L. B. and Geckler, L. H. (1950). The nature and relationships of taxonomic entities in the North American euoenotheras. *Indiana Univ. Publ. Sci. Ser.* **16**, 218–254.

Cobb, F. and Bartlett, H. H. (1919). On Mendelian inheritance in crosses between mass-mutating and non-mass-mutating strains of *Oenothera pratincola*. *J. Wash. Acad. Sci.* **9**, 462–483.

Columna, F. (1628 (1949)) *in* Hernandez F. "Rerum medicarum novae hispaniae thesaurus; historiae animalium et mineralium novae hispaniae liber unicus." Roma.

Correns, C. (1917). Ein Fall experimenteller Verschiebung des Geschlechtverhältnisses. *Sitzber. Preuss. Akad. Wiss.* **1917**, 685–717.

Correns, C. (1922). Vererbungsversuche mit buntolättrigen Sippen VI. Einige neue Fälle von Albomaculatio. VII. Ueber die *peraurea*-Sippe der *Urtica urens*. *Sitzber. Preuss. Akad. Wiss.* **1922**, 460–486.

Crowe, L. K. (1955). The evolution of incompatibility in species of *Oenothera*. *Heredity* **9**, 293–322.

Darlington, C. D. (1929). Ring formation in *Oenothera* and other genera. *J. Heredity* **20**, 345–369.

Darlington, C. D. (1931). The cytological theory of inheritance in *Oenothera*. *J. Genet.* **24**, 405–474.

Darlington, C. D. (1933). The behaviour of interchange heterozygotes in *Oenothera*. *Proc. Nat. Acad. Sci. Wash.* **19**, 101–103.

Darlington, C. D. (1936). The limitation of crossing over in *Oenothera*. *J. Genet.* **32**, 343–352.

Davis, B. M. (1909). Cytological studies on *Oenothera* I. Pollen development of *Oenothera grandiflora*. *Annals Bot.* **23**, 551–571.

Davis, B. M. (1910). Cytological studies on *Oenothera* II. Reduction divisions of *Oenothera biennis*. *Annals. Bot.* **24**, 631–651.

Davis, B. M. (1911). Genetical studies on *Oenothera* II. Some hybrids of *Oenothera biennis* and *O. grandiflora* that resemble *Oenothera lamarckiana*. *Am. Nat.* **45**, 193–233.

Davis, B. M. (1912a). Genetical studies on *Oenothera* III. Further hybrids of *Oenothera biennis* and *O. grandiflora* that resemble *O. lamarckiana*. *Am. Nat.* **46**, 377–427.

Davis, B. M. (1912b). Was Lamarck's evening primrose (*Oenothera lamarckiana* Seringe) a form of *Oenothera grandiflora* Solander? *Bull. Torrey Bot. Club.* **39**, 519–533.

Davis, B. M. (1915). The test of a pure species of *Oenothera*. *Proc. Am. Philos. Soc.* **54**, 226–245.

Davis, B. M. (1916a). Hybrids of *Oenothera biennis* and *Oenothera franciscana* in the first and second generations. *Genetics* **1**, 197–251.

Davis, B. M. (1916b). *Oenothera neo-lamarckiana*, hybrid of *O. franciscana* Bartlett × *O. biennis* Linnaeus. *Am. Nat.* **50**, 688–696.

Davis, B. M. (1917). A criticism of the evidence for the mutation theory of de Vries from the behavior of species of *Oenothera* in crosses and selfed lines. *Proc. Nat. Acad. Sci. Wash.* **3**, 704–710.

Davis, B. M. (1924). The behavior of *Oenothera neo-lamarckiana* in selfed line through seven generations. *Proc. Am. Philos. Soc.* **63**, 239–278.

Davis, B. M. (1926). The history of *Oenothera biennis* Linnaeus, *Oenothera grandiflora* Solander, and *Oenothera lamarckiana* of de Vries in England. *Proc. Am. Philos. Soc.* **65**, 349–378.

Davis, B. M. (1927). Lamarck's evening primose (*Oenothera lamarckiana* Seringe) was a form of *Oenothera grandiflora* Solander. *Proc. Am. Philos. Soc.* **66**, 319–355.

Davis, B. M. (1937). The segregation of sulfur and dwarf from crosses involving *Oenothera franciscana* and certain hybrid derivatives. *Proc. Am. Philos. Soc.* **77**, 99–160.

Deam, C. C. (1940). "Flora of Indiana." Department of Conservation, State Library, Indianapolis.

de Vries, H. (1889). "Intracelluläre Pangenesis." G. Fischer, Jena. (English translation, Open Court Publ. Co., Chicago, 1910.)

de Vries, H. (1895). Sur l'introduction de l'*Oenothera lamarckiana* dans les Pays-Bas. *Nederl. Kruidkundig Arch.* II. Ser. **VI**, 579–589.

de Vries, H. (1900a). Das Spaltungsgesetz der Bastarde. *Ber. Deut. bot. Ges.* **18**, 83–90. (Also in English: The law of separation of characters in crosses. *J. Roy. Hort. Soc.* **25**, 243–248. 1901.)

de Vries, H. (1900b). Sur les unités des caractères spécifiques et leur application à l'étude des hybrides. *Rev. gén. Bot.* **12**, 257–271.

de Vries, H. (1900c). Sur la loi de disjonction des hybrides. *C. R. Acad. Sci. Paris* **130**, 845–847.

de Vries, H. (1900d). Sur l'origine expérimentale d'une espèce vegetale. *C. R. Acad. Sci. Paris* **131**, 124–126.

de Vries, H. (1900e). Ueber Erbungleiche Kreuzungen (vorlauf. Mit.). *Ber. Deut. bot. Ges.* **18**, 435–443. (Also in English: On crosses with dissimilar heredity. *J. Roy. Hort. Soc.* **25**, 249–255. 1901.)

de Vries, H. (1900f). Sur la mutabilité de l'*Oenothera lamarckiana*. *C. R. Acad. Sci.* **131**, 561–563.

de Vries, H. (1901). Recherches expérimentales sur l'origine des espèces. *Rev. Gén. Bot.* **13**, 5–17.

de Vries, H. (1901–3). "Die Mutationstheorie." Vol. I (1901), Vol. II (1903). Von Veit, Leipzig. (English translation Vol. I (1909), Vol. II (1910). Open Court Publ. Co., Chicago.)

de Vries, H. (1903a). On atavistic variation in *Oenothera cruciata. Bull. Torrey Bot. Club.* **30**, 75–82.

de Vries, H. (1903b). Fécondation et hybridité. *Arch. Néerl. des Sci. Exactes et Naturelles* **8**, VIII–XVIII.

de Vries, H. (1903c). "Befruchtung und Bastardierung." 62 p. Leipzig.

de Vries, H. (1903d). Anwendung der Mutationslehre auf dis Bastardierungsgesetze. *Ber. Deut. bot. Ges.* **21**, 45–52.

de Vries, H. (1905a). Ueber die Dauer der Mutationsperiode bei *Oenothera lamarckiana. Ber. Deut. bot. Ges.* **23**, 382–387.

de Vries, H. (1905b). "Species and varieties, their origin by mutation: lectures delivered at the University of California." Ed. D. T. MacDougal. 847 p. Open Court Publ. Co., Chicago.

de Vries, H. (1907). On twin hybrids. *Bot. Gaz.* **44**, 401–407.

de Vries, H. (1908). Ueber die Zwillingsbastarde von *Oenothera nanella. Ber. Deut. bot. Ges.* **26a**, 667–676.

de Vries, H. (1909). On triple hybrids. *Bot. Gaz.* **47**, 1–8.

de Vries, H. (1911). Ueber doppelreziproke Bastarden von *Oenothera biennis* L. und *O. muricata.* L. *Biol. Centralbl.* **31**, 97–104.

de Vries, H. (1913). "Gruppenweise Artbildung." 365 p. Gebr. Borntraeger, Berlin.

de Vries, H. (1914a). L'*Oenothera grandiflora* de l'herbier des Lamarck. *Rev. gén. Bot.* **25**, 151–166.

de Vries, H. (1914b). The probable origin of *Oenothera lamarckiana. Ser. Bot. Gaz.* **17**, 345–361.

de Vries, H. (1915). The coefficient of mutation in *Oenothera biennis* L. *Bot. Gaz.* **59**, 169–196.

de Vries, H. (1916a). New dimorphic mutants of the oenotheras. *Bot. Gaz.* **62**, 249–280.

de Vries, H. (1916b). Gute, harte und leere Samen von *Oenothera. Zeitschr. ind. Abst.- u. Vererb.* **16**, 239–292.

de Vries, H. (1917a). *Oenothera lamarckiana* mut. *velutina. Bot. Gaz.* **63**, 1–24.

de Vries, H. (1917b). Halbmutanten und Zwillingsbastarde. *Ber. Deut. bot. Ges.* **35**, 128–135.

de Vries, H. (1918a). Mass mutations and twin hybrids of *Oenothera grandiflora* Ait. *Bot. Gaz.* **65**, 377–422.

de Vries, H. (1918b). Kreuzungen von *Oenothera lamarckiana* mut. *velutina. Zeitschr. ind. Abst.- u. Vererb.* **19**, 1–38.

de Vries, H. (1918c). Phylogenetische und gruppenweise Artbilding. *Flora* **111**, 208–226.

de Vries, H. (1918d). Mutations of *Oenothera suaveolens* Desf. *Genetics* **3**, 1–26.

de Vries, H. (1919a). *Oenothera rubrinervis*, a half mutant. *Bot. Gaz.* **67**, 1–26.

de Vries, H. (1919b). *Oenothera lamarckiana erythrina*, eine neue Halbmutante. *Zeitschr. ind. Abst.- u.Vererb.* **21**, 91–118.

de Vries, H. (1919c). *Oenothera lamarckiana* mut. *simplex. Ber. Deut. bot. Ges.* **37**, 65–73.

de Vries, H. (1923a). Ueber die Mutabilität von *Oenothera lamarckiana* mut. *simplex. Zeitschr. ind. Abst.- u. Vererb.* **31**, 313–351.

de Vries, H. (1923b). Ueber sesquiplex-Mutanten von *Oenothera lamarckiana*. *Zeitschr. Bot.* **15**, 368–408.

de Vries, H. (1923c). Ueber die Entstehung von *Oenothera lamarckiana* mut. *velutina*. *Biol. Centralbl.* **43**, 213–224.

de Vries, H. (1924a). Ueber Scheinbastarde. *Die Naturwiss.* **12**, 161–165.

de Vries, H. (1924b). Die Mutabilität von *Oenothera lamarckiana gigas*. *Zeitschr. ind. Abst.- u. Vererb.* **35**, 197–237.

de Vries, H. (1925a). On physiological chromomeres. *La Cellule* **35**, 5–17.

de Vries, H. (1925b). Androlethal factors in *Oenothera*. *J. Gen. Physiol.* **8**, 109–113.

de Vries, H. (1925c). Brittle races of *Oenothera lamarckiana*. *Bot. Gaz.* **80**, 262–275.

de Vries, H. (1925d). Die latente Mutabilität von *Oenothera biennis* L. *Zeitschr. ind. Abst.- u. Vererb.* **38**, 141–199.

de Vries, H. (1929). Ueber das Auftreten von Mutanten aus *Oenothera lamarckiana*. *Zeitschr. ind. Abst.- u. Vererb.* **52**, 121–191.

de Vries, H. (1935). Ueber semirecessive Anlagen in *Oenothera lamarckiana*. *Zeitschr. ind. Abst.- u. Vererb.* **57**, 222–256.

de Vries, H. and Bartlett H. H. (1912). The evening primroses of Dixie Landing, Alabama. *Science* **35**, 599–601.

de Vries, H. and Boedijn, K. (1923). On the distribution of mutant characters among the chromosomes of *Oenothera lamarckiana*. *Genetics* **8**, 233–238.

de Vries, H. and Boedijn, K. (1924). Die Gruppierung der Mutanten von *Oenothera lamarckiana*. *Ber. Deut. bot. Ges.* **42**, 174–178.

de Vries, H. and Gates, R. R. (1928). A survey of the cultures of *Oenothera lamarckiana* at Lunteren. *Zeitschr. ind. Abst.- u. Vererb.* **47**, 275–286.

Diers, L. and Schötz, F. (1965). Ueber den Feinbau pflanzlichen Mitochondrien. *Zeitschr. Pflanzenphysiol.* **53**, 334–343.

Diers, L. and Schötz, F. (1966). Ueber die dreidimensionale Gestaltung des Thylakoidsystems in den Chloroplasten. *Planta* **70**, 322–343.

Diers, L. and Schötz, F. (1968). Räumliche Beziehungen zwischen osmiophilen Granula und Thylakoiden. *Zeitschr. Pflanzenphysiol.* **58**, 252–265.

Diers, L. and Schötz, F. (1969). Über ring- und schalenförmige Thylakoidbildungen in den Plastiden. *Zeitschr. Planzenphysiol.* **60**, 187–210.

Diers, L., Schötz, F. and Bathelt, H. (1968). Zur Frage des Aufbaus von Thylakoiden aus Vesikeln. *Planta* **80**, 211–226.

Dolzmann, P. (1968). Photosynthese-Reaktionen einiger Plastom-Mutanten von *Oenothera*. III. Strukturelle Aspekte. *Zeitschr. Planzenphysiol.* **58**, 300–309.

Don, G. (1832). "General history of dichlamydeous plants." London.

Dulfer, H. (1926). Die Erblichkeitserscheinungen der *Oenothera lamarckiana semigigas*. *Rec. trav. botan. Néerl.* **23**, 1–72.

East, E. M. (1940). The distribution of self-sterility in the flowering plants. *Proc. Am. Philos. Soc.* **82**, 449–518.

Emerson, S. H. (1931a). Genetic and cytological studies on *Oenothera* II. Certain crosses involving *Oe. rubricalyx* and *Oe. "franciscana sulfurea"*. *Zeitschr. ind. Abst.- u. Vererb.* **59**, 381–394.

Emerson, S. H. (1931b). The inheritance of certain characters in *Oenothera* hybrids of different chromosome configurations. *Genetics* **16**, 325–348.

Emerson, S. H. (1932). Chromosome rings in *Oenothera*, *Drosphila* and maize. *Proc. Nat. Acad. Sci. Wash.* **18**, 630–632.

Emerson, S. H. (1935). The genetic nature of de Vries' mutations in *Oenothera lamarckiana*. *Am. Nat.* **69**, 545–559.

Emerson, S. H. (1936). Trisomic derivatives of *Oenothera lamarckiana*. *Genetics* **21**, 200–224.

Emerson, S. H. (1938). The genetics of incompatibility in *Oenothera organensis*. *Genetics* **23**, 190–202.

Emerson, S. H. (1940). Growth of incompatible pollen tubes in *Oenothera organensis*. *Bot. Gaz.* **101**, 890–911.

Emerson, S. H. and Sturtevant, A. H. (1931). Genetical and cytological studies in *Oenothera* III. The translocation hypothesis. *Zeitschr. ind. Abst.- u. Vererb.* **59**, 395–419.

Emerson, S. H. and Sturtevant, A. H. (1932). The linkage relations of certain genes in *Oenothera*. *Genetics* **17**, 393–412.

Ford, C. E. (1936). Nondisjunction in *Oenothera* and the genesis of trisomics. *J. Genet.* **33**, 275–303.

Fork, D. C. and Heber, U. W. (1968). Studies on the electron-transport reactions of photosynthesis in plastome mutants of *Oenothera*. *Pl. Physiol.* **43**, 606–612.

Fork, D. C., Heber, U. W. and Michael-Wolwertz, M. (1969). Studies on the photosynthesis of plastome mutants of *Oenothera*. *Carnegie Inst. Year Book.* **67**, 503–505.

Freese, E. B. (1967). Der Chemismus der urethaninduzierten Mutationen, 25 Jahre nach ihrer Entdeckung. *Molec. Gen. Genet.* **100**, 150–158.

Gardella, C. (1953). Studies on the chromosome structure of *Oenothera* (abstract). *Dissertation Abstr.* **13**, 957.

Gates, R. R. (1907a). Pollen development in hybrids of *Oenothera lata* × *O. lamarckiana* and its relation to mutation. *Bot. Gaz.* **43**, 81–115.

Gates, R. R. (1907b). Hybridization and germ cells of *Oenothera* mutants. *Bot. Gaz.* **44**, 1–21.

Gates, R. R. (1908a). The chromosomes of *Oenothera*. *Science* **27**, 193–195.

Gates, R. R. (1908b). A study of reduction in *Oenothera rubrinervis*. *Bot. Gaz.* **46**, 1–34.

Gates, R. R. (1909a). Further studies of *Oenothera* cytology. *Science* **29**, 269.

Gates, R. R. (1909b). Behavior of chromosomes in *Oenothera lata* × *O. gigas*. *Bot. Gaz.* **48**, 179–199.

Gates, R. R. (1909c). Stature and chromosomes of *Oenothera gigas* de V. *Arch. Zellf.* **3**, 527–552.

Gates, R. R. (1910). The earliest description of *Oenothera lamarckiana*. *Science* **31**, 425–426.

Gates, R. R. (1911a). Early historico-botanical records of the oenotheras. *Proc. Iowa Acad. Sci.* **1910**, 85–124.

Gates, R. R. (1911b). Mutation in *Oenothera*, *Am. Na* **15**, 577–606.

Gates, R. R. (1911c). The mode of chromosome reduction. *Bot. Gaz.* **51**, 321–344.

Gates, R. R. (1913). Tetraploid mutants and chromosome mechanisms. *Biol. Centrlbl.* **33**, 92–99, 113–150.

Gates, R. R. (1914a). Breeding experiments which show that hybridization and mutation are independent phenomena. *Zeitschr. ind. Abst.- u. Vererb.* **11**, 209–279.

Gates, R. R. (1914b). Some oenotheras from Cheshire and Lancashire. *Ann. Missouri Bot. Gar.* **1**, 383–400.

Gates, R. R. (1915). "The mutation factor in evolution." MacMillan, London.

Gates, R. R. (1922). Some points on the relation of cytology and genetics. *J. Heredity* **13**, 75–76.

Gates, R. R. (1924). Meiosis and crossing over. *J. Heredity* **15**, 237–240.

Gates, R. R. (1925). "Present problems of *Oenothera* research." Mem. Publ. in honor of 100th birthday of J. G. Mendel. Czechoslov. Eugenics Soc. pp. 135–145.

Gates, R. R. (1928a). The relation of cytology to genetics in *Oenothera*. *Zeitschr. ind. Abst.- u. Vererb.* (Verb. V. Intern. Kongr. Vererb. 1927), 749–758.

Gates, R. R. (1928b). The cytology of *Oenothera*. *Bibliographia Genetica* **IV**, 401–492.

Gates, R. R. (1933). The general bearings of recent research in *Oenothera*. *Am. Nat.* **67**, 352–364.

Gates, R. R. (1936). Genetical and taxonomic investigations in the genus *Oenothera*. *Philos. Trans. Roy. Soc. London* **B226**, 239–355.

Gates, R. R. (1939). Self-sterility in *Oenothera*. *Nature* **143**, 245.

Gates, R. R. and Goodwin, K. M. (1931). Meiosis in *Oenothera purpurata* and *Oe. blandina*. *Proc. Roy. Soc. London* **B109**, 149–164.

Gates, R. R. and Nandi, H. K. (1935). The cytology of trisomic mutations in a wild species of *Oenothera*. *Philos. Trans. Roy. Soc. London* **B225**, 227–254.

Gates, R. R. and Thomas, H. (1914). A cytological study of *Oenothera* mut. *lata* and *Oe.* mut. *semilata* in relation to mutation. *Quart. J. Microscop. Sci.* **59**, 523–571.

Geckler, L. H. (1950). The cytogenetics and phylogenetic relationships of certain races of *Euoenothera* from northeastern North America. *Indiana Univ. Publ. Sci. Ser.* **16**, 160–217.

Geerts, J. M. (1907). Ueber die Zahl der Chromosomen von *Oenothera lamarckiana*. *Ber. Deut. bot. Ges.* **25**, 191–195.

Geerts, J. M. (1909). Beiträge zur Kenntnis der Cytologie und der partieller Sterilität von *Oenothera lamarckiana*. *Rec. trav. botan. Néerl.* **5**, 93–208.

Gerhard, K. (1929). Genetische und zytologische Untersuchungen an *Oenothera grandiflora* Ait. *Jenaische Zeitschr. Naturwiss.* **4**, 283–338.

Goldschmidt, R. (1912). Die Merogonie der *Oenothera* Bastarde und die doppel-reziproken Bastarde von de Vries. *Arch. Zellf.* **9**, 331–334.

Goldschmidt, R. (1958). Genic conversion in *Oenothera*? *Am. Nat.* **92**, 93–104.

Göpel, G. (1970). Plastomabhängige Pollensterilität bei *Oenothera*. *Theoret. and Appl. Genetics* **40**, 111–116.

Gray, A. (1950). "Manual of Botany", 8th ed. American Book Co.

Hagen, C. W., Jr., (1950). A contribution to the cytogenetics of the genus *Oenothera*, with special reference to certain forms from South America. *Indiana Univ. Publ. Sci. Ser.* **16**, 305–348.

Håkansson, A. (1926). Ueber das Verhalten der Chromosomen bei der heterotypischen Teilung schwedischer *Oenothera lamarckiana* und einige ihrer Mutanten und Bastarde. *Hereditas* **8**, 225–304.

Håkansson, A. (1928). Die Reduktionsteilung in den Samenanlagen einiger Oenotheren. *Hereditas* **11**, 129–181.

Håkansson, A. (1930a). Die Chromosomenreduktion bei einigen Mutanten und Bastarden von *Oenothera lamarckiana*. *Jahrb. wiss. Bot.* **72**, 385–402.

Håkannson, A. (1930b). Zur Zytologie trisomatischer Mutanten aus *Oenothera lamarckiana*. *Hereditas* **14**, 1–32.

Hallier, U. W. (1966). Untersuchungen des Photosynthesesapparates bei Plastommutanten von Oenothera. *Ber. Deut. bot. Ges.* **79**, (69)–(71).

Hallier, U. W. (1967). On the use of ^{14}C and ^{32}P in locating genetically caused defects in photosynthesis of some plastid mutants. *In* "Isotopes in Plant Nutrition and Physiology", pp. 235–245. Intern. Atom. Energy Agency Vienna.

Hallier, U. W. (1968). Photosynthese-Reaktionen einiger Plastom-Mutanten von Oenothera II. Die Bildung von ATP und NADPH. *Zeitschr. Pflanzenphysiol.* **58**, 289–299.

Hallier, U. W., Heber, U. and Stubbe, W. (1968). Photosynthese-Reaktionen einiger Plastom-Mutanten von Oenothera I. Der reduktive Pentosephosphat-zyklus. *Zeitschr. Planzenphysiol.* **58**, 222–239.

Harte, C. (1942). Meiosis und Crossing-Over. Weitere Beiträge zur Zytogenetik von *Oenothera. Zeitschr. Bot.* **38**, 65–147.

Harte, C. (1948). Zytologisch-genetische Untersuchungen an spaltenden Oeno-theren-Bastarden. *Zeitschr. Vererb.* **82**, 495–640.

Harte, C. (1950). Dominanzwechsel bei *Oenothera* als genetisches und entwicklungs-physiologisches Problem. *Zeitschr. Vererb.* **83**, 318–323.

Harte, C. (1958a, b, c). Untersuchungen über die Gonenkonkurrenz in der Samen-anlage bei *Oenothera* unter Verwendung der Letalfaktoren als Markierungsgene. I. Die individuelle Komponente der Variabilität. *Zeitschr. Vererb.* **89**, 473–496. II. Die Umweltkomponente der Variabilität. *Zeitschr. Vererb.* **89**, 497–507. III. Die genetische Komponente der Variabilität. *Zeitschr. Vererb.* **89**, 715–728.

Haselwarter, A. (1937). Untersuchungen zur Physiologie der Meiosis V. *Zeitschr. Bot.* **31**, 273–328.

Hecht, A. (1944). Induced tetraploids of a self-sterile *Oenothera. Genetics* **29**, 69–74.

Hecht, A. (1950). Cytogenetic studies of *Oenothera* subgenus *Raimannia. Indiana Univ. Publ. Sci. Ser.* **16**, 255–304.

Heribert-Nilsson, N. (1912). Die Variabilität der *Oenothera lamarckiana* und das Problem der Mutation. *Zeitschr. ind. Abst.- u. Vererb.* **8**, 89–231.

Heribert-Nilsson, N. (1920a). Zuwachsgeschwindigkeit der Pollenschläuche und gestorte Mendelzahlen bei *Oenothera lamarckiana. Hereditas* **1**, 41–67.

Heribert-Nilsson, N. (1920b). Kritische Betrachtungen und faktorielle Erklärung der *laeta-velutina* Spaltung bei *Oenothera. Hereditas* **1**, 312–342.

Herzog, G. (1940). Genetische und cytologische Untersuchungen über 15-chromo-somige Mutanten von *Oenothera biennis* und *O. lamarckiana. Flora* **134**, 377–432.

Hoeppener, E. and Renner, O. (1928). Genetische und zytólogische Oenotheren-studien I. Zur Kenntnis der *Oenothera ammophila* Focke. *Zeitschr. ind. Abst.-u. Vererb.* **49**, 1–25.

Hoeppener, E. and Renner, O. (1929). Genetische und zytólogische Oenotheren-studien II. Zur Kenntnis von *Oe. rubrinervis, deserens, lamarckiana-gigas, biennis-gigas, franciscana, hookeri, suaveolens, lutescens. Bot. Abhandl.* **15**, 3–86.

Hoff, Victor (1962). An analysis of outcrossing in certain complex-heterozygous euoenotheras I. Frequency of outcrossing. *Am. J. Bot.* **49**, 715–724.

Hoff, Victor (1966). Cytogenetic observations on some euoenotheras (Onagraceae) of Northwestern Arkansas. *Southwestern Nat.* **11**, 217–222.

Honing, J. A. (1911). Die Doppelnatur der *Oenothera lamarckiana. Zeitschr. ind. Abst.- u. Vererb.* **4**, 227–278.

Ishikawa, M. (1918). Studies on the embryo sac and fertilization in *Oenothera. Annals Bot.* **32**, 279–317.

Japha, B. (1939). Die Meiosis von *Oenothera* II. *Zeitschr. Bot.* **34**, 321–369.

Jean, R., Lambert, A. M. and Linder, R. (1966). Analyse cytogénétique de l'*Oenothera nuda* Renner. *Bull. Soc. Bot. Nord France* **19**, 6–26.

Jávorka, S. (1925). "Magyar Flora (Flora Hungarica)." Budapest.

Kisch, R. (1937). Die Bedeutung der Wasserversorgung für den Ablauf der Meiosis (Untersuchungen zur Physiologie der Meiosis VI). *Jahrb. wiss. Bot.* **85**, 450–484.

Kowalewicz, R. (1956). Entwicklungsgeschichtliche Studien an normalen und cruciaten Blüten von *Epilobium* und *Oenothera*. *Planta* **46**, 569–603.

Krumholz, G. (1925). Untersuchungen über Scheckung der Oenotherenbastarde, insbesondere über die Möglichkeit der Entstehung von Periklinal-Chimëren. *Jenaische Zeitschr. Naturwiss.* **62**, 187–260.

Krumholz, G. (1930). Ueber Verschiendenheiten in den Embryonengrösse einiger Oenotheren und ihrer reziproken Bastarde. *Zeitschr. ind. Abst.- u. Vererb.* **56**, 383–392.

Langendorf, J. (1930). Zur Kenntnis der Genetik und Entwicklungsgeschichte von *Oenothera fallax, rubririgida*, and *hookeri-albata*. *Bot. Archiv.* **29**, 474–530.

Lewis, D. (1942). The physiology of incompatibility in plants. I. The effect of temperature. *Proc. Roy. Soc. London* **B131**, 13–26.

Lewis, D. (1948). Structure of the incompatibility gene. I. Spontaneous mutation rate. *Heredity* **2**, 219–236.

Lewis, D. (1949). Structure of the incompatibility gene. II. Induced mutation rate. *Heredity* **3**, 339–355.

Lewis, D. (1951). Structure of the incompatibility gene. III. Types of spontaneous and induced mutation. *Heredity* **5**, 399–414.

Lewis, D. (1952). Serological reactions of pollen incompatibility substances. *Proc. Roy. Soc. London* **B140**, 127–135.

Lewis, D. (1960). Genetic control of specificity and activity of the S-antigen in plants. *Proc. Roy. Soc. London* **B151**, 468–477.

Lewitsky, G. A. (1931). The morphology of chromosomes (in English). *Bull. Appl. Bot., Genet. and Pl. Breed. USSR* **27**, 103–174.

Linder, R. (1957). Les *Oenothera* récemment reconnus en France. *Bull. Soc. Bot. France* **104**, 515–525.

Linnaeus, C. (1737). "Hortus Cliffortianus." Amsterdam.

Linnaeus, C. (1753). "Species Plantarum." Stockholm.

Linskens, H. F. (1958). Zur Frage der Entstehung der Abwehr-Körper bei der Inkompatibilitäts-Reaktion von *Petunia* I. *Ber. Deut. bot. Ges.* **71**, 3–10.

Linskens, H. F. (1959). Zur Frage der Entstehung der Abwehr-Körper bei der Inkompatibilitäts-Reaktion von *Petunia* II. *Ber. Deut. bot. Ges.* **72**, 84–92.

Linskens, H. F. (1960). Zur Frage der Entstehung der Abwehr-Körper bei der Inkompatibilitäts-Reaktion von *Petunia* III. *Zeitschr. Bot.* **48**, 126–135.

Lotsy, J. P. (1914). On the origin of species. *Proc. Linnaean Soc. London* **126**, 73–89.

Lutz, A. M. (1907). Preliminary note on the chromosomes of *Oenothera lamarckiana* and one of its mutants, *Oe. gigas*. *Science* **26**, 151–152.

Lutz, A. M. (1908). Chromosomes of somatic cells of the oenotheras. *Science* **27**, 335.

Lutz, A. M. (1909). Notes on the first-generation hybrid of *Oenothera lata* ♀ × *gigas* ♂. *Science* **29**, 263–267.

MacDougal, D. T., Vail, A. M., Shull, G. H. and Small, J. K. (1905). Mutants and hybrids of the oenotheras. *Carnegie Institution Wash. Publ.* **24**, 1–57.

MacDougal, D. T., Vail, A. M. and Shull, G. H. (1907). Mutations, variations and relationships of the oenotheras. *Carnegie Institution Wash. Publ.* **81**, 1–92.

Mäkinen, Y. L. A. and Lewis, D. (1962). Immunological analysis of incompatibility

(S) proteins, and of cross reacting material in a self-compatible mutant of *Oenothera organensis*. *Genet. Res., Cambr.* **3**, 352–363.

Mansfeld, R. (1940). Verzeichnis der Farn- und Blütenpflanzen des Deutschen Reiches. *Ber. Deut. bot. Ges.* **58a**, 12–323.

Marquardt, H. (1937). Die Meiosis von *Oenothera* I. *Zeitschr. Zellf. u. Mikroscop. Anat.* **27**, 159–210.

Marquardt, H. (1948). Das Verhalten röntgeninduzierter Viererringe mit grossen interstitialen Segmenten bei *Oenothera hookeri*. *Zeitschr. ind. Abst.- u. Vererb.* **82**, 415–429.

McClintock, B. (1941). The stability of broken ends of chromosomes in *Zea mays*. *Genetics* **26**, 234–282.

Michaelis, P. (1930). Ueber experimentell erzeugte, heteroploide Pflanzen von *Oenothera hookeri*. *Zeitschr. Bot.* **23**, 288–308.

Michaelis, P. (1955). Modellversuche zur Plastiden- und Plasmavererbung. *Der Züchter* **25**, 209–221.

Mickan, M. (1936). Zur Kenntnis der *Oenothera argillicola* MacKenzie. Genetische und Zytologische Untersuchungen. *Flora* **130**, 1–20.

Morison, P. (1669). "Hortus Regius Blesensis." London.

Morison, P. (1680). "Plantarum Historiae Universalis Oxoniensis." Oxford.

Morrison, Roger, B. (1965). Quaternary geology of the Great Basin. *In* "The Quarternary of the United States" (Wright and Frey), pp. 265–285.

Mühlethaler, K. and Bell, P. R. (1962). Untersuchungen über die Kontinuität von Plastiden und Mitochondrien in der Eizelle von *Pteridium aquilinum* (L) Kuhn. *Naturwiss.* **49**, 63–64.

Muller, H. J. (1917). An *Oenothera*-like case in *Drosophila*. *Proc. Nat. Acad. Sci. Wash.* **3**, 619–626.

Muller, H. J. (1918). Genetic variability, twin hybrids and constant hybrids, in a case of balanced lethal factors. *Genetics* **3**, 422–499.

Munz, P. A. (1965). Onagraceae. *North American Flora.* Ser. II, pt. **5**, 1–231. N.Y. Botanical Garden.

Nanney, D. L. (1963). Aspects of mutual exclusion in *Tetrahymena*. *In* "Biological Organization at Cellular and Supercellular Levels", pp. 91–109. Academic Press.

Newcomb, M. and Cleland, R. E. (1946). Aseptic cultivation of excised plant embryos. *Science* **104**, 329–330.

Noack, K. L. (1925). Weitere Untersuchungen über das Wesender Buntblättrigkeit bei *Pelergonium*. *Verh. Phys. Med. Ges. Würzburg.* **50**, 47–97.

Oehlkers, F. (1921). Vererbungsversuche an Oenotheren I. *Zeitschr. ind. Abst.- u. Vererb.* **26**, 1–31.

Oehlkers, F. (1924). Sammelreferat über neuere experimentelle Oenotheren-arbeiten. *Zeitschr. ind. Abst.- u. Vererb.* **34**, 259–283.

Oehlkers, F. (1926). Erblichkeit und Zytologie einiger Kreuzungen mit *Oenothera strigosa*. *Jahrb. wiss. Bot.* **65**, 401–446.

Oehlkers, F. (1930a). Studien zum Problem der Polymerie und des multiplen Allelomorphismus I. *Zeitschr. Bot.* **22**, 473–537.

Oehlkers, F. (1930b). Studien zum Problem der Polymerie und des multiplen Allelomorphismus II. *Zeitschr. Bot.* **23**, 967–1003.

Oehlkers, F. (1933). Crossing over bei *Oenothera* (Vererbungsversuche an Oeno-theren V). *Zeitschr. Bot.* **26**, 385–430.

Oehlkers, F. (1935a). Studien zum Problem der Polymerie und des multiplen

Allelomorphismus III. Die Erblichkeit der Sepalodie bei *Oenothera* und *Epilobium*. *Zeitschr. Bot.* **28**, 161–222.

Oehlkers, F. (1935b). Untersuchungen zur Physiologie der Meiosis I, *Zeitschr. Bot.* **29**, 1–53.

Oehlkers, F. (1936). Untersuchungen zur Physiologie der Meiosis III. *Zeitschr. Bot.* **30**, 253–276.

Oehlkers, F. (1938). Ueber die Erblichkeit des *cruciata*-Merkmals bei den Oenotheren; eine Erwiderung. *Zeitschr. ind. Abst.- u. Vererb.* **75**, 277–297.

Oehlkers, F. (1940). Meiosis und crossing over. Zytogenetische Untersuchungen an Oenotheren. *Zeitschr. ind. Abst.- u. Vererb.* **78**, 157–186.

Oehlkers, F. (1943). Die Auslösung von Chromosomenmutationen in der Meiosis durch Einwirkung von Chemikalien. *Zeitschr. ind. Abst.- u. Vererb.* **81**, 313–341.

Oehlkers, F. (1949). Mutationsauslösung durch Chemikalien. *Sitz. Heidelberg Akad. Wiss., Math-Naturw. Kl.* **9**, 3–40.

Oehlkers, F. and Bergfeld-Gaertner, H. (1962). Mutationsauslösung durch Radioisotope. *Zeitschr. Vererb.* **93**, 264–279.

Oehlkers, F. and Linnert, G. (1949). Neue Versuche über die Wirkungsweise von Chemikalien bei der Auslösung von Chromosomenmutationen. *Zeitschr. ind. Abst.- u. Vererb.* **83**, 136–156.

Oehlkers, F. and Linnert, G. (1951). Weitere Untersuchungen über die Wirkungsweise von Chemikalien bei der Auslösung von Chromosomenmutationen. *Zeitschr. ind. Abst.- u. Vererb.* **83**, 429–438.

Parkinson, John (1629). "Paradisi in sole Paradisus terrestris." London.

Pohl, J. (1895). Ueber Variationsweit der *Oenothera lamarckiana*. *Österr. Bot. Zeitschr.* **45**, 166–169, 205–212.

Preer, L. B. (1950). A cytogenetic study of certain broad-leaved races of *Oenothera*. *Indiana Univ. Publ. Sci. Ser.* **16**, 82–159.

Raven, Peter, H. (1964). The generic subdivision of *Onagraceae*, tribe *Onagreae*. *Brittonia* **16**, 276–288.

Raven, Peter, H. (1968). *Oenothera* L. *In* "Flora Europaea", Vol. II.

Ray, John, (1686). "Historia Plantarum." London.

Renner, O. (1913). Ueber angebliche Merogonie der Oenotherabastarde. *Ber. Deut. bot. Ges.* **31**, 334–335.

Renner, O. (1914). Befruchtung und Embryobilding bei *Oenothera lamarckiana* und einiger. verwandten Arten. *Flora* **107**, 115–150.

Renner, O. (1917a). Die tauben Samen der Oenotheren. *Ber. Deut. bot. Ges.* **34**, 858–869.

Renner, O. (1917b). Artbastarde und Bastardarten in der Gattung *Oenothera*. *Ber. Deut. bot. Ges.* **35**, (21)–(26).

Renner, O. (1917c). Versuche über die gametische Konstitution der Oenotheren. *Zeitschr. ind. Abst.- u. Vererb.* **18**, 121–294.

Renner, O. (1918a). Bemerkungen zu der Abhandlung von Hugo de Vries. Kreuzungen von *Oenothera lamarckiana* mut. *velutina*. *Ber. Deut. bot. Ges.* **36**, 446–456.

Renner, O. (1918b). Weitere Vererbungsstudien an Oenotheren. *Flora* **111–112**, 641–667.

Renner, O. (1919a). Ueber Sichtbarwarden der Mendelschen Spaltung in Pollen von *Oenothera*bastarden. *Ber. Deut. bot. Ges.* **37**, 129–135.

Renner, O. (1919b). Zur Biologie und Morphologie der männlichen Haplonten einiger Oenotheren. *Zeitschr. Bot.* **11**, 305–380.

Renner, O. (1921a). Heterogamie im weiblichen Geschlecht und Embryosackentwicklung bei den Oenotheren. *Zeitschr. Bot.* **13**, 609–621.

Renner, O. (1921b). Das Rotnervenmerkmal der Oenotheren. *Ber. Deut. bot. Ges.* **39**, 264–270.

Renner, O. (1924). Die Scheckung der Oenotherenbastarde. *Biol. Zentrabl.* **44**, 309–336.

Renner, O. (1925). Untersuchungen über die faktorielle Konstitution einiger komplexheterozygotischer Oenotheren. *Bibliotheca Genetica* **9**, 1–168.

Renner, O. (1928). Ueber Koppelungswechsel bei *Oenothera. Zeitschr. ind. Abst.-u. Vererb.* (Verhandl. 5th Intern. Kongr. Vererb.) **2**, 1216–1220.

Renner, O. (1933). Zur Kenntnis der Letalfaktoren und des Koppelungswechsels der Oenotheren. *Flora* **27**, 215–250.

Renner, O. (1936). Zur Kenntnis der nichtmendelenden Buntheit der Laubblätter. *Flora* **30**, 218–290.

Renner, O. (1937a). Zur Kenntnis der Plastiden- und Plasmavererbung. *Cytologia* (Fujii Jubil. Vol.), 643–653.

Renner, O. (1937b). Ueber *Oenothera atrovirens* Sh. et Bart. und über somatische Konversion im Erbgang der *cruciata*-Merkmals der Oenotheren, *Zeitschr. ind. Abst.- u. Vererb.* **74**, 91–124.

Renner, O. (1938a). Kurze Mitteilungen über Oenothera II. Zu den Chromoso-menformeln der Komplexe *albicans, curvans, flectens, gaudens (rubens), rigens. Flora* **132**, 319–324.

Renner, O. (1938b). Alte und neue Oenotheren in Norddeutschland. *Feddes Repertor.* Beih. **100** (Bornmüller-Festschr.), 94–105.

Renner, O. (1939). Kurze Mitteilungen über *Oenothera* III. Ueber *gigas*- und hemigigas-Formen und ihre Verwendung zur Untersuchung des *cruciata*-Merkmals. *Flora* **133**, 215–238.

Renner, O. (1940a). Kurze Mitteilungen über *Oenothera* IV. Ueber die Beziehungen zwischen Heterogamie und Embryosackentwicklung und über diplarrhene Verbindungen. *Flora* **134**, 145–158.

Renner, O. (1940b). Zur Kenntnis der 15-chromosomigen Mutanten von *Oenothera lamarckiana. Flora* **134**, 257–310.

Renner, O. (1941). Ueber die Entstehung homozygotischer Formen aus Komplex-heterozygotischen Oenotheren. *Flora* **135**, 201–238.

Renner, O. (1942a). Ueber das Crossing-over bei *Oenothera. Flora* **136**, 117–214.

Renner, O. (1942b). Europäische Wildarten von *Oenothera. Ber. Deut. bot. Ges.* **60**, 448–466.

Renner, O. (1942c). Beiträge zur Kenntnis des *cruciata*-Merkmals der Oenotheren IV. *Gigas*-Bastarde, Labilität und Konversibilität der *Cr*-Gene. *Zeitschr. ind. Abst.- u. Vererb.* **80**, 590–611.

Renner, O. (1943a). Ueber die Entstehung homozygotischer Formen aus Kom-plexheterozygotischen Oenotheren II. Die Translokationshomozygoten. *Zeitschr. Bot.* **39**, 49–105.

Renner, O. (1943b). Zur Kenntnis des Pollenkomplexes *flectens* der *Oenothera atrovirens* Sh. et. Bart. *Zeitschr. ind. Abst.- u. Vererb.* **81**, 391–483.

Renner, O. (1943c). Kurze Mitteilungen über *Oenothera* V. Zur Kenntnis von *O. silesiaca* n. sp., *parviflora* L., *ammophila* Focke, *rubricaulis* Kleb. *Flora* **136**, 324–334.

Renner, O. (1943d). Kurze Mitteilungen über *Oenothera* VI. Ueber die 15-Chromo-somigen Mutanten *dependens, incana, scintillans, glossa, tripus. Flora* **137**, 216–229.

Renner, O. (1946). Artbildung in der Gattung *Oenothera. Naturwiss.* **33**, 211–218.

Renner, O. (1948). Die cytologischen Grundlagen des Crossing-over bei *Oenothera*. *Zeitschr. Naturforsch.* **3b**, 186–196.

Renner, O. (1949). Die 15-Chromosomigen Mutanten der *Oenothera lamarckiana* und ihrer Verwandten. *Zeitschr. ind. Abst.- u. Vererb.* **83**, 1–25.

Renner, O. (1950). Europäische Wildarten von *Oenothera* II. *Ber. Deut. bot. Ges.* **63**, 129–138.

Renner, O. (1952). Zur Genetik von *Oenothera* und *Epilobium*. *Zeitschr. Naturforsch.* **7b**, 368–371.

Renner, O. (1953). Ueber *Oenothera* hybrida mut. *helix*. *Planta* **42**, 30–41.

Renner, O. (1956). Europäische Wildarten von *Oenothera* III. *Planta* **47**, 219–254.

Renner, O. (1957). Ueber den Erbgang des *cruciata*-Merkmals der Oenotheren V. *Planta* **48**, 343–392.

Renner, O. (1958a). Ueber den Erbgang des *cruciata*-Merkmals der Oenotheren VI. Verbindung der *Oenothera lamarckiana* mut. *blandina*. *Zeitschr. ind. Abst.- u. Vererb.* **89**, 14–35.

Renner, O. (1958b). Ueber den Erbgang des *cruciata*-Merkmals der Oenotheren VII. Verbindungen der *Oenothera hookeri, Oe. franciscana* und *Oe. purpurata*. *Flora* **145**, 339–373.

Renner, O. (1958c). Ueber den Erbgang des *cruciata*-Merkmals der Oenotheren VIII. Verbindungen der *Oenothera atrovirens*, und Rückblick. *Zeitschr. Vererb.* **89**, 377–396.

Renner, O. (1959a). Somatic conversion in the heredity of the *cruciata* character in *Oenothera*. *Heredity* **13**, 283–288.

Renner, O. (1959b). Paralipomena zur Genetik von *Oenothera*. Mutanten von *Oe. hookeri*, reziproke Bastarde, Samengrösse, sublethale Kombinationen, Analyse von *hookericurva*. *Zeitschr. Vererb.* **90**, 132–147.

Renner, O. and Cleland, R. E. (1933). Zur Genetik und Zytologie der *Oenothera chicaginensis* und ihrer Abkömmlinge. *Zeitschr. ind. Abst.- u. Vererb.* **66**, 275–318.

Renner, O. and Hirmer, U. (1956). Zur Kenntnis von *Oenothera*. I. Ueber *Oe. conferta* n. sp. II. Ueber künstliche Polyploidie. *Biol. Zentralbl.* **75**, 513–531.

Renner, O. and Sensenhauer, R. (1942). Versuche über den Erbgang des *cruciata*-Merkmals der Oenotheren. III. Weitere Belege für somatische Konversion. *Zeitschr. ind. Abst.- u. Vererb.* **80**, 570–589.

Rossmann, G. (1963a). Analyse der *Oenothera coronifera* (Renner). *Flora* **153**, 451–468.

Rossmann, G. (1963b). Ueber den Erbgang des *Sepalodie*- und des *cruciata*-Merkmals der Oenotheren. I. Die Entstehung des *Sepalodie*-Merkmals und seine Verwandten bei Kreuzung mit normalkronigen Arten. *Zeitschr. Vererb.* **94**, 19–23.

Rostanski, K. (1965). The species of the genus *Oenothera* L. in Silesia (in Polish with English abstract). *Fragmenta Flor. et Geobotanica* **11**, 491–497.

Rostanski, K. (1968). Neophytism of species of the genus *Oenothera* L. occurring in Europe (in Polish with English abstract). *Materialy zakhadu fitosocjologii Stosowanej U.W.* Nr. **25**, 67–79.

Rudloff, C. F. (1929). Zur Kenntnis der *Oenothera purpurata* Klebahn und *O. rubricaulis* Klebahn. *Zeitschr. ind. Abst.- u. Vererb.* **52**, 191–235.

Rudloff, C. F. and Stubbe, H. (1935). Mutationsversuche mit *Oenothera hookeri*. *Flora* **129**, 347–362.

Scholz, H. S. (1956). Die Oenothera-Arten in Berlin und Umgebung. *Wiss. Zeitschr. Pädagog. Hochschule Potsdam* **2**, 205–209.

Schötz, F. (1954). Ueber Plastidenkonkurrenz bei *Oenothera*. *Planta* **43**, 182–240.

Schötz, F. (1958a). Beobachtungen zur Plastidenkonkurrenz bei *Oenothera* und Beiträge zum Problem des Plastidenvererbung. *Planta* **51**, 173–185.

Schötz, F. (1958b). Periodische Ausbleichungserscheinungen des Laubes bei *Oenothera*. *Planta* **52**, 351–392.

Schötz, F. (1962a). Zur Kontinuität der Plastiden. *Planta* **58**, 333–336.

Schötz, F. (1962b). Pigmentanalytische Untersuchungen an *Oenothera*. I. Vorversuche und Analyse der Blätter und Blüten von *Oenothera suaveolens* Desf., mutante "Weissherz." *Planta* **58**, 411–434.

Schötz, F. (1965a). Elektronenmikroskopische Untersuchungen an den Plastiden eines Oenotheren-Bastardes mit disharmonischer Genom- und Plastom Kombination. *Ber. Deut. bot. Ges.* **77**, 372–378.

Schötz, F. (1965b). Zur Frage der Vermehrung der Thylakoidschichten in den Chloroplasten. *Planta* **64**, 376–380.

Schötz, F. (1968). Über Plastidenkonkurrenz bei *Oenothera* II. *Biol. Zentralbl.* **87**, 33–61.

Schötz, F. and Bathelt, H. (1964a). Pigmentanalytische Untersuchungen an *Oenothera* II. Der *albivelutina*-Typ. (I Teil). *Planta* **63**, 213–232.

Schötz, F. and Bathelt, H. (1964b). Pigmentanalytische Untersuchungen an *Oenothera*. III. Der *albivelutina*-Typ. (II Teil). *Planta* **63**, 233–252.

Schötz, F. and Bathelt, H. (1965). Pigmentanalytische Untersuchungen an *Oenothera*. IV. Der *pictirubata*-Typ. *Planta* **64**, 330–362.

Schötz, F. and Diers, L. (1965). Electronenmikroskopische Untersuchungen über die Abgabe von Plastidenteilen ins Plasma. *Planta* **66**, 269–292.

Schötz, F. and Diers, L. (1968). Beeinflussung der Thylakoidbildung durch Disharmonie zwischen Genom und Plastom. *Protoplasma* **65**, 335–348.

Schötz, F., Diers, L. and Bathelt, H. (1968). Ueber den Einfluss einer Genom-Plastom-Disharmonie auf die Thylakoidanordnung. *Zeitschr. Naturforsch.* **23b**, 1247–1252.

Schötz, F., Diers, L. and Rüffer, P. (1971). Abgabe von Plastidenteilen ins Cytoplasma. Eine vergleichend lichtmikroskopische und electronenmikroskopische Untersuchung. *Ber. Deutsch. bot. Gesell.* **84**, 41–51.

Schötz, F. and Heiser, F. (1969). Ueber Plastidenkonkurrenz bei *Oenothera* III. Zahlenverhältnisse in Mischzellen. *Wiss. Zeitschr. Pädagog. Hochschule Potsdam* **13**, 65–89.

Schötz, F. and Senser, F. (1964). Untersuchungen über die Chloroplastenentwicklung bei *Oenothera*. III. Der *pictirubata*-Typ. *Planta* **63**, 191–212.

Schötz, F. and Senser, F. (1965). Untersuchungen über die Plastidenentwicklung und Pigmentausstattung der Oenotheren. Der *rubiflexa*-Typ. *Zeitschr. Vererb.* **96**, 250–266.

Schötz, F., Senser, F. and Bathelt, H. (1966). Untersuchungen über die Plastidenentwicklung und Pigmentausstattung der Oenotheren. II. Mutierte *biennis*-Plastiden. *Planta* **70**, 125–154.

Schwemmle, J. (1924). Vergleichend zytologische Untersuchungen an Onagraceen. *Ber. Deut. bot. Ges.* **42**, 238–243.

Schwemmle, J. (1927). Der Bastard *Oenothera Berteriana* × *Onagra* (*muricata*) und seine Zytologie. *Jahrb. wiss. Bot.* **66**, 579–595.

Seitz, F. W. (1935). Zytologische Untersuchungen an tetraploiden Oenotheren. *Zeitschr. Bot.* **28**, 481–542.

Senser, F. (1962). Zur Plastidenentwicklung bei *Oenothera*. *Diss. Naturwiss.-Fac. Univ. München*.

Senser, F. and Schötz, F. (1964). Untersuchungen über die Chloroplastenentwicklung bei *Oenothera*. II. Der *albivelutina*-Typ. *Planta* **62**, 171–190.

Sheffield, F. M. L. (1927). Cytological studies of certain meiotic stages in *Oenothera*. *Annals Bot.* **41**, 779–811.

Shull, G. H. (1921). Three new mutations in *Oenothera lamarckiana*. *J. Hered.* **12**, 354–363.

Shull, G. H. (1923a). Linkage with lethal factors the solution of the *Oenothera* problem. *Eugen., Genet. and the Family* **1**, 86–99.

Shull, G. H. (1923b). Further evidence of linkage with crossing over *Oenothera*. *Genetics* **8**, 154–167.

Shull, G. H. (1925). The third linkage group in *Oenothera*. *Proc. Nat. Acad. Sci. Wash.* **11**, 715–718.

Shull, G. H. (1927). Crossing over in the third linkage group in *Oenothera*. *Proc. Nat. Acad. Sci. Wash* **13**, 21–24.

Shull, G. H. (1928a). *Oenothera* cytology in relation to genetics. *Am. Nat.* **62**, 97–114.

Shull, G. H. (1928b). A new gene mutation (mut. *bullata*) in *Oenothera lamarckiana* and its linkage relations. *Zeitschr. ind. Abst.- u. Vererb.* (Verhandl. 5th Intern. Kongr. Vererb) **2**, 1322–1342.

Shull, G. H. (1930). The first two cases of crossing-over between old-gold and *bullata* factors in the third linkage group of *Oenothera*. *Proc. Nat. Acad. Sci. Wash.* **16**, 106–109.

Sikka, S. M. (1939). A study of chromosome catenation in *Oenothera* hybrids. *J. Genet.* **39**, 309–334.

Sloatman, R. J. (1953). A cytogenetic study of certain Euoenotheras from western New York. *Am. J. Bot.* **40**, 835–836.

Smith, J. E. (1806). *Oenothera biennis. English Botany*, 1st ed. **22**, 1534.

Sonneborn, T. M. (1957). Breeding systems, reproductive methods and species problems in Protozoa. *In* "The Species Problem", Amer. Assn. Adv. Sci. Wash.

Steiner, E. E. (1951). Phylogenetic relationships of certain races of *Euoenothera* from Mexico and Guatemala. *Evolution* **5**, 265–272.

Steiner, E. E. (1952). Phylogenetic studies in *Euoenothera*. *Evolution* **6**, 69–80.

Steiner, E. E. (1955). A cytogenetic study of certain races of *Oenothera elata*. *Bull. Torrey Bot. Club.* **82**, 292–297.

Steiner, E. E. (1956). New aspects of the balanced lethal mechanism in *Oenothera*. *Genetics* **41**, 486–500.

Steiner, E. E. (1957). Further evidence of an incompatibility allele system in the complex-heterozygotes of *Oenothera*. *Am. J. Bot.* **44**, 582–585.

Steiner, E. E. (1961). Incompatibility studies in *Oenothera*. *Zeitschr. Vererb.* **92**, 205–212.

Steiner, E. E. (1964). Incompatibility studies in *Oenothera*: The distribution of S_I alleles in *biennis* 1 populations. *Evolution* **18**, 370–378.

Steiner, E. E. and Schultz, M. S. (1958). Distribution of incompatibility alleles among the complex-heterozygotes of *Oenothera*. *Science* **127**, 516–517.

Stern, C. (1936). Somatic crossing over and segregation in *Drosophila melanogaster*. *Genetics* **21**, 625–730.

Stinson, H. T. (1953). Cytogenetics and phylogeny of *Oenothera argillicola* MacKenz. *Genetics* **38**, 389–406.

Stinson, H. T. and Steiner, E. (1955). Phylogenetic studies in *Oenothera*: Further analysis of plants from the south-eastern United States. *Am. J. Bot.* **42**, 905–911.

Stomps, T. (1913). Das *cruciata*-Merkmal. *Ber. Deut. bot. Ges.* **31**, 166–172.

Stubbe, W. (1953). Genetische und zytologische Untersuchungen an verschiedenen Sippen von *Oenothera suaveolens*. *Zeitschr. ind. Abst.- u. Vererb.* **85**, 180–209.

Stubbe, W. (1955). Erbliche Chlorophylldefekte bei *Oenothera*. *Photo. Wiss.* **4**, 3–8.

Stubbe, W. (1957). Dreifarbenpanaschierung bei *Oenothera*. I. Entmischung von drei in der Zygote vereinigten Plastidomen. *Ber. Deut. bot. Ges.* **70**, 221–226.

Stubbe, W. (1958). Dreifarbenpanaschierung bei *Oenothera*. II. Wechselwirkungen zwischen Geweben mit zwei erblich verschiedenen Plastidensorten. *Zeitschr. Vererb.* **89**, 189–203.

Stubbe, W. (1959). Genetische Analyse des Zusammenwirkens von Genom und Plastom bei *Oenothera*. *Zeitschr. ind. Abst.- u. Vererb.* **90**, 288–298.

Stubbe, W. (1960). Untersuchungen zur genetischen Analyse des Plastoms von *Oenothera*. *Zeitschr. Bot.* **48**, 191–218.

Stubbe, W. (1962). Sind Zweifel an der genetischen Kontinuität der Plastiden berechtigt? Eine Stellungnahme zu den Ansichten von Mühlethaler und Bell. *Zeitschr. Vererb.* **93**, 175–176.

Stubbe, W. (1963a). Die Rolle des Plastoms in der Evolution der Oenotheren. *Ber. Deut. bot. Ges.* **76**, 154–167.

Stubbe, W. (1963b). Extrem disharmonische Genom-Plastom-Kombinationen und väterliche Plastidenvererbung bei *Oenothera*. *Zeitschr. Vererb.* **94**, 392–411.

Stubbe, W. (1966). Die plastiden als Erbträger. *In* "Probleme der biologischen Reduplikation." P. Sitte, Heidelberg.

Stubbe, W. (1970). Das *falcifolia*-Syndrom der Oenotheren. *Molec. Gen. Genet.* **106**, 213–227.

Stubbe, W. and von Wettstein D. (1955). Zur Structure erblich verschiedener Chloroplasten von *Oenothera*. *Protoplasma* **45**, 241–250.

Sturtevant, A. H. (1931). Genetic and cytological studies on *Oenothera*. I. Nobska, Oakesiana, Ostreae, Shulliana, and the inheritance of old-gold flower color. *Zeitschr. ind. Abst.- u. Vererb.* **59**, 367–380.

Tournefort, J. (1694). "Élémens de Botanique." Paris.

Tournefort, J. (1700). "Institutiones Rei Herbariae." Paris.

Van Beneden, E. (1883). Recherches zur la maturation de l'oeuf, la fecondation et la division cellulaire. *Arch. Biol.* **4**, 265–648.

van Overeem, C. (1922). Ueber Formen mit abwichender Chromosomenzahl bei *Oenothera*. *Bot. Centralbl. Beih.* **39**, 1–80.

Weier, T. E. (1930). A comparison of meiotic prophase in *Oenothera lamarckiana* and *O. hookeri*. *La Cellule* **39**, 271–305.

Wilson, E. B. (1925). "The cell in development and heredity." MacMillan.

Winkler, H. (1930). "Die Konversion der Gene." G. Fischer, Jena.

Winkler, H. (1932). Konversions-Theorie und Austausch-Theorie. *Biol. Zentralbl.* **53**, 165–189.

Wisniewska, E. (1935). Zytologische Untersuchungen an *Oenothera hookeri* de Vries (in Polish, German summary). *Acta Soc. Bot. Polon.* **12**, 113–164.

Zürn, K. (1937a). Untersuchungen zur Physiologie der Meiosis IV. *Zeitschr. Bot.* **30**, 577–603.

Zürn, K. (1937b). Untersuchungen zur Physiologie der Meiosis IX. Die Bedeutung der Plastiden für den Ablauf der Meiosis. *Jahrb. wiss. Bot.* **85**, 706–731.

Author Index

Numbers in italics indicate the page in the References section where the reference is given in full.

A

Alpinus, Prosper, 314, *344*

B

Baerecke, M., 106, 340, *344*
Bartlett, H. H., 17, 34, 35, 130, 192, 219, *344*, *347*, *350*
Bathelt, H., 183, *350*, *359*
Bauhin, C., 314, *344*
Baur, E., 170, *344*
Bell, P. R., 185, *355*
Belling, J., 57, 58, *344*
Bergfeld-Gaertner, H., 216, *356*
Bhaduri, D. N., 73, 74, *344*
Blakeslee, A. F., 57, 59, 60, 230, 312, 330, 333, 340, *344*, *347*
Boedijn, K., 194, 195, 196, 197, *344*, *350*
Borbás, V., 226, *344*
Bridges, C. B., 58, *344*
Brittingham, W. H., 49, 53, 120, 212, *344*, *347*
Brown, W. V., *344*

C

Catcheside, D. G., 65, 83, 84, 95, 106, 110, 123, 124, 125, 199, 200, 207, 208, 210, 211, 212, 339, 340, *344*, *345*
Chrometzka, P., 105, 118, *345*
Clapham, A. R., 226, *345*
Cleland, R. E., 44, 48, 49, 53, 54, 59, 60, 62, 65, 79, 80, 84, 86, 95, 96, 107, 120, 123, 160, 173, 195, 197, 211, 227, 230, 232, 234, 287, 312, 323, 330, 331, 332, 333, 334, 335, 336, 337, 338, 339, 340, *344*, *345*, *346*, *347*, *355*, *358*
Cobb, F., 35, *347*
Columna, F., *347*
Correns, C., 170, *347*
Crowe, L. K., 160, *347*

D

Darlington, C. D., 58, 73, 77, 83, 113, 121, 210, 260, *347*

Davis, B. M., 24, 43, 191, 219, 220, 221, 224, *347*, *348*
Deam, C. C., 228, *348*
de Vries, H., 3, 4, 5, 8, 9, 10, 11, 14, 15, 16, 17, 18, 19, 20, 22, 23, 24, 25, 31, 33, 34, 37, 119, 120, 130, 168, 192, 193, 194, 195, 196, 197, 198, 200, 209, 211, 213, 218, 219, 220, 221, 222, 223, 313, *348*, *349*, *350*
Diers, L., 184, 185, *350*, *359*
Dolzmann, P., 185, *350*
Don, G., 130, *350*
Dulfer, H., 199, *350*

E

East, E. M., 160, *350*
Emerson, S. H., 49, 53, 59, 60, 61, 95, 105, 108, 109, 111, 114, 116, 120, 121, 133, 160, 161, 199, 340, *350*, *351*

F

Ford, C. E., 207, *351*
Fork, D. C., 185, *351*
Freese, E. B., 216, *351*

G

Gardella, C., 75, 76, 77, 78, *351*
Gates, R. R., 43, 54, 160, 192, 314, 316, *350*, *351*, *352*
Geckler, L. H., 65, 258, 279, 336, 337, 338, 339, *347*, *352*
Geerts, J. M., 43, 100, *352*
Gerhard, K., 29, *352*
Goldschmidt, R., 27, 137, *352*
Goodwin, K. M., 54, *352*
Göpel, G., 187, *352*
Gray, A., 228, *352*

H

Hagen, C. W., Jr., 160, *352*
Håkansson, A., 59, 85, 209, *352*

Subject Index